Generative AI for Communications Systems

IEEE Press

445 Hoes Lane

Piscataway, NJ 08854

Generative AI for Communications Systems

Fundamentals, Applications, and Prospects

*Edited by Diep N. Nguyen, Nam H. Chu, Dinh Thai Hoang,
Octavia A. Dobre, Dusit Niyato, and Petar Popovski*

IEEE Press
WILEY

Published by John Wiley & Sons, Inc., Hoboken, New Jersey.

Library of Congress Cataloging-in-Publication Data has been applied for:

Hardback ISBN: 9781394293902
ePDF ISBN: 9781394293926
ePub ISBN: 9781394293919
oBook ISBN: 9781394293933

Cover Design: Wiley
Cover Image: © Vertigo3d/Getty Images

For my family and my students.

—Diep N. Nguyen

To my family—your patience and support throughout this project have been invaluable. The countless hours spent coordinating with contributors, reviewing manuscripts, and writing my own chapter often meant time away from you. Thank you for understanding the demands of this work and for encouraging me when the finish line seemed distant. I am deeply grateful.

—Nam H. Chu

To my family and my students.

—Dinh Thai Hoang

I would like to thank my coauthors and collaborators for their shared vision and insightful discussions, and my family for their continued support and inspiration throughout this journey.

—Octavia A. Dobre

For my family.

—Dusit Niyato

I would like to thank my coauthors for their diligent work on this project.

—Petar Popovski

Contents

List of Contributors

Asmaa Abdallah received the Ph.D. degree in electrical engineering from the American University of Beirut, Beirut, Lebanon, in 2020. She is currently a research scientist at King Abdullah University of Science and Technology (KAUST).

Abdulrahman Alabbasi (Abbasi) joined Ericsson in 2018 and working as a master researcher. His focus is on enabling AI in RAN via solidifying a unified effort across Ericsson to address this area. This includes all aspects of AI application for RAN parameters optimization and the innovation of AI-native RAN. He envisions a Telecom future where AI deployment in the network (and UE nodes) enables the realization of the 6G-driven use-cases and corresponding KPIs. Abbasi holds a master's degree in electronics engineering, Tokyo, Japan, and a Ph.D. degree in electrical engineering from KAUST in Saudi Arabia.

Pegah Alizadeh is a senior researcher at Ericsson Research in Paris, specializing in reinforcement learning, generative decision-making, optimization, and large-scale artificial intelligence systems. Her work bridges offline and multi-agent reinforcement learning, decision transformers, and diffusion models, with applications to telecommunications, natural language processing, and autonomous systems. Before joining Ericsson, she was an Assistant Professor at the De Vinci Research Center (Paris), leading work on meta-learning, learning-augmented optimization, and reinforcement learning for sequential decision-making and urban mobility. She holds a Ph.D. in Computer Science from Sorbonne Paris Nord University (LIPN), where she developed methods for planning in Markov Decision Processes with unknown rewards. Author of more than 30 peer-reviewed publications and several patents on AI-driven network management, she focuses on developing robust, adaptive, safe, and explainable learning systems that connect optimization, reasoning, and real-world decision-making.

Samad Ali received the Ph.D. degree in wireless communications engineering from the University of Oulu, Finland. He is currently a Senior Staff Research Specialist at Nokia and an Adjunct Professor (Docent) at the University of Oulu. His main research interests include the applications of AI/ML in wireless communication networks.

Ender Ayanoglu received his Ph.D. degree in electrical engineering from Stanford University in 1986. He was with Bell Laboratories from 1996to 1999. From 1999 to 2002, he was a systems architect with Cisco Systems Inc. Since 2002, he has been a professor in the Department of Electrical Engineering and Computer Science, University of California, Irvine. He was a recipient of the IEEE Communications Society Stephen O. Rice Prize Paper Award in 1995, the IEEE Communications Society Best Tutorial Paper Award in 1997, and the IEEE Communications Society Joseph L. LoCicero Award in 2023. He has been an IEEE Fellow since 1998.

Eren Balevi received the Ph.D. degree in Electrical and Electronics Engineering from Middle East Technical University, Ankara, Turkey, in 2016. He was employed as an Adjunct Professor and Post-doctoral Research Scholar in the Department of Electrical Engineering at the University of South Florida in 2017. He worked as a Post-doctoral Research Scholar in the Department of Electrical and Computer Engineering at The University of Texas at Austin between 2018 and 2020. He also served as an Adjunct Professor for the Fall 2019 semester at the University of Texas at Austin. He started working at the Qualcomm Research Center in 2020. He was an Assistant Professor at the Middle East Technical University between 2021 and 2024. He is now an Editor in the *IEEE Transactions on Wireless Communications*. His current research interests lie mainly in the intersection between machine learning and communication theory. He is also interested in distributed edge artificial intelligence (AI).

Abdulkadir Celik received the Ph.D. degrees in electrical engineering and computer engineering (double major) from Iowa State University, Ames, IA, USA, in 2016. He is currently an Associate Professor at the University of Sharjah.

Nam H. Chu received the B.E. degree in electronics and telecommunications engineering from Hanoi University of Science and Technology, Hanoi, Vietnam, in 2009, the master's degree in software engineering from the University of Canberra, Canberra, ACT, Australia, in 2014, and the Ph.D. degree in electrical and computer engineering from the University of Technology Sydney, Sydney, NSW, Australia, in

2023. He is currently with the School of Information Technology, Crown Institute of Higher Education, Australia. He is also with the School of Electrical and Data Engineering, University of Technology Sydney, Australia and the Department of Telecommunications Engineering, University of Transport and Communications, Vietnam. His research interests include applying advanced machine learning and optimization methods for enhancing and securing wireless communications. He was the co-recipient of the Computer Networks 2021 Best Paper Award. He serves as a TPC and reviewer for top venues, such as IEEE JSAC, IEEE TWC, IEEE TMC, IEEE IoTJ, IEEE GLOBECOM, IEEE ICC, and IEEE WCNC.

Kemal Davaslioglu is a principal scientist at Nexcepta Inc., where he leads R&D efforts in wireless communications, wireless networking, radar signal processing, generative AI, and computer vision. He received his Ph.D. degree in electrical and computer engineering from the University of California, Irvine. He previously worked at Broadcom, Intelligent Automation/BlueHalo, and University Technical Services. His research focuses on AI/ML for communications, radar, computer vision, and anomaly detection with emphasis on AI security and robust learning. He is a senior IEEE member and recipient of the IEEE HST Best Paper Award.

Kaushik Dey is a Senior Research Manager at Ericsson Research, leading the AI/ML research unit in India. With more than two decades of experience at Ericsson and IBM, he has worked across AI research and product development, driving innovations in Zero-Touch Networks, multi-agent systems, and energy-efficient AI. His research interests include Reinforcement Learning, Autonomous Agents, and AI-driven communication networks. He holds 20+ international patents and a Ph.D. in Computer Science from Indian Institute of Technology, Kharagpur.

Zhi Ding holds the position of Distinguished Professor in the Department of Electrical and Computer Engineering at the University of California, Davis. He received his Ph.D. degree from Cornell University in 1990. His major research interests cover wireless networking, signal processing, multimedia, and learning. He is a Fellow of IEEE. He served as an Associate Editor for IEEE Trans. Signal Processing from 1994 to 1997 and 2001 to 2004, and for IEEE Signal Processing Letters from 2002 to 2005. He was the General Chair of the 2016 IEEE Int'l. Conf. Acoustics, Speech, and Signal Processing and the Technical Program Chair of the IEEE Globecom 2006. He is a coauthor of the textbook: *Modern Digital and Analog Communication Systems, 5th edition*, Oxford University Press, 2019. He received the IEEE Communication Society's WTC Award in 2012 and the IEEE Communication Society's Education Award in 2020.

Ahmed M. Eltawil received the Ph.D. degree in electrical engineering from the University of California, Los Angeles, CA, USA, in 2003. He is currently a Full Professor at King Abdullah University of Science and Technology (KAUST).

Tugba Erpek is a Principal Scientist at Nexcepta, Inc. Prior to Nexcepta, Dr. Erpek was a Research Associate Professor with Virginia Tech National Security Institute, Blacksburg; a Lead Scientist and Network Communications Technical Area Lead with the Intelligent Automation, Inc., Rockville, MD, USA; and a Senior Communications Systems Engineer with the Shared Spectrum Company, Vienna, VA, USA. She received the Ph.D. degree in electrical and computer engineering from Virginia Tech, Blacksburg, VA, USA, in 2019. Her research interests include wireless communications and networking, 5G and beyond, dynamic spectrum sharing, cognitive radio, radar systems, IoT, security, machine learning, adversarial machine learning, optimization, network protocol design, and implementation. Dr. Erpek has published extensively in these areas. She has served as a TPC member and reviewer for major IEEE conferences and journals.

Paria Mohammadzadeh Hesar received the B.Sc. degree in Electrical Engineering from Urmia University, Urmia, Iran, in 2015, and the M.Sc. degree in Electrical Engineering (Communication Systems) from the University of Tabriz, Tabriz, Iran, in 2018. She is currently pursuing the Ph.D. degree in Electrical Engineering (Communication Systems) at York University, Toronto, ON, Canada, where she is a Research Assistant with the Next Generation Wireless Networks (NGWN) Lab, supervised by Dr. Hina Tabassum. Her research interests include generative AI for wireless communications, radio-map reconstruction, and electromagnetic field (EMF) exposure modelling in next-generation networks.

Dinh Thai Hoang (M'16, SM'22) is currently a faculty member at the School of Electrical and Data Engineering, University of Technology Sydney, Australia. He received the Ph.D. degree in Computer Science and Engineering from Nanyang Technological University, Singapore, in 2016. His research interests include emerging wireless communications and networking topics, especially machine learning applications in networking, edge computing, and cybersecurity. He has received several prestigious awards, including the Australian Research Council Discovery Early Career Researcher Award, IEEE TCSC Award for Excellence in Scalable Computing for Contributions on "Intelligent Mobile Edge Computing Systems" (Early Career Researcher), IEEE Asia-Pacific Board (APB) Outstanding Paper Award 2022, and IEEE Communications Society Best Survey Paper Award 2023. He is currently an Editor of IEEE TMC, IEEE TWC, IEEE TCOM, and IEEE TNSE.

Nguyen Van Huynh received the Ph.D. degree in Electrical and Computer Engineering from the University of Technology Sydney (UTS), Australia, in 2022. He is currently a Lecturer at the School of Computer Science and Informatics, University of Liverpool (UoL), United Kingdom. His research interests include cybersecurity, mobile computing, 5G/6G, IoT, and machine learning. He currently serves as an Editor for the IEEE Transactions on Communications and as an Associate Editor for the IEEE Transactions on Mobile Computing (IEEE TMC) and IEEE Transactions on Vehicular Technology (IEEE TVT).

Thien Huynh-The received the Ph.D. degree in Computer Science and Engineering from Kyung Hee University (KHU), South Korea, in 2018. From September 2018 to May 2022, he was a Postdoctoral Researcher with the ICT Convergence Research Center, Kumoh National Institute of Technology, South Korea. He is currently an Assistant Professor in the Department of Computer and Communication Engineering at the Ho Chi Minh City University of Technology and Education, Vietnam. His current research interests include digital image processing, radio signal processing, computer vision, wireless communications, machine learning, and deep learning.

Athanasios Karapantelakis is a principal researcher in AI at Ericsson. He holds a Ph.D. from KTH Royal Institute of Technology, Stockholm, Sweden. His research interests include automated planning and reasoning and their connection to machine learning. Throughout his career, he has held diverse engineering roles, ranging from software design to cloud and AI research. He has coauthored a book on AI and contributed to over 100 patent families, along with 18 peer-reviewed scientific conference publications.

Brian Kim is a Post-doctoral Research Associate at the DEVCOM US Army Research Laboratory, Maryland. He was also a Post-doctoral Research Associate at GENESYS Lab, Department of Electrical and Computer Engineering, Northeastern University, directed by Prof. Kaushik Roy Chowdhury. He received his Ph.D. degree under the supervision of Prof. Sennur Ulukus from the Department of Electrical and Computer Engineering, University of Maryland, College Park, in 2022. Prior to this, he received the B.S. and M.S. degrees in electrical engineering from Korea Advanced Institute of Science and Technology (KAIST), Daejeon, South Korea, in 2014 and 2016, respectively.

Matti Latva-aho (IEEE Fellow) received his M.Sc., Lic.Tech. and Dr. Tech (Hons.) degrees in Electrical Engineering from the University of Oulu, Finland in 1992, 1996, and 1998, respectively. From 1992 to 1993, he was a Research Engineer at

Nokia Mobile Phones, Oulu, Finland, after which he joined the Centre for Wireless Communications (CWC) at the University of Oulu. He was Director of CWC during the years 1998–2006 and as Head of the Department for Communication Engineering until August 2014. He is currently a Professor of Wireless Communications at the University of Oulu and served as the Director of the National 6G Flagship Programme until 6/2025. He was an Academy of Finland Professor from 2017 to 2021 and was appointed as a Global Fellow at the University of Tokyo in 2021. He is currently the Vice President for Research at the University of Oulu. He has published over 600 conference and journal papers in the field of wireless communications. He received the Nokia Foundation Award in 2015 for his achievements in mobile communications research.

Mehdi Letafati is a doctoral researcher at the Center for Wireless Communications, University of Oulu, Finland. His research interests lie at the intersection of generative AI, data science, and semantic communications. He has received the Best Paper Award at the 2024 IEEE Wireless Communication and Networking Conference (WCNC'24). His doctoral thesis has been supported by multiple research grants from Elisa Research Foundation (HPY:n Tutkimussäätiö), the Finnish Foundation for Technology Promotion (Tekniikan edistämissäätiö), and the Jenny and Antti Wihuri Foundation (Jenny ja Antti Wihurin rahasto). He served as a distinguished lecturer at the 2024 COST INTERACT Training School on AI/ML for 6G in Barcelona and has contributed to several national and EU-funded projects, including Hexa-X-II, 6GARROW, and 6GFlagship.

Amirhossein Mohammadi received the B.Sc. degree in Electrical Engineering from the University of Tehran, Tehran, Iran, in 2023. He is currently an M.Sc. candidate in Electrical and Computer Engineering at York University, Toronto, ON, Canada, and a Research Assistant with the Next Generation Wireless Networks (NGWN) Lab, supervised by Dr. Hina Tabassum. His research interests include machine learning and representation learning for time-series data.

Diep N. Nguyen received the M.E. degree in electrical and computer engineering from the University of California at San Diego (UCSD), La Jolla, CA, USA, in 2008, and the Ph.D. degree in electrical and computer engineering from the University of Arizona (UA), Tucson, AZ, USA, in 2013. He is currently the Head of 5G/6G Wireless Communications and Networking Lab, Director of Agile Communications and Computing group, Faculty of Engineering and Information Technology, University of Technology Sydney (UTS), Sydney, Australia. Before joining UTS, he was a DECRA Research Fellow with Macquarie University, Macquarie Park, NSW, Australia, and a Member of the Technical Staff with Broadcom Corporation, CA, USA, and ARCON Corporation, Boston, MA, USA, and consulting the Federal Administration of Aviation on turning detection of UAVs and aircraft, and the US Air Force Research Laboratory, USA, on anti-jamming. His research interests include computer networking, wireless communications, and machine learning application, with emphasis on systems' performance and security/privacy. Dr. Nguyen received several awards from US Congress, the US National Science Foundation, and the Australian Research Council. He has served on the Editorial Boards of the *IEEE Transactions on Mobile Computing, IEEE Communications Surveys & Tutorials (COMST)*, and *IEEE Open Journal of the Communications Society*.

Alexandros Nikou received his degree in electrical and computer engineering in 2014. He holds a Ph.D. degree from KTH Royal Institute of Technology in 2019. Since 2019, he has been with Ericsson and has contributed to the fields of network automation, reinforcement learning, and agentic AI-based networks.

Quoc-Viet Pham is currently an Assistant Professor in Networks and Distributed Systems at the School of Computer Science and Statistics, Trinity College Dublin, Ireland. He earned his Ph.D. degree (with Best Ph.D. Dissertation Award) in Telecommunications Engineering from Inje University, South Korea, in 2017. His research interests include wireless AI, edge computing, Internet of Things (IoT), and distributed machine learning. He was a recipient of Clarivate Highly Cited Researcher Award in 2024 and IEEE ComSoc Best Young Researcher Award for EMEA in 2023, in recognition of his research contributions to the Society.

Petar Popovski is a Professor at Aalborg University, where he heads the Connectivity Section and serves as the Director of the Center for Classical Communication in the Quantum Era (CLASSIQUE). He also holds a Visiting Excellence Chair at the University of Bremen. He received his Dipl.-Ing and M. Sc. degrees in communication engineering from the *University of Sts. Cyril and Methodius* in Skopje and his Ph.D. from Aalborg University in 2005. He is a Fellow of the IEEE and a past Chair of the IEEE Communication Theory Technical Committee. He is currently the Editor in-Chief of the *IEEE Journal on Selected Areas in Communications*. He is the author of the book *Wireless Connectivity: An Intuitive and Fundamental Guide*. His research interests lie in the broad area of communication theory.

Yalin E. Sagduyu is the chief technology officer at Nexcepta. He oversees research in wireless communications, security, machine learning, and next-generation networks. He received his Ph.D. in electrical and computer engineering from the University of Maryland, College Park, and previously served as a research professor at Virginia Tech and director of networks and security at Intelligent Automation/BlueHalo. He has published over 200 papers, received an IEEE best paper award, and served as an Editor for *IEEE Transactions on Communications* and *IEEE Transactions on Cognitive Communications and Networking.*

Yi Shi is a Research Associate Professor at the Commonwealth Cyber Initiative (CCI), Virginia Tech. He is an IEEE Fellow. Before joining Virginia Tech, Dr. Shi was a Senior Lead Scientist at Intelligent Automation, a BlueHalo company. He holds three patents on wireless communications and social networks. Dr. Shi's research focuses on machine learning, algorithm design, and optimization for 5G and NextG networks, satellite networks, cognitive radio networks, MIMO and cooperative communication networks, sensor networks, ad hoc networks, and social networks. His work has been published in leading IEEE and ACM journals and presented at top-tier international conferences. He received a number of awards, including the Test of Time Paper Award at IEEE INFOCOM 2023 and the Best Paper Award at IEEE INFOCOM 2008. Dr. Shi currently serves as a Workshop Chair for IEEE INFOCOM 2026, a distinguished lecturer for IEEE Vehicular Technology Society, an Editor for *IEEE Communications Surveys and Tutorials and IEEE Transactions on Cognitive Communications and Networking*. He has also served as a Co-Chair for various IEEE and ACM Workshops, Conference Tracks, and Symposia, as well as a TPC member for many top international conferences.

Hina Tabassum (Senior Member, IEEE) received the Ph.D. degree from King Abdullah University of Science and Technology (KAUST). She is currently an Associate Professor at the Lassonde School of Engineering, York University, Canada, where she joined as an Assistant Professor in 2018. In 2023, she was appointed as the York Research Chair in 5G/6G-enabled mobility and sensing applications for a five-year term. In 2024, she was also appointed as a Visiting Faculty at the University of Toronto. She has been selected as an IEEE Communications Society Distinguished Lecturer during 2025–2026. Dr. Tabassum has been recognized among Stanford University's list of the World's Top Two-Percent Researchers from 2021 to 2024. Her accolades include the Lassonde Innovation Early-Career Researcher Award (2023) and the N2Women Rising Stars in Computer Networking and Communications Award (2022). She has co-authored more than 120 peer-reviewed articles in leading IEEE journals, magazines, and conferences. Dr. Tabassum was the Founding Chair of the IEEE Communications Society Radio Communications Committee's Special Interest Group on THz Communications. She has served as an Associate Editor for *IEEE Communications Letters, IEEE Open Journal of the Communications Society*, and *IEEE Transactions on Green Communications and Networking* from 2019 to 2023. Currently, she serves as an Area Editor for *IEEE Open Journal of the Communications Society* and as an Associate Editor for *IEEE Transactions on Communications, IEEE Transactions on Mobile Computing, IEEE Transactions on Wireless*

Communications, and *IEEE Communications Surveys & Tutorials*. Her research interests include multi-band 6G wireless communications and sensing networks, connected and autonomous systems, and AI-enabled network mobility and resource management solutions.

Sennur Ulukus is a Distinguished University Professor and the Anthony Ephremides Professor in Information Sciences and Systems in the Department of Electrical and Computer Engineering at the University of Maryland at College Park, where she also holds a joint appointment with the Institute for Systems Research (ISR). Prior to joining UMD, she was a Senior Technical Staff Member at AT&T Labs-Research. Her research interests include information theory, wireless communications, machine learning, signal processing and networks, with recent focus on age of information, private information retrieval, blockchain consensus, quantum information, machine learning for wireless, federated learning, and large language models. She is a Fellow of the IEEE, and a Distinguished Scholar-Teacher of the University of Maryland. She has received many awards, including the 2003 IEEE Marconi Prize Paper Award in Wireless Communications, the 2020 IEEE Communications Society Technical Committee on Green Communications and Computing (TCGCC) Distinguished Technical Achievement Recognition Award, and the 2011 ISR Outstanding Systems Engineering Faculty Award. She was a Distinguished Lecturer of the IEEE Information Theory Society during 2018–2019. She is a Senior Editor for the *IEEE Transactions on Green Communications and Networking*. She served as the TPC Chair of IEEE Globecom 2021, and as a TPC Co-Chair of IEEE Globecom 2024, IEEE DySPAN 2024, IEEE MILCOM 2023, IEEE ITW 2019, IEEE ISIT 2017, IEEE Globecom 2016, IEEE PIMRC 2014, and IEEE CTW 2011.

Songyang Zhang received the Ph.D. degree in Department of Electrical and Computer Engineering from the University of California at Davis, Davis, CA, USA, in 2021, where he was a Post-doctoral Research Associate from August 2021 to July 2023. He is currently an Assistant Professor with the Department of Electrical and Computer Engineering in University of Louisiana at Lafayette, Lafayette, LA, USA. His current research interests include machine learning, signal processing, IoT intelligence, and wireless communication. He received the Best Paper Finalist in the 2020 IEEE Int'l. Conf. Image Processing and was recognized as the Best TCSVT Reviewer of 2022 by IEEE Circuits and System Society.

Preface

The paradigm of generative artificial intelligence (GenAI) refers to a branch of artificial intelligence (AI) that focuses on machine learning models that can generate new and original content, such as text, images, videos, music, or even applications. Unlike traditional/discriminative AI systems typically aim to perform specific tasks, such as classification or prediction, GenAI aims to produce new content/knowledge that resembles the training data it was exposed to. Recently, GenAI has achieved remarkable success in generating realistic images, such as faces, landscapes, objects, and even videos and training data for numerous applications. GenAI has thus been leveraged in various areas, e.g., in industry, education, and entertainment. Within the communications and networking industry, GenAI has been considered an enabler of the future systems across the PHY, MAC, NET, and Application layers.

The release of ChatGPT by OpenAI in 2022 and other following applications such as DeepSeek and Claude draws paramount attention to GenAI due to its enormous potential. For instance, Hyundai employs GenAI-based design software to explore various design options and optimize them for performance, cost, and sustainability. IBM uses GenAI to identify and test computer chip materials. All these examples foretell a future where GenAI holds immense potential for numerous domains, serving as a powerful catalyst for innovation. However, fully unleashing the potential of GenAI in practice faces various communications and networking challenges.

The characteristics of future communications are anticipated to include advancements and improvements in several key areas, such as higher data rates, ultra-low latency, massive device connectivity, enhanced energy efficiency, improved reliability and security, and seamless mobility. To achieve that, future communications will likely integrate various cutting-edge technologies (e.g., 6G and beyond 6G technologies) and fully incorporate GenAI for network design, optimization, and management. However, various challenges need to be addressed to realize such potential and benefits of GenAI in future communications and networking. First, training a GenAI model requires a diverse and representative dataset, which can be challenging due to the dynamic nature of the environment and limited access to real-world data. Second, GenAI models, e.g., Generative Adversarial Networks (GANs), Variational Autoencoders (VAEs), and Generative Diffusion Models (GDMs), often require significant computational resources and training time. In wireless communications, where real-time or near real-time processing is crucial, addressing the computational complexity is another challenge. Third, ensuring that GenAI models can generalize unseen scenarios is a challenge that needs to be addressed. This is because wireless communication systems operate in various environments and scenarios, including different channel conditions, network topologies, and user distributions. Fourth, since wireless channels are prone to noise, interference (both advertent and inadvertent), and fading, the received signals likely are distorted. GenAI models in communications, hence, should be robust to noisy and incomplete data to provide reliable and accurate outputs. Moreover, it is worth noting that GenAI models (which are often based on Pretrained Foundation Models (PFMs) such as BERT and ChatGPT) are imperfect and may sometimes produce flawed, biased, or inappropriate outputs. Therefore, ensuring that GenAI models do not generate malicious or deceptive content and safeguarding user privacy is a critical challenge in communications systems.

Besides the potential of GenAI in solving problems of future communications systems, the growing interest in applications of GenAI to various areas, ranging from industry to education, entertainment or even creative arts also calls for novel network architectures/solutions to enable/facilitate GenAI as a Service. According to a report by McKinsey, "Generative AI is now being used by 79% of workers and 22% use it regularly to complete their day-to-day work tasks. Generative AI is also being used by around 40% of organizations." Due to the large-scale computation modeling parameters (in the order of billions or even trillions), the distributed nature of data sources, the demanding computational power of GenAI applications/services, the extremely low-latency requirement, communications/networking systems must be revisited.

Given the above, this book titled *Generative AI for Communications Systems: Fundamentals, Applications, and Prospects* aims to lay a systematic foundation for the research community and industry who is interested in GenAI for communications and networking, covering recently proposed applications of GenAI to communications and networking, challenges in applying GenAI for future communications systems, and potential solutions. Additionally, to enable and facilitate "Generative AI as a Service," this book aims to explore novel communications/networking architecture/solutions to meet the computing/communications challenges/demands of large-scale GenAI models. In particular, the book will start by providing a fundamental background on GenAI, future communications systems, and AI-empowered communications systems, emphasizing the significance of GenAI in this domain, including networking architectures, protocols, key challenges, and research trends. Then, it will delve into the applications of GenAI across various layers, including the PHY layer, MAC layer, Network layer, and Application layer. New concepts, ideas like Trainable Radios, Native AI, Mobile AI-generated content, implications of GenAI to wireless and cybersecurity are introduced and discussed in depth. Novel communications/networking solutions/architectures to support GenAI as a service are then explored. Finally, it will discuss the cybersecurity implications of GenAI and the crucial challenges (e.g., training data availability, computational complexity, generalization to various scenarios, robustness to noisy and incomplete data, real-time adaptation and learning, as well as ethics and privacy consideration) of adopting GenAI in future communications and networking. The book will then provide insights and potential solutions to tackle these challenges, including open/future research directions. The key features of this book are as follows:

- Introduce a cutting-edge research topic focused on the applications of GenAI in future communications and networking. The book will highlight the potential impact and promising applications of this technology.
- Provide a concise overview of GenAI and its role in future communications and networking development, fundamental concepts, architectural aspects, and the significance of integrating GenAI in this domain.
- Provide a comprehensive review of the state-of-the-art research and development in various aspects of GenAI in future communications and networking to enable GenAI as a service. This will encompass advancements, methodologies, and innovative approaches in the field.
- Highlight the significant challenges and open issues that arise when employing GenAI in communications and networking. This may include training data availability, computational complexity, generalization to various scenarios, robustness to noisy and incomplete data, real-time adaptation and learning, ethics, and privacy consideration.
- Explore potential research directions to leverage the latest advances of GenAI for future communications and networking. This book will delve into how GenAI can contribute to the development and enhancement of communications and networking.

To achieve these objectives, the book comprises **three main parts** as follows.

Part I: Fundamental Background

In Part I, we will first introduce the future communications systems, including the networking protocols and architectures, key challenges and research trends in this domain. Then, we provide the basics of GenAI and explain the significance of GenAI in communications and networking. Finally, we review several case studies highlighting applications of GenAI to communications systems.

Part II: Applications of Generative AI to Communications and Networking and Generative AI as a Service

In Part II, we will provide in-depth discussion of GenAI applications to address various problems in communications and networking, such as signal detection, channel modeling, resource allocation, signal processing. To present these applications logically, we group them into different layers, i.e., the PHY layer, MAC layer, Network layer, and Application layer. Then, at each layer, we will focus on critical problems together with potential applications of GenAI techniques. Comparisons and detailed discussions are also provided to help the readers have a comprehensive view of the advantages as well as current limitations of using GenAI to solve problems in networks.

Additionally, given the growing potential of offering GenAI as a Service, we will discuss the emerging applications with different data modalities (pictures/photos, sounds, 3D models, videos/animations) and the challenges that must be addressed by future network architectures. This is due to the large-scale computation modeling parameters, the distributed nature of data sources, the demanding computational power of GenAI applications/services. For example, with billions of parameters, it is difficult to deploy GenAI models at the edge servers due to their limited computing, communication, and storage resources. However, the deployment of these models at core networks/servers can be prohibited by the large-scale distributive data, the excessive communications overhead/latency. It is thus very much desirable to propose novel network architectures, solutions to facilitate the GenAI services.

Part III: Challenges and Future Opportunities for Generative AI in Communications Systems

In Part III, the book will discuss critical issues and challenges associated with GenAI applications. Specifically, Part III will first provide a comprehensive investigation of the implications of GenAI on wireless and cybersecurity. We plan to cover both the risks and potential applications of GenAI to the security of communications systems. It then discusses the inherent challenges of GenAI when applied to communications systems, e.g., training data availability, required computational complexity, and the network architecture to support GenAI (e.g., centralized vs. distributed, fog/edge computing). The book will then explore potential solutions to these challenges, along with highlighting future research directions in the field.

Acronyms

3D	Three Dimensional
3GPP	3rd Generation Partnership Project
3P	Third-party
4G	Fourth Generation of Mobile Networks
5G	Fifth Generation
5G-LDPC	5G Low Density Parity Check
6G	Sixth Generation
64-QAM	64-Quadrature Amplitude Modulation
AAC	Authentication and Access Control
AC-WGAN	Auxiliary Classifier Wasserstein GAN
ACM	Association for Computing Machinery
Adam	Adaptive Moment Estimation
ADMM	Alternating Direction Method of Multipliers
AE	Autoencoder
AEC	Autoencoder Communication
AGOD	AI-Generated Optimal Decision
AI	Artificial Intelligence
AI Act	Artificial Intelligence Act
AI-RAN	AI-based Radio Access Network
AIGC	AI-Generated Content
AIGCaaS	AIGC-as-a-Service
ALBERT	A Lite BERT
AMD	Advanced Micro Devices
AMF	Access Management Function
AML	Adversarial Machine Learning
AoC	Age of Cache
API	Application Programming Interface
AR	Augmented Reality

ARQ	Automatic Repeat Request
ASIC	Application-Specific Integrated Circuit
ASP	AIGC Service Provider
ATSSS	Access Traffic Steering, Switching, and Splitting
AUROC	Area Under the Receiver Operating Characteristic Curve
AV	Autonomous Vehicle
AWGN	Additive White Gaussian Noise
B5G	Beyond Fifth Generation
Bard	Bard (Google's AI chatbot)
BER	Bit Error Rate
BERT	Bidirectional Encoder Representations from Transformers
BFE-CGAN	Bispectrum Feature Extraction-Conditional GAN
BiGANs	Bidirectional GANs
C2PA	Coalition for Content Provenance and Authenticity
CBRS	Citizen Broadband Radio Service
CDL	Clustered Delay Line
CDMA	Code Division Multiple Access
cGANs	conditional Generative Adversarial Network
ChatGPT	Chat Generative Pre-trained Transformer
CIFAR	Canadian Institute For Advanced Research
CNN	Convolutional Neural Network
CNNs	Convolutional Neural Networks
COLM	Conference on Language Modeling
CommGAN	Communication GAN
Copilot	Microsoft Copilot
CoT	Chain-of-Thought
CPU	Central Processing Unit
CRN	Cognitive Radio Network
CRS	Congressional Research Service
CS	Cosine Similarity
CSI	Channel State Information
CSINet	Channel State Information Network
CSS	Cascading Style Sheets
CTDE	Centralized Training with Decentralized Execution
CSV	Comma-Separated Values
CVPR	Conference on Computer Vision and Pattern Recognition
cVAEs	Conditional VAEs
DAC	Digital-to-Analog Converter
DAI	Discriminative AI
DALL-E	DALL-E (OpenAI's text-to-image model)

DCGANs	Deep Convolutional GANs
DDIMs	Denoising Diffusion Implicit Models
DDoS	Distributed Denial of Service
DDPG	Deep Deterministic Policy Gradient
DDPM	Denoising Diffusion Probabilistic Model
DeepJSCC	Deep Learning-based Joint Source Channel Coding
DEVCOM	Development Command
DGM	Deep Generative Model
DHCP	Dynamic Host Configuration Protocol
DIL	Disconnected, Intermittent, and Limited
DL	Deep Learning
DLP	Data Loss Prevention
D-MIMO	Distributed Multiple-input Multiple-output
DMPM	Deep Multi-agent reinforcement learning-based Power Management
DNA	Deep Neighbor Adaptation
DNN	Deep Neural Network
DOA	Direction of Arrival
DoS	Denial of Service
DP	Differential Privacy
DQN	Deep Q-Networks
DR-VAE	Deep Recurrent Variational Autoencoder
DRL	Deep Reinforcement Learning
DSA	Dynamic Spectrum Access
DSM	Denoising Score Matching
DTs	Digital Twins
DVFS	Dynamic Voltage and Frequency Scaling
EAI	Edge AI
EDMA-RM	Event-Driven and Mobility-Aware Resource Management
ELBO	Evidence Lower Bound
EM	Electromagnetic
eMBB	enhanced Mobile Broadband
EM-GM-AM	EPxpectation-Maximization Gaussian-Mixture Approximate Message Passing
EMA	Exponential Moving Average
eNodeB	evolved Node B
ESP	Edge Service Provider

EU	European Union
F-AnoGAN	Fast AnoGAN
FA	FedAvg
FDD	Frequency-Division Duplex
FDSC	Full-Duplex Semantic Communications
FedAvg	Federated Averaging
FedGAN	Federated Generative Adversarial Network
FedProx	Federated Proximal
FFN	Feed-Forward Network
FGMs	Flow-based Generative Models
FID	Fréchet Inception Distance
FL	Federated Learning
FLOP	Floating Point Operation
FLOPS	Floating Point Operations Per Second
FM	FedMix
FMARL	Federated Multi-Agent Reinforcement Learning
FMNIST	Fashion Modified National Institute of Standards and Technology
FP	Forward Pass
FVSBNs	Fully Visible Sigmoid Belief Networks
GAI	Generative AI
GAIaaMS	GenAI-as-a-Mobile Service
GAMs	Generative Autoregressive Models
GANs	Generative Adversarial Networks
GCE	Generative Channel Estimate
GDPR	General Data Protection Regulation
GenAI	Generative AI
GEO	Geostationary Earth Orbit
GFLOPS	Giga Floating Point Operations Per Second
GMs	Generative Models
gNB	gNodeB
gNodeB	Next Generation NodeB
Google	Google LLC
GPT	Generative Pre-trained Transformer
GPU	Graphics Processing Unit
GRUs	Gated Recurrent Units
GS	Gram-Schmidt
GT	Ground-Truth
GTMs	Generative Transformer Models
HAPs	High Altitude Platforms

HDT	Human Digital Twin
HF	High Frequency
HIPAA	Health Insurance Portability and Accountability Act
HPF	High-Pass Filter
HRLLC	Hyper-Reliable and Low Latency Communication
HTML	HyperText Markup Language
HTTP	Hypertext Transfer Protocol
HTTPS	HTTP Secure
IAB	Integrated Access and Backhaul
IAF	Inverse Autoregressive Flow
ICLR	International Conference on Learning Representations
IEEE	Institute of Electrical and Electronics Engineers
IMT	International Mobile Telecommunications
InfoGANs	Information Maximizing GANs
IoE	Internet of Everything
IoT	Internet of Things
IoV	Internet of Vehicles
IP	Internet Protocol
IPFS	InterPlanetary File System
IRS	Intelligent Reflecting Surface
ISAC	Integrated Sensing and Communication
ISO	International Organization for Standardization
ITS	Intelligent Transportation System
ITU	International Telecommunication Union
ITU-R	International Telecommunication Union's Radiocommunication Sector
JCAS	Joint Communication and Sensing
JSON	JavaScript Object Notation
JSCC	Joint Source-Channel Coding
KDE	Kernel Density Estimation
KL	Kullback-Leibler
KNN	*k*-Nearest Neighbors
KPIs	Key Performance Indicators
KV	Key Value
KYC	Know Your Customer
LAION	Large-scale Artificial Intelligence Open Network
LC	Least Cost
LDM	Latent Diffusion Model
LDPL	Log-Distance Path Loss
LEO	Low Earth Orbit
Llama	Large Language Model Meta AI

LLMs	Large Language Models
LO	Local Oscillator
LoRA	Low-Rank Adaptation
LoRa	Long Range
LoRaWAN	Long Range Wide Area Network
LOS	Line of Sight
LS	Least-Squares
LS-GAN	Least-Squares GAN
LSTM	Long Short-Term Memory
LTE	Long-Term Evolution
MAC	Medium Access Control
MADE	Masked Autoencoder for Distribution Estimation
MAF	Masked Autoregressive Flow
MARL	Multi-Agent Reinforcement Learning
MBI	Model-Based Interpolation
MC-AE	Multi-Carrier Auto-Encoder
MCS	Mobile Crowdsourcing
MEC	Multi-access Edge Computing
MEO	Medium Earth Orbit
Meta	Meta Platforms
MHSAM	Multi-Head Self-Attention Mechanism
Microsoft	Microsoft Corporation
MIMO	Multiple-Input Multiple-Output
MitM	Man-in-the-Middle
MITM-RIS	Man-in-the-Middle Malicious RIS
ML	Machine Learning
MLOps	Machine Learning Operations
mmWave	Millimeter Wave
mMTC	Massive Machine Type Communication
MNIST	Modified National Institute of Standards and Technology
MO	Manifold Optimization
MR	Mixed Reality
MS-SSIM	Multi-Scale Structural Similarity Index Measure
MSE	Mean Squared Error
MUI	Multi-User Interference
NADE	Neural Autoregressive Distribution Estimation
NAS	Neural Architecture Search
NAS	Non-Access Stratum
NB-IoT	Narrowband Internet of Things

NCSN	Noise-Conditioned Score Network
NEF	Network Exposure Function
NeurIPS	Conference on Neural Information Processing Systems
NextG	Next Generation
NFs	Network Functions
NFV	Network Function Virtualization
NICE	Nonlinear Independent Components Estimation
NLP	Natural Language Processing
NLOS	Non-Line of Sight
NM	Naivemix
NMSE	Normalized Mean Square Error
NNs	Neural Networks
non-IID	non-Independent and Identically Distributed
NR	New Radio
NSFW	Not Safe for Work
NTN	Non-Terrestrial Network
NTIA	National Telecommunications and Information Administration
NVIDIA	NVIDIA Corporation
OFDM	Orthogonal Frequency Division Multiplexing
OMP	Orthogonal Matching Pursuit
OpenAI	OpenAI
ORAN	Open Radio Access Network
PA	Power Amplifier
PCC	Policy and Charging Control
PCF	Policy Control Function
PDCP	Packet Data Convergence Protocol
PDE	Partial Differential Equation
PDF	Probability Density Function
PDU	Protocol Data Unit
PEFT	Parameter-Efficient Fine-Tuning
PFL-GAN	Personalized Federated Learning GAN
PFM	Pre-trained Foundation Model
PGC	Professionally Generated Content
PHY	Physical Layer
PIGL	Physics-Inspired Generative Learning
PL-SKG	Physical Layer Secret Key Generation
PLS	Physical Layer Security
PRB	Physical Resource Block

ProGANs	Progressive GANs
PRS	Positioning Reference Signal
PS	PS-FedGAN
PS	Phase Shifter
PS-FedGAN	Partially Shared Federated Generative Adversarial Network
PSD	Power Spectral Density
PSNR	Peak Signal-to-Noise Ratio
PWSCI	Public Wireless Supply Chain Innovation
QLoRA	Quantized LoRA
QoE	Quality of Experience
QoPE	Quality of Personal Experience
QoS	Quality of Service
Qualcomm	Qualcomm Incorporated
RAG	Retrieval-Augmented Generation
RAN	Radio Access Network
RATs	Radio Access Technologies
RCS	Radar Cross-Section
RealNVP	Real-valued Non-Volume Preserving
ReLU	Rectified Linear Unit
REPL	Read-Eval-Print Loop
RF	Radio Frequency
RFMLS	Radio Frequency Machine Learning Systems
RGB	Red, Green, Blue
RIS	Reconfigurable Intelligent Surface
RL	Reinforcement Learning
RLC	radio link control
RLHF	Reinforcement Learning from Human Feedback
RME	Radiomap Estimation
RME-GAN	Radiomap Estimation Generative Adversarial Network
RMSProp	Root Mean Square Propagation
RNNs	Recurrent Neural Networks
ROC	Receiver Operating Characteristic
ROHC	Robust Header Compression
RRC	Radio Resource Control
RSS	Radio Signal Strength
RSRP	Reference Signal Received Power
RSSI	Received Signal Strength Indicator
RTX	Real-Time Ray Tracing
Rx	Reception
SAGA	Spectrum Augmentation/Adaptation with GAN
SBA	Service-Based Architecture

SCAFFOLD	Stochastic Controlled Averaging for Federated Learning
SD	SCAFFOLD
SDAP	Service Data Adaptation Protocol
SDE	Stochastic Differential Equation
SDN	Software-Defined Networking
SEC	Securities and Exchange Commission
SEI	Specific Emitter Identification
SemCom	Semantic Communication
SFC	Service Function Chaining
SGM	Score-based Generative Model
SHF	Super-High Frequency
SL	SDA-FL
SLA	Service-Level Agreement
SLO	Service-Level Objective
SMF	Service Management Function
SNR	Signal-to-Noise Ratio
SOC	Security Operations Center
SOI	Signal of Interest
SON	Self-Organizing Network
SPADE	Spatially Adaptive Normalization
SSL	Self-Supervised Learning
SSIM	Structural Similarity Index Measure
SVHN	Street View House Numbers
SVM	Support Vector Machine
SVD	Singular Value Decomposition
TAI	Trustworthy AI
TAN	Terahertz Airborne Network
TCP	Transmission Control Protocol
TDD	Time-Division-Duplex
TDL	Tapped Delay Line
TDMA	Time-Division Multiple Access
THz	Terahertz
TOPS	Tera Operations Per Second
TPU	Tensor Processing Unit
TSMC	Taiwan Semiconductor Manufacturing Company
TV	Total Variation
Tx	Transmission
UAVs	Unmanned Aerial Vehicles
UDP	User Datagram Protocol

UE	User Equipment
UFed-GAN	Unsupervised Federated GAN
UGC	User-Generated Content
UHF	Ultra-High Frequency
ULA	Uniform Linear Array
U-Net	U-shaped Network
UPF	User Plane Function
URL	Uniform Resource Locator
URLLC	Ultra-reliable Low-latency Communications
USENIX	USENIX Association
V2I	Vehicle-to-Infrastructure
V2V	Vehicle-to-Vehicle
V2X	Vehicle-to-Everything
VAE	Variational Autoencoder
VAEAC	VAE with Arbitrary Conditioning
VAEs	Variational Autoencoders
VANETs	Vehicular Ad Hoc Networks
VPU	Vision Processing Unit
VQ-VAE	Vector-Quantized Variational Autoencoder
VQ-VAEs	Vector Quantized VAEs
VRAM	Video Random Access Memory
vRAN	virtualized Radio Access Network
VR	Virtual Reality
WAEs	Wasserstein Autoencoders
WGAN-GP	Wasserstein GAN with Gradient Penalty
WGANs	Wasserstein GANs
WiFi	Wireless Fidelity
xAI	Explainable AI
XL-MIMO	Extremely Large-scale Multiple-Input-Multiple-Output
XLSX	Excel Open XML Spreadsheet
XML	eXtensible Markup Language
XR	Extended Reality
YOLO	You Only Look Once

1

Future AI-empowered Communications Systems

Nguyen Van Huynh[1,*], *Thien Huynh-The*[2], *and Quoc-Viet Pham*[3]

[1] *School of Computer Science and Informatics, University of Liverpool, Liverpool, United Kingdom*
[2] *Department of Computer and Communications Engineering, Ho Chi Minh City University of Technology and Education, Ho Chi Minh City, Vietnam*
[3] *School of Computer Science and Statistics, Trinity College Dublin, Dublin, Ireland*

*Corresponding author: huynh.nguyen@liverpool.ac.uk

1.1 Fundamental Background of Future Communications Systems

1.1.1 Overview of Future Communications Systems

Over the last few years, we have witnessed the rapid development of various emerging applications that are expected to change every aspect of our lives, such as the Internet of Everything (IoE), augmented/virtual reality (AR/VR), and autonomous vehicles. These applications can create a massive volume of traffic with diverse requirements on security, latency, and robustness. A report from Ericsson forecasts that the global mobile data traffic can reach 473 exabytes per month in 2030 [1]. Unfortunately, current communication systems (e.g., 5G) may be unable to accommodate these requirements effectively. For example, the next generation of AR/VR, such as holographic teleportation, is expected to have Tbps-level data rate and microsecond-level latency demands to ensure its Quality of Service (QoS) and Quality of Experience (QoE) [2]. In addition, as we are moving toward the era of automated and intelligent ecosystems, existing communication systems are unable to fulfill their requirements for the convergence of communication, sensing, control, computing, and intelligence capabilities [3, 4]. It is, therefore, essential to develop new communication systems that can meet the demanding requirements of these emerging applications in the future.

Generative AI for Communications Systems: Fundamentals, Applications, and Prospects,
First Edition. Edited by Diep N. Nguyen et al.

5G Advanced and 6G have recently been introduced as promising solutions to fulfill these emerging requirements. In particular, 3GPP started working on 5G Advanced from Release 18 with the main focus of further expanding and extending the current 5G capabilities and use cases [5, 6]. In 5G Advanced, the communication performance will be significantly improved by utilizing massive MIMO systems with enhanced demodulation reference symbols. Moreover, AI/ML will be the essential component of future communication systems for intelligent management services to optimize the system performance in terms of energy saving, load balancing, network slicing, and mobility optimization. Moreover, AI/ML will play an important role in several emerging applications such as AR/VR, indoor positioning, and ubiquitous NR IoT coverage, which are expected to be supported in 5G Advanced. Finally, NTN will also be integrated into 5G Advanced to further enhance its communication coverage and enable more exciting applications [7]. Beyond 5G Advanced, 6G aims to make the communication performance even better with a wide range of new applications for the future. In November 2023, the International Telecommunication Union's Radiocommunication sector (ITU-R) introduced IMT-2030 Framework [8], which is expected to be a fundamental milestone for a global 6G standard. Specifically, the IMT-2030 Framework identifies 15 capabilities for 6G technology, including enhanced 5G capabilities such as security and resilience, high data rate, and spectrum efficiency. In addition, it also envisions several new capabilities for 6G such as coverage, sustainability, sensing, AI-native, and interoperability. Various use cases for 6G are also introduced in this framework, including immersive communication, hyper-reliable and low-latency communication, ubiquitous connectivity, massive communication, and integrated sensing and communication [9]. To accommodate these visions and use cases, 5G Advanced and 6G are expected to utilize several promising technologies and designs such as Terahertz band communications, extremely large multiple-input-multiple-output (MIMO), non-terrestrial network (NTN), quantum internet, intelligent reflecting surface (IRS), and open radio access network (ORAN), which will be discussed in detail in the following.

1.1.1.1 Terahertz Band Communications

To meet the exponential growth in data traffic of emerging applications, Terahertz (THz) band communication, ranging from 0.1 to 10 THz, has been considered a promising technology for the next decade [2]. As illustrated in Figure 1.1, the THz band has a massive amount of available bandwidth, which can be up to several THz and much higher than that of mmWave frequencies. As a result, the THz band can offer data rates of up to several terabits per second, allowing it to handle the massive data traffic of future applications with ease [10]. Although operating at very high frequencies, the THz waves can still penetrate thin objects, enabling non-line-of-sight communications. Moreover, the THz band has been widely used

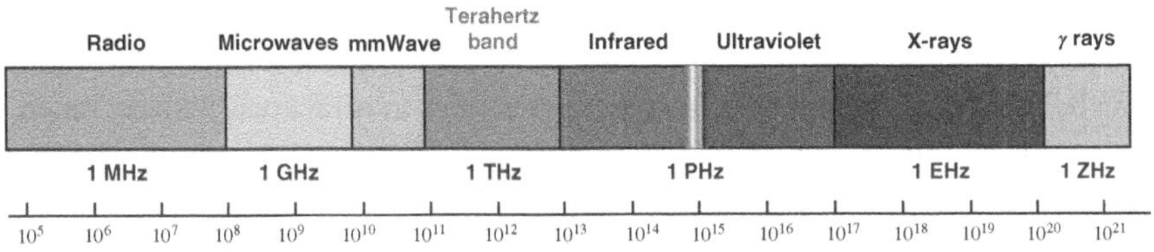

Figure 1.1 Terahertz band provides more spectrum resources for wireless communications [2] / IEEE / CC BY 4.0.

for applications such as imaging, positioning, and sensing [11, 12], making it well fit with the vision of future communication systems where multiple functionalities will be equipped besides communications. With these promising characteristics, the THz band is expected to be a key technology in future communication systems (e.g., 6G).

1.1.1.2 Massive and Extremely Large-scale Multiple-input-multiple-output

MIMO and massive MIMO technologies have been studied extensively to improve the capacity and reliability of wireless systems by using a large number of antennas at base stations and/or wireless users [13]. However, as mentioned, the next-generation wireless networks, such as 5G advanced and 6G, are envisioned to meet the increasing demand of massive data traffic, ultra-low latency, and extremely low bit error rate [14]. Existing MIMO technologies may not effectively accommodate these requirements due to their design limitations [15]. To this end, extremely large-scale MIMO (XL-MIMO) has emerged as a promising technology. The key idea of XL-MIMO is to employ very large antenna arrays which contain at least several hundreds or even thousands of antennas [15]. There are several antenna array designs for XL-MIMO that have been investigated, such as holographic MIMO [16], continuous aperture MIMO [17], and extremely large antenna array [18]. By using a large number of antennas, XL-MIMO not only can significantly improve beamforming gain and spectral efficiency but also can enhance the array spatial resolution. Due to these advantages, XL-MIMO is expected to be an enabler for a wide range of applications such as augmented/virtual/mixed reality, holographic display, and high-accuracy localization and sensing [15].

1.1.1.3 Non-terrestrial Networks

The next generation of communication systems is expected to provide global coverage, enabling seamless and reliable communications in remote regions or during disaster events where existing communication systems are unable to be established due to their terrestrial infrastructures and lack of flexibility [19]. To fulfill this vision, non-terrestrial communication stations such as unmanned aerial vehicles (UAVs), high altitude platforms (HAPs), and satellites (including geostationary earth orbit (GEO), medium earth orbit (MEO), and low earth orbit (LEO)) can be used to quickly provide Internet connections to rural areas without relying on terrestrial infrastructures [20], as illustrated in Figure 1.2. For example, UAVs can be deployed to build temporary communication networks to support rescue missions in disaster situations while HAPs and satellites can provide reliable connections to a massive number of users on Earth with high bandwidth and low latency. By integrating with terrestrial communications systems, NTN can ensure

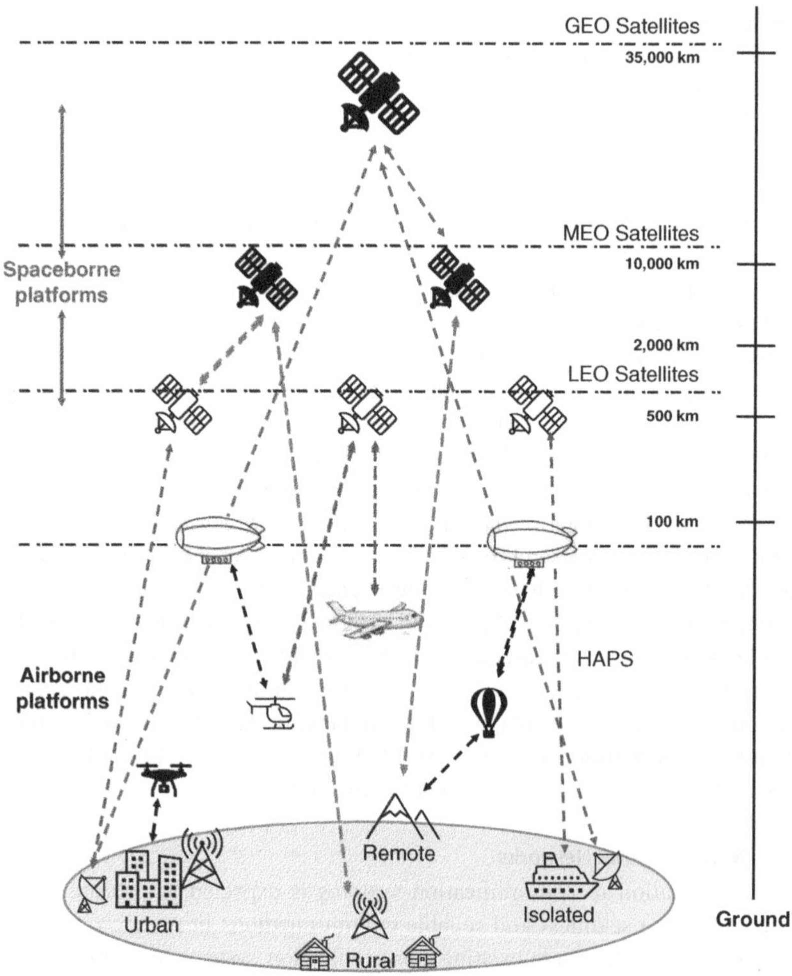

Figure 1.2 Overview of NTN. Adapted from [21].

the service continuity, service ubiquity, and service scalability of future communication systems [22].

1.1.1.4 Quantum Internet

Quantum computing has the great potential to solve problems that are intractable for conventional computing systems, such as climate change, chemical simulations, financial modeling, and cryptography [23, 24]. The key fundamental of quantum computing is the superposition and entanglement principles of quantum bits (qubits) where they can be in multiple states simultaneously and have correlations with each other despite their distances. Future communication systems can benefit from quantum computing by connecting multiple quantum computers to solve larger-scale problems distributed on quantum clusters [25]. To make quantum computers work together, qubits are required to be transmitted or "teleported" from one to another. Unfortunately, current communication infrastructures are not capable of exchanging qubits. As such, *quantum internet* has been proposed to realize the vision of quantum communications [23, 26, 27].

As constrained by physical laws, qubits cannot be copied or amplified [24, 28]. Hence, it is impossible to physically transmit the particles storing qubits. To overcome this challenge, *quantum internet* leverages the entanglement principle of quantum computing to transmit or "teleport" quantum states from the source to the destination, as illustrated in Figure 1.3. In particular, an entanglement generator can be used to generate two entangled qubits (i.e., particle A and particle B) and distribute each of them to the source and the destination. When the source wants to transmit a particular qubit state, it entangles this qubit with particle A and performs the Bell-state measurement on them. This results in two information bits, which will be sent to the destination over a classical channel. Finally, the received measurement will be processed with particle B to reconstruct the selfsame state. This *quantum internet* architecture has been widely studied recently and is expected to become popular in future communication systems, allowing us

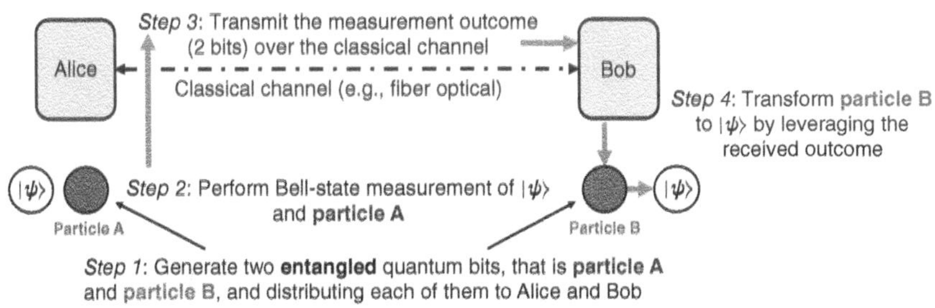

Figure 1.3 Overview of quantum internet. Adapted from [23].

to better leverage the power of quantum computers in many aspects of our lives [23, 26, 27, 28].

1.1.1.5 Intelligent Reflecting Surface

Another technology that can help to improve communication performance in terms of data rate and robustness is intelligent reflecting surface (IRS). Unlike existing communication systems where channel environments are uncontrollable and modeled by probability [29], IRS can enable intelligent communication networks with controllable channel environments. As illustrated in Figure 1.4, a typical IRS is a planar antenna array that consists of low-cost passive reflecting elements. These elements can be controlled by the control circuit board to intelligently change the phase, frequency, amplitude, and polarization of incident signals to enhance the signal strength, suppress interference, and address blockage issues [29–31]. Consequently, IRS can significantly improve can significantly improve the transmission rate, reliability, and robustness of future communication systems. Moreover, the deployment cost of IRS systems is low since they use low-cost and lightweight passive reflecting elements. More importantly, IRS has high compatibility with existing communication networks with the only change in the network protocol [29], making it a promising technique for communication systems in the near future.

1.1.1.6 Open Radio Access Network

Open radio access network (ORAN) has recently emerged as a promising technology to revolutionize the future telecommunication ecosystem. In particular, ORAN aims to disaggregate the radio access network into components which are

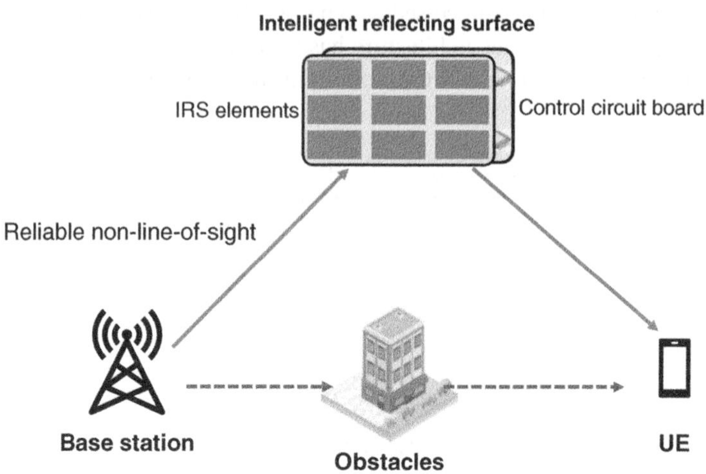

Figure 1.4 Overview of intelligent reflecting surface systems. Adapted from [32].

interconnected by open and interoperable interfaces, as depicted in Figure 1.5. In this way, the telecommunication system can be built with parts from multiple vendors. With ORAN, network operators have more flexibility in choosing vendors for components in the RAN network instead of using a single vendor for end-to-end supply [33]. This can help to enable low-cost and flexible deployments, increase the resilience and reconfigurability of the RAN, and avoid vendor lock-in to select best-of-breed components from different suppliers. By integrating artificial intelligence (AI)/machine learning (ML) into ORAN, mobile network operators can further enhance and optimize the performance of future communication systems, enabling the development of self-organizing and fully autonomous networks [34].

1.1.2 Key Challenges and Research Trends

Although having great potential to realize the vision of future communication systems, these technologies and designs face several challenges that must be fully addressed to ensure their readiness and effectiveness in practice.

- **Spectrum Scarcity:** As mentioned, future communication systems are expected to serve a huge number of wireless devices, leading to congestion in the available radio frequency spectrum. For that, novel dynamic spectrum

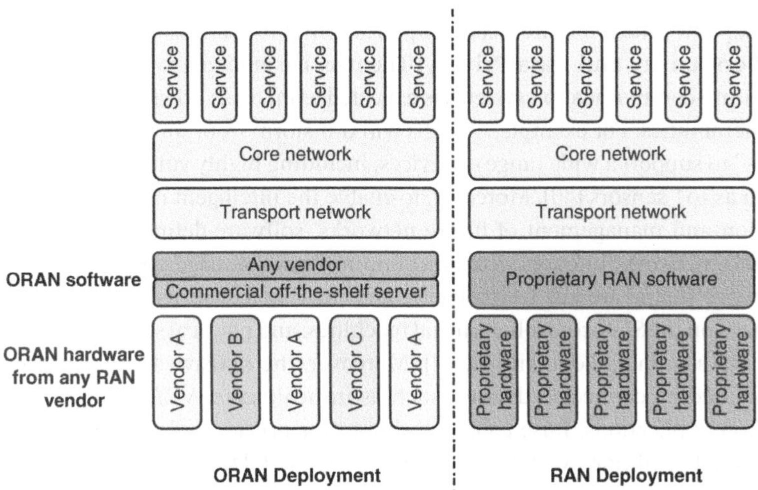

Figure 1.5 Architecture of ORAN [35] / IEEE / CC BY NC ND 4.0.

access schemes are required to intelligently utilize and allocate the existing spectrum resources to different types of users and applications. In addition, ultra-high frequency bands like THz can offer a huge available bandwidth to accommodate these emerging services. However, their unique characteristics introduce new and pressing challenges that have never been encountered in lower frequencies, such as severe spreading loss, high channel sparsity, molecular absorption loss, and limited advancements in THz band digital processors and frequency synchronizers [36]. This demands new spectrum access approaches designed specifically for THz band communications, for example, multi-band-based spectrum allocation scheme [37].

- **Complexity:** Integrating various technologies, such as 6G, satellite communications, quantum internet, and IRS, makes future communication systems complex and heterogeneous. This poses intricate challenges to network operators to ensure seamless coordination and communication between different entities [3, 38]. Moreover, with a vast number of devices having diverse requirements (e.g., IoT sensors, autonomous vehicles, and VR/AR devices) and diverse spectrum bands (e.g., sub-6GHz, millimeter wave (mmWave), and THz), network management will be a challenging task in future communications systems [39]. Several advanced technologies have been proposed to deal with these issues such as software-defined networking, network slicing, and network function virtualization [38]. However, further efforts are required to enable autonomous and intelligent network management to better support future communication applications and services.
- **Cybersecurity and Data Privacy:** The massive emergence of connections in future communication systems is expected to increase privacy and security vulnerabilities. For example, 6G cells will transform from small cells to "tiny cells" to support a wide range of devices, including highly vulnerable devices such as IoT sensors [40]. Moreover, to enable the intelligent network orchestration and management of future networks, software-defined networking (SDN), network function virtualization, multi-access edge computing, and network slicing are key technologies, and thus their security issues such as attacks on the SDN controller, virtual machines, and physical security threats, remain a primary concern [40, 41]. Moreover, the massive amount of data generated by connected devices, such as in healthcare, VR/AR, and localization applications, raise concerns about user privacy and data protection [41–43]. In addition, with the advancements in digital technologies like quantum computing and AI, adversaries will become more powerful in performing intelligent cyber attacks. Unfortunately, post-quantum cryptography and

quantum internet are still in their early stage, and thus more research efforts are needed to deal with these emerging security and privacy threats in future communication systems.

- **Energy Efficiency:** The gigantic increase in future network traffic and the deployment of a massive number of cells, edge computing nodes, and connected devices are expected to introduce an enormous computational burden on data processing systems, resulting in extremely high energy consumption. In addition, potential key technologies in future communication networks, such as massive and XL-MIMO and THz communication, can consume a significant amount of energy to operate due to their complex design and unique characteristics [3, 44, 45]. To meet sustainability goals and ensure a green future for the next generations, balancing performance requirements with the demand to reduce greenhouse gas emissions and minimize the ecological footprint of future network infrastructures is an essential task.
- **Latency and Reliability:** Although future communication systems are envisioned to provide ultra-low latency and high-reliability communications to diverse applications, there is a need for new designs and techniques to realize this vision in practice. Specifically, achieving ultra-low latency and high-reliability metrics across heterogeneous environments and conditions with a massive number of devices is a significant technical challenge that current optimization and management approaches may not be effective in dealing with [46]. Moreover, efficient coordination of resources (e.g., radio, computing, and storage) across terrestrial, aerial, and satellite networks is vital to guarantee the end-to-end performance of future applications [47]. This demands lightweight yet effective and intelligent resource allocation and operation coordination approaches. It is worth noting that future communication systems are expected to operate based on various advanced technologies to meet the pressing demands of future applications and services. However, existing communication and networking approaches may not be fully applicable to these new technologies and designs to deliver ultra-low latency and high-reliability performance, especially in the physical layer. For instance, massive and XL-MIMO systems will need new methods for their operation such as precoding, channel estimation, and signal detection due to their complex designs and unique characteristics that have not been seen in existing communication systems [15]. NTNs will also need novel management and coordination mechanisms to ensure high communication performance while taking into account the special features and challenges of satellite communications.

1.2 AI-powered Communication Enablers

1.2.1 Deep Learning-based Approaches

Deep learning (DL) has revolutionized the field of communication systems, offering a data-driven approach to address complex challenges. By exploiting the power of neural networks (NNs), DL enables the development of intelligent systems capable of learning from data and making adaptive decisions. Besides introducing the foundations of DL in communication systems, this part discusses its applications in physical layer optimization, resource allocation, network performance enhancement, and integrated sensing and communication (ISAC).

1.2.1.1 Foundations of Deep Learning in Communication Systems

Deep learning, a subset of AI and ML, has revolutionized the design and optimization of communication systems. DL leverages multi-layered NNs to extract complex, high-dimensional patterns from raw data, thus making it highly effective for addressing the inherent complexities of modern communication systems. Especially, by learning representations directly from raw data, DL outperforms traditional model-driven approaches, accordingly enabling the design of systems that are adaptive, scalable, and capable of handling non-linear relationships in communication tasks [48].

The fundamental building blocks of DL are NNs, which consist of interconnected layers of artificial neurons. Each neuron performs a weighted sum of its inputs followed by a non-linear activation function, which allows the network to approximate complex functions. In communication systems, these networks are leveraged for tasks such as channel estimation, modulation classification, and signal detection. Some advanced architectures like convolutional neural networks (CNNs) and recurrent neural networks (RNNs) further enhance the capability of DL by incorporating spatial and temporal correlations, respectively [49]. CNNs, originally designed for image processing and computer vision, have found extensive applications in communication systems for their ability to comprehensively capture spatial features. For example, CNNs can process spectrograms or other time-frequency representations of signals to perform modulation classification or spectrum sensing. By applying convolutional operations, CNNs effectively reduce the dimensionality of the data while preserving essential features to be suitable for resource-constrained environments. RNNs and their variants, such as long short-term memory (LSTM) networks, are well-suited for tasks involving sequential data, such as decoding and predicting signal patterns. These networks maintain a memory of previous states for consequently seizing temporal dependencies in time-series data. For instance, RNNs are widely used for channel prediction and adaptive coding schemes in scenarios where signal behavior changes over time.

Recent advancements in DL have introduced transformer models, which deploy self-attention mechanisms to process sequences of data in parallel. Transformers have presented some remarkable results in tasks such as natural language processing and are now being adapted for communication systems, particularly in scenarios that require large-scale processing, such as network traffic analysis and resource allocation [50]. In a nutshell, compared to traditional ML approaches, DL provides several key advantages in communication systems:

- **Data-driven modeling:** Unlike traditional methods that rely on predefined mathematical models, DL learns directly from data, allowing it to handle complex, dynamic environments.
- **Adaptability:** DL models can be retrained and fine-tuned to adapt to changing network conditions or new use cases.
- **Scalability:** DL architectures can be scaled to process large datasets and accommodate the high-dimensional nature of modern communication scenarios.

In fact, the adoption of DL in communication systems is a transformative step by offering the potential to optimize system performance and support the demands of next-generation networks [51]. However, it also reveals some essential challenges, such as high computational complexity and the need for large labeled datasets, which are actively being addressed by ongoing research.

1.2.1.2 Applications in Physical Layer Optimization

DL has proven to be a powerful tool for optimizing the physical layer (PHY) of communication systems, which includes tasks such as channel estimation, signal detection, and waveform design. By exploiting the capability of DL to model complex, non-linear relationships, these applications address the limitations of traditional signal processing techniques, particularly in environments with high noise, interference, or rapidly changing conditions.

Channel Estimation and Equalization: One of the key areas where DL has shown significant potential is in channel estimation and equalization. Conventional approaches often rely on predefined models that are limited in flexibly adapting to varying channel conditions, such as multipath fading or non-line-of-sight (NLoS) scenarios. DL-based methods, such as CNNs and autoencoders, learn directly from raw channel data, thus allowing for robust estimation and equalization. For example, NNs trained on simulated or real channel data can predict channel state information (CSI) with high accuracy, even in challenging scenarios like massive MIMO or mmWave communications [52].

Beamforming and MIMO System Design: DL enhances beamforming and MIMO system design by automatically selecting optimal beam patterns and antenna configurations. RNNs and deep reinforcement learning (DRL) have been considered for dynamically optimizing predefined beamforming strategies in real-time conditions. These learning-based methods outperform traditional optimization techniques by adapting to changes in user mobility, interference, and other environmental factors, thereby improving spectral efficiency and reducing energy consumption. In Ref. [53], a novel data-driven DL framework is proposed for spectral-efficient broadband multi-user hybrid beamforming in aerial massive MIMO-OFDM systems, enabling joint optimization of all key transmission modules with implicit CSI, reduced pilot/feedback overhead, and adaptation to practical low-resolution phase shifters through transfer learning.

Waveform Design and Modulation Recognition: DL facilitates the design of waveforms for specific applications, in which the trade-offs between spectral efficiency and robustness against noise and interference are balanced effectively. Generative adversarial networks (GANs) and deep unfolding techniques have been exploited to create optimized waveforms to perform effectively and robustly under varying communication environments. Similarly, modulation recognition tasks, important to cognitive radios and dynamic spectrum sensing, benefit from CNNs and transformers, which automatically identify modulation schemes with high accuracy using spectrograms or other signal representations, such as in-phase and quadrature data. For instance, MONet [54], a 3D convolutional neural network, leverages cuboidal filters to effectively capture intra- and inter-antenna correlations in high-dimensional data arrays derived from MIMO-OFDM signals, thus enabling robust and accurate modulation classification under various channel conditions.

Signal Detection and Error Correction: In signal detection, DL-based detectors, such as those employing LSTMs or gated recurrent units (GRUs), have demonstrated resilience in handling inter-symbol interference and complex noise profiles. Furthermore, DL models are being used to optimize error correction codes, such as turbo and low-density parity-check codes, by learning decoding algorithms that achieve near-optimal performance while reducing computational complexity. For example, in Ref. [55], a turbo-style multi-carrier auto-encoder (MC-AE) for OFDM systems is developed to flexibly integrate with any channel coding scheme while achieving iteration gains, by independently designing the MC-AE and leveraging its bit-dependency for extrinsic information exchange.

1.2.1.3 Resource Allocation and Scheduling

Efficient resource allocation and scheduling are critical to optimize communication system performance, particularly in the context of next-generation networks like 5G and 6G. These tasks involve managing resources such as spectrum, power, time slots, and antennas to maximize network efficiency, reliability, and user satisfaction. Traditional optimization techniques, while effective, are often restricted by the high complexity and poor adaptation to modern communication environments. DL has offered advanced data-driven solutions to overcome these challenges.

Spectrum and Power Allocation: DL models have recently been exploited to monitor and allocate spectrum and power dynamically in multi-user and multi-cell networks. Some architectures with specific learning mechanisms such as CNNs and LSTMs incorporated with reinforcement learning (RL) enable systems to learn optimal allocation policies directly from raw data, consequently adapting to variations in user demands, interference, and propagation conditions. Notably, DL-based spectrum allocation can intelligently predict underutilized or unused frequency bands to reduce interference and enhance spectral efficiency. In Ref. [56], an efficient deep model for spatiotemporal spectrum usage prediction in dynamic spectrum allocation is introduced by leveraging CNNs and ResNets for accurate predictions with minimal sensing cost by combining DL with nearest neighbor interpolation to effectively handle sparse sensor data and outperform existing CNN and ConvLSTM models. Similarly, RL algorithms optimize power allocation to obtain a good trade-off between energy efficiency and QoS.

Time Slot Scheduling: In time-division multiple access (TDMA) or similar systems, scheduling time slots for data transmission is a complicated issue due to the dynamic nature of traffic patterns. RNNs and transformers, which are proficient in capturing temporal dependencies, are leveraged to accurately estimate traffic loads and accordingly allocate time slots. Ref. [57] proposes an effective DL-based real-time scheduler for stochastic edge-cloud environments using an asynchronous-advantage-actor-critic framework integrated with residual recurrent neural network architecture, which allows decentralized learning across multiple agents and efficiently captures temporal workload patterns for adaptive scheduling to significantly improve energy consumption, response time, service-level agreement adherence. Based on the real-time scheduling decisions made by these DL-based methods, the throughput and minimize latency of systems are improved robustly.

Network Slicing and Virtualization: As a key feature of 5G and beyond, network slicing, wherein resources are partitioned to create virtualized network slices tailored for specific applications (e.g., IoT, autonomous vehicles, and high-bandwidth streaming), is upgraded comprehensively thanks to the capability of DL in learning complex patterns of resource requirements. DL-based models dynamically allocate resources to various slices based on predicted demand, thus ensuring QoS while maintaining overall network efficiency. Some learning techniques like federated learning (FL) and split learning further enhance this process by facilitating decentralized resource management across distributed edge devices. For instance, the work [58] studies a cutting-edge DL-based intelligent network application for network slicing in ORAN, with an intelligent and automated slicing architecture illustrated in Figure 1.6, thus enabling intelligent and automated network slicing for IoT services with guaranteed service level agreements through a near real-time RAN intelligent controller.

1.2.1.4 Network Performance Enhancement

The rapidly expanding complexity of modern communication networks, driven by the increasing number of connected devices and diverse application requirements, has necessitated network performance enhancement. In this context, DL offers innovative solutions to address challenges in traffic prediction, load balancing, fault management, and anomaly detection, enabling more intelligent, adaptive, and efficient networks.

Traffic Prediction and Management: Accurate traffic prediction is essential for proactive resource allocation and congestion management. DL models with such RNN and LSTM architectures are competent to capture temporal dependencies in network traffic patterns. These models are employed to potentially forecast traffic loads in real-time conditions, thus allowing operators to optimize resource utilization, reduce latency, and ensure a high QoS. For example, in dense urban networks, the work [59] studies a novel DL method for cellular traffic prediction that addresses limitations in existing approaches by incorporating a non-local

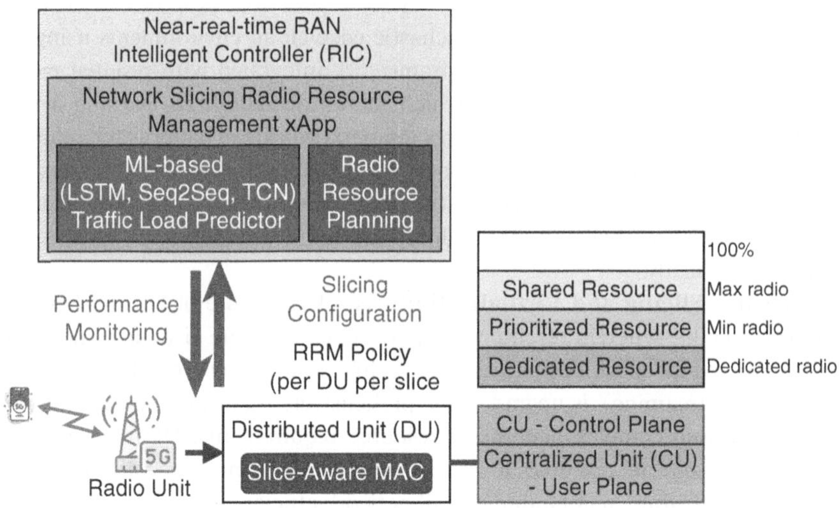

Figure 1.6 An overview of intelligent and automated RAN slicing architecture deployed in Ref. [58]. Adapted from [58].

graph-based attention mechanism model to capture dynamic non-local spatial correlations and a calibration layer to effectively fuse and utilize spatio-temporal features with varying granularities.

Dynamic Load Balancing: DL algorithms are recently deployed for dynamic load balancing strategies in multi-access edge computing (MEC) and core networks. By analyzing real-time data, NNs can identify underutilized and overburdened nodes to redistribute tasks to maintain balanced workloads. CNNs and autoencoders are usually deployed to detect imbalances and recommend optimal routing paths, accordingly improving network stability and efficiency. These approaches are particularly beneficial in scenarios with varying user densities, such as during large-scale events or in disaster-stricken areas.

Fault Detection and Anomaly Management: Maintaining network reliability requires rapid identification and resolution of faults and anomalies. DL models, particularly autoencoders in combination with unsupervised learning techniques, present some remarkable achievements in detecting subtle anomalies that may cause network failures. In Ref. [60], an integrated learning approach of a modified adversarial autoencoder for data augmentation and a cost-sensitive support vector machine for classification is deployed to improve detection performance by mitigating bias and considering the varying costs of misclassification. The whole sleeping cell detection of this learning approach is illustrated in Figure 1.7, which includes measurement, data processing, and sleeping cell detection. These models learn patterns in network behavior and flag deviations, thus allowing operators to address issues proactively. By deploying some cutting-edge deep architectures for anomaly detection models, hardware failures, or cyberattacks can be predicted early and accurately to reduce downtime and enhance security accordingly.

Intelligent QoS Provisioning: Ensuring QoS in heterogeneous and dynamic networks is a challenging task. By utilizing predictive analytics and classification techniques, recent DL models can adaptively allocate resources to meet diverse QoS requirements for specific applications like real-time video streaming, IoT communications, and AR. By simultaneously learning historical and real-time data, these models can prioritize critical traffic and dynamically adjust policies to meet user expectations [61].

1.2.1.5 Integrated Sensing and Communication

ISAC has emerged as a transformative paradigm in next-generation wireless networks, particularly 6G, by combining communication and sensing functionalities into a unified system. This integration enables the shared use of spectrum, hardware, and computational resources, thus significantly improving system efficiency and reducing costs. ISAC potentially revolutionizes various applications ranging

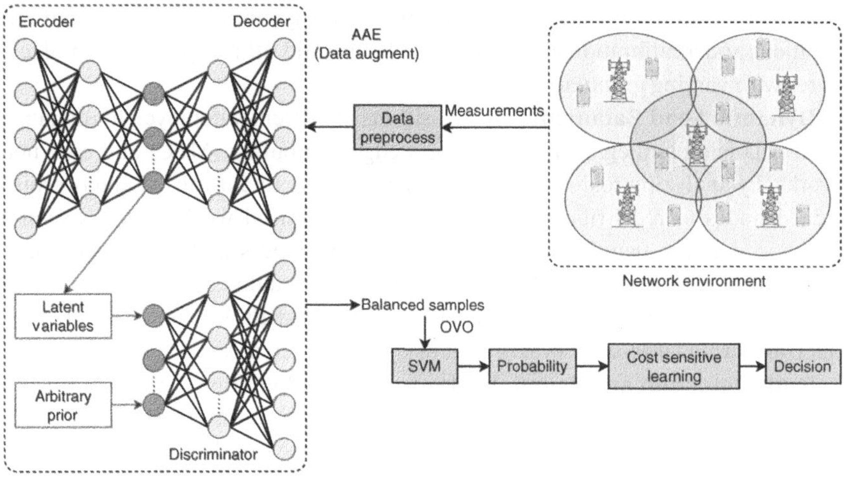

Figure 1.7 An overview of the sleeping cell detection in Ref. [60] / IEEE.

from autonomous vehicles and smart cities to healthcare and industrial automation by providing simultaneous high-speed data transfer and precise environmental sensing [62].

Principles of ISAC: Fundamentally, ISAC leverages a shared infrastructure to simultaneously perform communication and sensing tasks. Unlike traditional systems that handle these functions independently, ISAC systems design waveforms, beamforming patterns, and resource allocation strategies to optimize both functionalities. For instance, a single transmitted signal in an ISAC system should provide high-quality communication while capturing useful information about the environment, such as object detection or motion tracking. This dual-purpose capability is achieved through advanced signal processing techniques and the integration of advanced learning algorithms.

Key Techniques Empowered by DL Used in ISAC: DL has become a cornerstone for advancing ISAC systems, providing tools to address challenges and optimize both sensing and communication functionalities [63]. The following are some of the most prominent DL-based techniques used in ISAC:

- **Signal Classification and Target Detection:** DL models, particularly CNNs and RNNs, have shown noteworthy performance in signal identification and modulation classification tasks, thus allowing ISAC systems to differentiate between communication signals and environmental reflections. These models are also deployed for radar-based target detection and object classification to offer precise situational awareness.

- **Beamforming Optimization:** Advanced NNs and RL models dynamically optimize beamforming patterns to further achieve efficient energy consumption while maintaining high sensing accuracy and communication quality. These techniques are particularly beneficial in dynamic environments like vehicular and UAV-aided wireless networks.
- **Waveform Design:** GANs and autoencoders can be considered for designing dual-purpose waveforms that simultaneously meet the demands of sensing and communication functionalities. The ultimate goal of waveform design for ISAC systems is to balance trade-offs like spectral efficiency and detection accuracy.
- **Resource Allocation:** Advanced DRL frameworks can optimize the allocation of shared resources, such as spectrum and power, between sensing and communication tasks to dynamically adapt to real-time network conditions, thereby minimizing interference and maximizing overall network efficiency.
- **Environmental Sensing and Localization:** RNNs and LSTM are commonly studied to effectively analyze sequential data, hence facilitating continuous environmental sensing and user localization. Especially, by processing temporal patterns, these models are capable of improving the accuracy of object tracking and position estimation, which are essential components of autonomous systems and IoT applications.
- **Interference Mitigation:** By leveraging autoencoders and adversarial training mechanisms, the interference between sensing and communication signals can be separated and mitigated significantly. In addition to signal clarity enhancement, DL can improve the overall robustness of the system.

1.2.2 Reinforcement Learning-based Approaches

1.2.2.1 Basics of Reinforcement Learning for Communications

In principle, RL is an ML paradigm designed for sequential decision-making tasks, where an agent learns optimal policies by interacting with an environment. In the context of communication systems, RL is particularly effective in solving complex and dynamic problems that require adaptive and real-time decision-making. Unlike supervised learning, which relies on labeled data, RL enables an agent to learn through trial-and-error, ruled by a reward signal that reflects the success of its actions [64]. In a typical RL framework, some key concepts should be taken into consideration:

- **Agent, Environment, and Actions:** In RL, the agent represents the decision-making entity, such as a communication node, base station, or user device, while the environment includes all external factors affecting the agent, such as network traffic, channel conditions, and interference. The agent interacts with the environment by taking actions, such as allocating power, selecting a frequency band, or scheduling transmissions.

- **States and Rewards:** The environment's state is a representation of its current condition, such as the channel quality or user demand. The agent observes the state and receives a reward signal based on the immediate effect relying on its actions. For example, a successful packet delivery or reduced interference may result in a positive reward.
- **Policy and Value Functions:** A policy defines the agent's strategy for choosing appropriate actions in a given state. Value functions should estimate the expected long-term reward of states (or state-action pairs), regulating the agent to maximize cumulative rewards over time.
- **Exploration Versus Exploitation:** RL algorithms must balance exploration (trying new actions to discover their effects) and exploitation (choosing actions known to yield high rewards). This balance is important for communication systems, where dynamic environments necessitate continuous adaptation.

Based on their approach to learning and interacting with the environment, RL algorithms are broadly categorized into three groups as follows:

- **Model-free RL:** These algorithms, such as Q-learning and Deep Q-networks (DQN), do not require prior knowledge of the environment. They are widely used in scenarios where the system model is unknown or complex.
- **Model-based RL:** These algorithms build a model of the environment to predict outcomes of actions, which can improve efficiency in scenarios with predictable dynamics.
- **Multi-agent RL:** In multi-user or cooperative systems, MARL enables multiple agents to learn simultaneously, optimizing global network performance.

1.2.2.2 RL in Resource Allocation and Spectrum Management

As playing an important role in modern communication systems, resource allocation and spectrum management involve distributing limited resources, such as spectrum, power, and time slots—among users and applications to maximize overall system performance. RL provides an adaptive data-driven approach to address these challenges, especially in scenarios with high complexity and uncertainty.

Dynamic Spectrum Access and Allocation: In cognitive radio networks, RL is exploited to enable Dynamic Spectrum Access (DSA), where agents intelligently identify and manipulate underutilized frequency bands. Q-learning and DQN are popular choices for this task, wherein they allow the agent to learn optimal spectrum allocation policies by interacting with the environment. For instance, a novel distributed DSA framework [65] for multi-use MIMO networks using multi-agent RL is proposed, where multiple cells learn to optimize power allocation policies in a distributed manner to maximize average signal-to-noise ratio (SNR) while mitigating interference to incumbent networks, accordingly analyzing performance and demonstrating the efficacy of DQN and Deep Deterministic Policy Gradient (DDPG) algorithms with varying trade-offs in performance, complexity, and scalability.

Power Control and Interference Management: Recently, RL has presented some advantages in power control, where it enables communication nodes to adjust their transmission power dynamically to achieve energy efficiency while minimizing interference. Multi-agent reinforcement learning (MARL) is often employed in some specific scenarios with multiple users or devices competing for shared resources. By coordinating actions among agents, MARL algorithms can optimize power allocation to balance throughput, interference mitigation, and fairness. For instance, Ref. [66] introduces a novel deep multi-agent reinforcement learning-based power-management strategy, namely DMPM with the working diagram illustrated in Figure 1.8, for underwater wireless sensor networks that maximizes fair spatial reuse while maintaining network capacity and successfully demonstrating its effectiveness through numerical results and analysis of different reward functions.

Joint Optimization of Spectrum and Power: The collaboration between RL and DNNs allows for the joint optimization of spectrum and power allocation, addressing the intricate trade-offs between spectral efficiency, energy consumption, and QoS. For example, an innovative hierarchical multi-agent DRL-based

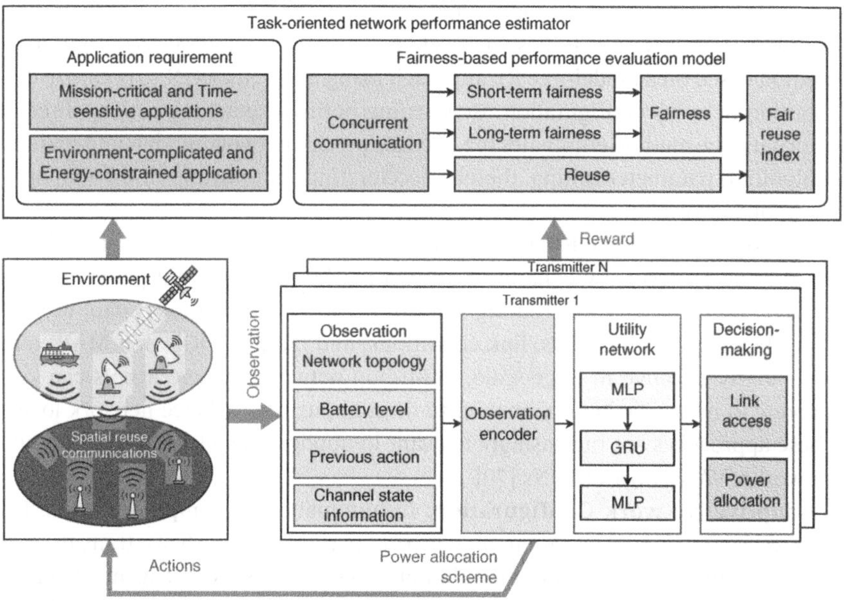

Figure 1.8 Working diagram of the fairness-based BMPM framework in Ref. [66] / IEEE.

joint partial spectrum sensing and power allocation scheme [67] is introduced for DSA, thus enabling secondary users to autonomously learn optimal SW and PA strategies in dynamic spectrum environments with reduced mutual interference, maximizing network throughput. Furthermore, proximal policy optimization and DDPG algorithms can be considered for optimizing resource allocation in complex environments, such as heterogeneous networks and vehicular communication systems [68].

Real-time Adaptation to Dynamic Environments: One of the major advantages of RL in resource allocation is its ability to adapt to dynamic and unpredictable network conditions. Traditional optimization techniques often require static assumptions or extensive computation, whereas RL models can learn and adjust policies in real time based on feedback from the environment. This capability allows RL to be particularly suitable for dynamic networks, where user mobility, interference, and channel conditions vary rapidly.

1.2.2.3 Network Optimization and Configuration

Network optimization and configuration are pivotal for ensuring efficient and reliable operations in modern communication systems. With the increasing complexity of networks in 5G and beyond, traditional heuristic and rule-based approaches struggle to handle dynamic traffic patterns, user demands, and environmental conditions. In this context, RL offers a promising alternative, leveraging its ability to learn optimal policies through interaction with the network environment. For example, the work [69] studies an effective distribution network reconfiguration method based on a model-free RL approach using NoisyNet DQN to optimize distribution network configuration. By learning optimal switching actions directly from data, the method eliminates the need for precise network models and manual exploration parameter tuning, thereby accelerating training and improving optimization performance.

Routing and Path Selection: RL has shown remarkable success in optimizing routing protocols by dynamically selecting the most efficient paths for data transmission. Unlike static routing algorithms, RL-based methods adapt to real-time changes in traffic loads, link conditions, and failures. DQN and MARL are particularly effective in large-scale, multi-hop networks, where agents can collaborate to minimize latency, maximize throughput, and balance network loads. These approaches are increasingly relevant in applications like vehicular ad hoc networks (VANETs) and SDNs [70].

Adaptive Network Configuration: In automatic and complex systems, RL enables adaptive configuration of network parameters, such as transmission power, channel selection, and modulation schemes. In SDN, RL agents dynamically adjust configurations based on traffic demands, interference levels, and QoS requirements. This flexibility enhances network efficiency and adaptability, especially in heterogeneous and dense environments [69]. Moreover, RL-based configuration is critical for next-generation wireless systems, where the networks need to self-optimize in real time to handle the diverse requirements of IoT, AR, and autonomous systems.

Self-healing Networks: One of the most promising applications of RL in network optimization is its role in self-healing networks. RL agents can detect faults, predict failures, and autonomously reconfigure network settings to maintain service continuity. For instance, autoencoders combined with RL can identify anomalies, accordingly, the RL agent can take corrective actions, such as rerouting traffic or reallocating resources, without human intervention. For instance, the work [71] proposes a novel distributed service function chaining (SFC) self-healing framework for edge-cloud infrastructures using a combination of RL-based SFC-to-datacenter association, iterative optimization of individual SFC healing decisions, and a combinatorial auction for resource allocation to ensure efficient and resilient service delivery across diverse stakeholder interests.

1.2.2.4 RL for Edge and Fog Computing

Edge and fog computing have emerged as essential paradigms for enabling low-latency, high-efficiency processing in modern communication systems. These frameworks facilitate edge computing, offloading computational tasks and data processing from centralized cloud servers to edge devices, thereby reducing network congestion and latency [72, 73]. However, managing the distributed and resource-constrained nature of edge and fog environments introduces significant challenges, such as task offloading, resource allocation, and energy efficiency. From this perspective, RL offers an adaptive, data-driven approach to optimize these systems in real time.

Task Offloading Optimization: In edge and fog computing, task offloading involves deciding whether to process a task locally, at the edge, or in the fog/cloud layer. RL algorithms, particularly DQN and proximal policy optimization, dynamically learn a complicated pattern of multiple offloading strategies that balance trade-offs between latency, energy consumption, and computational efficiency. For instance, an RL agent can optimize offloading decisions based on task size, network conditions, and resource availability, ensuring timely execution while minimizing energy use [74].

Resource Allocation and Scheduling: Efficient allocation of limited resources, such as CPU, memory, and bandwidth, is critical in edge and fog networks. MARL is particularly effective in these environments, where multiple devices or nodes operate collaboratively. RL agents can dynamically allocate resources to tasks based on priorities, user demands, and network conditions.

For example, in a smart transportation scenario, a novel vehicular fog computing resource allocation framework that combines a heuristic algorithm with RL to minimize service latency for vehicular applications by dynamically allocating resources to parked vehicles, accordingly leveraging real-time vehicle movement and parking data collected in a real world [75].

Mobility-aware Optimization: The dynamic nature of user and device mobility in edge and fog networks, thus demanding continuous adaptation to changing locations and connectivity conditions. RL-based approaches are aptly deployed to optimize task migration and resource reallocation in response to mobility. In particular, RL agents can predict user trajectories and pre-allocate resources at neighboring edge nodes, thus reducing handover latency and maintaining service quality. For instance, a novel EDMA-RM framework for green IoT-edge-fog-cloud networks [76], with an overall diagram presented in Figure 1.9, that addresses mobility challenges and optimizes resource allocation through event-driven mechanisms, thus demonstrating significant improvements in CPU usage, memory usage, delay, and jitter compared to existing approaches.

Energy Efficiency in Distributed Environments: Due to the limited power budgets, edge and fog devices require energy-efficient designs to ensure sustainable operation. Advanced RL algorithms like DDPG optimize resource utilization to reduce energy consumption while meeting QoS requirements. These methods dynamically adjust parameters like task scheduling, processing power, and data transmission rates to achieve energy-efficient operations.

1.2.2.5 Multi-agent and Cooperative RL in 6G

The intricate interplay of massive connectivity, dynamic spectrum allocation, and diverse applications in 6G networks demands sophisticated control mechanisms. In this regard, MARL and cooperative RL offer a promising approach to enable decentralized, collaborative decision-making among network entities. In MARL, multiple agents operate in a shared environment to optimize local and global

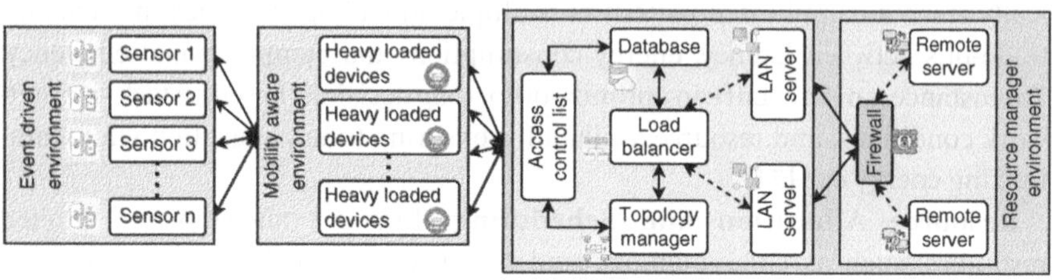

Figure 1.9 An overall diagram of an event-driven and mobility-aware resource management in [76]. Adapted from [76].

objectives via effective learning policies. Each agent interacts with the environment to receive localized observations and rewards, while also cooperating or competing with other agents. Cooperative MARL extends this framework by advancing the collaboration among agents to achieve shared goals, such as maximizing network efficiency, ensuring fairness, or minimizing latency [77].

Indeed, MARL offers a powerful framework for decentralized, cooperative solutions to complex 6G applications, including dynamic spectrum sharing, energy-efficient power control, vehicular communication, and resource allocation in heterogeneous networks.

- **Dynamic Spectrum Sharing:** MARL facilitates decentralized spectrum sharing among devices, thus allowing agents to dynamically negotiate and access underutilized frequency bands. Cooperative MARL ensures efficient spectrum utilization while minimizing interference in dense 6G environments, such as smart cities or industrial automation.
- **Energy-efficient Power Control:** In 6G networks, power control is critical for expanding device battery life and reducing energy consumption. In this context, MARL agents can collaboratively optimize power allocation across devices, balancing energy efficiency and QoS.
- **Vehicular Communication and Autonomous Systems:** MARL facilitates decentralized, cooperative decision-making among vehicles in 6G-aided autonomous systems, consequently resulting in efficient resource allocation, collision avoidance, and optimized traffic flow. Cooperative RL agents further optimize communication paths and spectrum usage to ensure the reliability and safety of vehicular networks.
- **Resource Allocation in Heterogeneous Networks:** In the theme of 6G networks incorporating heterogeneous nodes (such as small cells, drones, and satellites), MARL agents can dynamically allocate resources across these nodes, ensuring seamless connectivity and efficient resource utilization in complex scenarios.

Despite offering several highly innovative features for diverse 6G applications, MARL faces challenges in scalability, coordination, and privacy, which require innovative solutions to ensure efficient and secure operation.

- **Scalability and Computational Overhead:** As the number of agents increases, the complexity of learning and coordination grows exponentially. Advanced MARL techniques, such as Centralized Training with Decentralized Execution (CTDE), help mitigate this by leveraging centralized coordination during training and distributed decision-making during execution.
- **Coordination Among Agents:** Ensuring effective cooperation among agents is challenging, especially in non-stationary environments where

agents' policies evolve over time. Strategies like reward shaping and shared global objectives are used to align agent behaviors.

- **Privacy and Security Concerns:** Decentralized learning can expose sensitive data and make the network vulnerable to attacks. Federated MARL (F-MARL) addresses these concerns by enabling agents to collaboratively learn without sharing raw data, ensuring privacy and security.

1.2.3 Federated/Distributed Learning-based Approaches

1.2.3.1 Basics of Federated Learning

Federated Learning (FL) has emerged as a powerful learning paradigm to address critical challenges in conventional centralized learning approaches, such as DL and single-agent RL [78]. Unlike centralized learning, FL does not require distributed clients, such as mobile devices and wearable devices, to share their data with the central server for data processing and model training. In doing so, FL enables distributed devices to train distributed ML models where the data is generated, and thus it can avoid the need to share the data between devices and the server. As such, FL has the potential to protect the privacy of sensitive information contained in the data. In addition, FL can seamlessly adapt and scale with network growth, particularly in the context of large-scale networks and IoT systems, where each device can generate data and contribute its computing resources for model training, thus alleviating the pressure on the central server in conventional centralized learning methods.

As shown in Figure 1.10, a global communication round of vanilla FL is composed of three main steps, including model initialization, local training, and model aggregation. These steps are briefly presented below.

Step 1: Model Initialization: The server initializes a model weight $\boldsymbol{w}^0$ of the global model and broadcasts its value to the selected clients in the current communication round. Note that due to data heterogeneity (e.g., non-independent and identically distributed (non-IID) data and imbalanced data), limited resources, and channel conditions, there may be only a subset of clients to be selected to contribute to the FL process [79, 80]. Moreover, the server may need to specify the parameters of local models (e.g., accuracy of the local models and learning rate) and the parameters of the global model (e.g., accuracy of the global model).

Step 2: Local Training: At communication round t, each client k trains a local model using its raw dataset, denoted as $\mathcal{D}_k$. The objective of local training of client k is to minimize the loss function $f_k(\boldsymbol{w})$ as $\boldsymbol{w}_k^t = \arg\min f_k(\boldsymbol{w})$. Then, the selected clients shared the local model parameters with the server for model aggregation. It is worth noting that different loss functions may be applied to different learning tasks. For example, the linear regression model over the dataset $(\boldsymbol{x}, \boldsymbol{y})$ is identified by the loss function $f(\boldsymbol{w}) = \frac{1}{2}(\boldsymbol{x}^T\boldsymbol{w} - \boldsymbol{y})^2$.

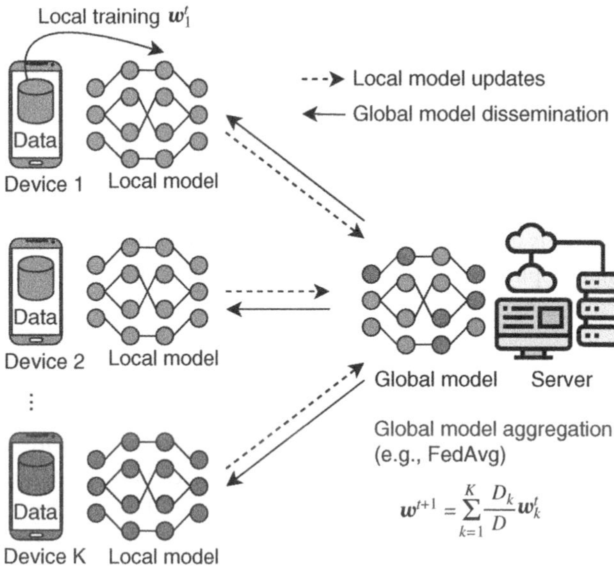

Figure 1.10 Overview of federated learning. Adapted from [81].

Step 3: Model Aggregation: Once the server receives the local model updates from the selected clients, the global model parameter can be obtained by minimizing the global loss function as $f(\mathbf{w}) = \sum_{k=1}^{K} \frac{D_k}{D} f_k(\mathbf{w})$. A popular approach to this problem is Federated Averaging (also known as FedAvg) [82]. In particular, the global model parameter can be updated as $\mathbf{w}^{t+1} = \sum_{k=1}^{K} p_k \mathbf{w}_k^t$, where $p_k = D_k/D$ is the ratio of the data size of the IoT device k to the total data. The FedAvg update indicates that the server takes the weighted average of the local models. In this regard, each client can update the local model parameter multiple times before sending the result to the server for model aggregation. According to Ref. [83], the number of local iterations can be specified by the learning hyperparameters and the required accuracy of local models.

The two steps of local training and model aggregation can be executed iteratively until the learning criteria are met. For example, the FL process may stop once the difference of the loss function in the two consecutive iterations is less than a threshold ε, that is, $|f(\mathbf{w}^{t+1}) - f(\mathbf{w}^t)| \leq \varepsilon$. Similar to the local model iteration, the number of global iterations is dependent on the learning hyperparameters, and the learning accuracy of the global model [83]. It is worth noting that the learning task in FL is typically generated by the cloud/server. However, FL clients are the model owners in many scenarios, where clients have a common interest in collaboratively building an AI model via knowledge sharing. In such cases, the server is employed to facilitate the FL process [84].

Since its successful deployment for the Google keyboard, FL has found great successes in many vertical areas, such as smart healthcare [85], smart transportation [86], and mobile communications and networking [87]. With recent advancements in FL and AI milestones, such as the deployment of Large Language Model (LLM)-based services like ChatGPT and Copilot and the emergence of federated foundation models [88], FL is expected to continue playing a key role in various mobile services and smart applications, as well as significantly shaping the development of AI-enabled future networks.

1.2.3.2 Applications of Federated Learning for Communications and Networking

FL has been recognized not only as a technique to enhance data protection in communications and networking but also as a powerful AI approach to improve mobile edge networks. This includes cyber detection, where FL can overcome the limitations of conventional DL algorithms that require sensitive data to achieve superior performance. For edge caching and computation offloading, FL enables clients to train their local models without exposing sensitive information, while the server can predict popular files for caching and offloading through a model aggregation approach. In the context of vehicular networks, each vehicle, typically equipped with computing resources, acts as a client, and the roadside unit can be leveraged as the server. Herein, the server is responsible for averaging the model updates from vehicles, updating the global model, and disseminating it to all the vehicles, thereby enhancing QoS.

From the IoT perspective, each client in FL is seen as an IoT device, such as smartphones in mobile networks and vehicles in intelligent transportation systems. Notably, a recent report by Statista showed that the number of IoT connections will continue increasing significantly in the next decade, jumping from around 18 billion in 2024 to approximately 39.6 billion in 2033 [89]. In this context, FL will be a key to the realization of beyond 5G and 6G network systems, where it will be a complement to centralized AI/ML techniques as their combination well aligns with the IoT-edge-cloud continuum architecture. The opportunities offered by distributed learning for IoT services, such as computation offloading, IoT localization, MCS, and IoT security and privacy, and for IoT applications, such as IoT-enabled healthcare, smart grids, autonomous vehicles, aerial IoT networks, and IoT-enabled smart industries, have been extensively explored in the literature.

As an example of the application of FL for privacy preservation in fingerprint-based localization applications, where database heterogeneity due to volatile client behaviors is often a major factor in performance degradation, the heterogeneous FedLoc algorithm with the area of convex hull-based aggregation is introduced [90]. As illustrated in Figure 1.11, the FedLoc scheme includes two phases: Phase I

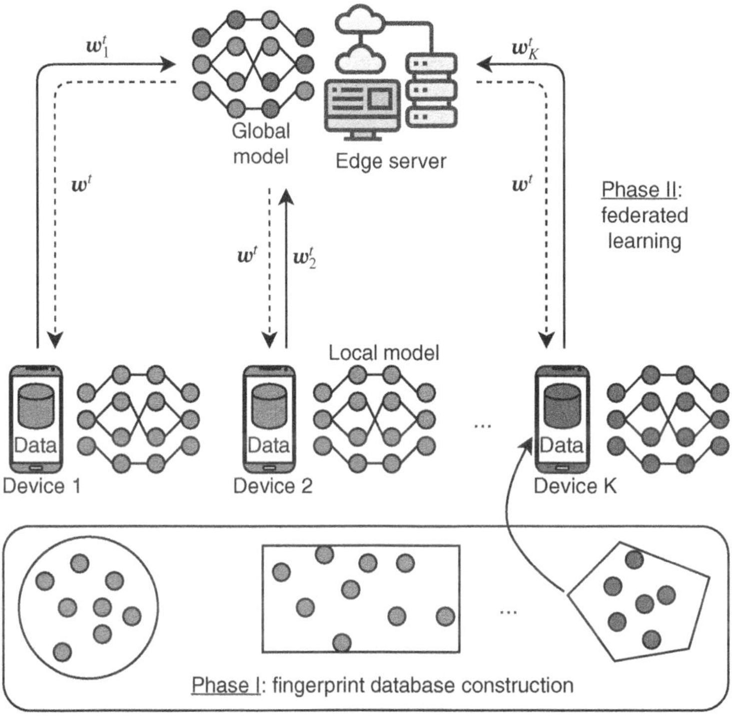

Figure 1.11 Illustration of FL-enabled IoT localization. Adapted from [90].

of fingerprint database construction and Phase II of FL. In the working principle, each IoT device moves around the area of interest to collect the Radio Signal Strength (RSS) fingerprint data. Then, IoT devices optimize their ML models based on the collected local dataset before uploading the updated models. At the server, the proportion weight of each IoT device is chosen in proportion to the area of the convex hull characterized by the local fingerprint data. Based on that aggregation rule, the server generates the new global model and transmits it back to the distributed nodes for executing the next learning round. Compared to the FedLoc framework in Ref. [91], the proposed heterogeneous FedLoc algorithms can obtain a significant location prediction gain of around 20%. FL can also be used to provide high-precision positioning correction for dynamic IoT environments. In this regard, by taking into account specific applications, an add-on FL-integrated transfer learning approach can greatly reduce individual location prediction errors in cooperative IoV systems [92]. Experimental datasets with sensor-rich vehicles show that the proposed framework is not affected by changes in the scale of the dataset but also offers a convergence rate lower than 10 epochs, much faster than using conventional deep neural networks.

1.2.4 Existing Challenges

Given the above, it is undeniable that AI has great potential to solve a wide range of issues in communication and networking. Unfortunately, existing AI-based approaches face several challenges in future communication systems.

- **Dependence on Labeled Data:** It is well-known that traditional AI approaches (e.g., DL) require a huge amount of labeled data to achieve good learning performance. Nevertheless, it may not always be the case in practice since collecting and labeling data are time-consuming, costly, and require human interventions [93]. It is expected that future communication systems will be self-organizing and fully autonomous. Without human interventions, it may be impossible to collect and label enough training data from a massive number of devices with different configurations and requirements co-existing in the same system.
- **Less Effective Under Dynamic Networks:** Theoretically, traditional AI approaches are trained on specific environments/scenarios with known training data. This makes them less effective in dealing with new system conditions that they have not trained on Ref. [93, 94]. Unfortunately, this is likely the case in future communication systems where network conditions are highly dynamic due to their unique characteristics such as massive heterogeneous connections and applications, use mobility between different network environments, and evolving user demands.
- **Performance Degradation with Complex and High-dimensional Data:** As discussed in Section 1.2, AI (and DL in particular) has great potential to extract useful information from a massive amount of data. Nevertheless, traditional AI techniques are mainly designed for discriminative tasks, and thus they may be less effective in complex problems with high-dimensional, heterogeneous, and biased training data [93]. Again, future communication systems, unfortunately, are heterogeneous, complex, and consist of diverse devices and application settings. Consequently, traditional AI-based methods that work well in current communication systems may no longer be effective in future communication systems.

1.2.5 Potential of Generative AI

Given the above, it is, therefore, essential to develop novel AI techniques that can effectively deal with heterogeneous, complex, and high-dimensional data generated by future communication systems to enable their intelligence and autonomy. Among various emerging AI techniques, Generative AI is a promising solution thanks to its unique features in knowledge generalization, learning complex data, and effective adaptability [93–95].

- **Generative Modeling and Data Augmentation:** Generative AI is designed to create new data, such as images, text, wireless signals, or network traffic, that is similar to the original training data by learning its underlying distribution. Moreover, Generative AI can produce data based on specific constraints or conditions, such as generating network traffic in particular regions or with unique characteristics. This is particularly helpful when the training data is inadequate or biased, which is expected to be the case in future communication systems.
- **Anomaly Detection:** Generative models can be used to detect anomalies by identifying deviations from expected patterns and provide probabilistic insights into their outputs, aiding applications that require security and uncertainty quantification in future communication systems such as sensitive applications in banking and industry.
- **Domain Adaptation:** Generative models can effectively generate data in a target domain while leveraging knowledge from a source domain. This can help enhance model performance in diverse environments, which discriminative models often struggle to achieve as discussed above.
- **Data Privacy and Anonymization:** Generative AI can support data privacy and anonymization through synthetic data generation that preserves key statistical properties of the original data while ensuring privacy and anonymity. This unique feature of Generative AI can help to develop novel AI-based approaches to address the security and data privacy challenge of future communication systems, which is highlighted in Section 1.1.2.
- **Variational Learning and Sampling:** Generative AI's key techniques like generative adversarial networks (GANs) and variational autoencoders (VAEs) provide effective probabilistic sampling features that can help to explore data distributions flexibly, offering more versatility than discriminative models in probabilistic sampling. This can help to further enhance the performance of AI-based approaches, especially in tasks like channel estimation, beamforming, joint source-channel coding, and resource allocation in future communication systems.

With these advantages of Generative AI, this book is dedicated to providing a comprehensive overview of Generative AI, its key techniques, and applications in various scenarios and settings of future communication systems. In particular, Chapter 2 will provide the background of Generative AI and its potential for future communications systems. Chapter 3 then introduces key study cases of Generative AI applications to communications systems. After that, applications of Generative AI at the PHY layer, MAC layer, Network layer, and Application layer will be discussed in detail in Chapters 4–7, respectively. Chapter 8 then highlights the implications of Generative AI on wireless and cybersecurity. Finally, challenges and future research directions of Generative AI in future communication systems will be presented in Chapters 9 and 10, respectively.

1.3 Conclusion

In this chapter, we have first provided an overview of future communication systems, including key designs and technologies such as THz band communications, non-terrestrial networks, massive and extremely large MIMO, intelligent reflecting surface, and quantum internet. Open challenges and research trends of future communication systems are then presented. Next, we have provided an overview of existing AI-based approaches in communication systems, including their applications in the physical layer, resource allocation, network optimization, and privacy-preserving distributed networks. The existing challenges of these traditional AI-based methods are then highlighted. Finally, we briefly discuss the potentials of Generative AI in addressing these challenges by highlighting the unique features of Generative AI such as generative modeling, anomaly detection, domain adaptation, data privacy and anonymization, and variational learning and sampling.

References

1 Ericsson. Mobile data traffic forecast – Ericsson mobility report. https://www.ericsson.com/en/reports-and-papers/mobility-report/dataforecasts/mobile-traffic-forecast, 2024. (Accessed on 13/12/2024).

2 Ian F. Akyildiz, Ahan Kak, and Shuai Nie. 6G and beyond: The future of wireless communications systems. *IEEE Access*, 8: 133995–134030, 2020.

3 Chamitha De Alwis, Anshuman Kalla, Quoc-Viet Pham, Pardeep Kumar, Kapal Dev, Won-Joo Hwang, and Madhusanka Liyanage. Survey on 6G frontiers: Trends, applications, requirements, technologies and future research. *IEEE Open Journal of the Communications Society*, 2: 836–886, 2021.

4 Syed Junaid Nawaz, Shree Krishna Sharma, Shurjeel Wyne, Mohammad N. Patwary, and Md Asaduzzaman. Quantum machine learning for 6G communication networks: State-of-the-art and vision for the future. *IEEE Access*, 7: 46317–46350, 2019.

5 3GPP. 3gpp release 18. 2023. https://www.3gpp.org/specifications-technologies/releases/release-18.

6 Xingqin Lin. An overview of 5G advanced evolution in 3GPP release 18. *IEEE Communications Standards Magazine*, 6(3): 77–83, 2022.

7 Ericsson. 5g advanced: Evolution towards 6g. 2023. https://www.ericsson.com/en/reports-and-papers/white-papers/5g-advanced-evolution-towards-6g.

8 ITU. Itu advances the development of imt-2030 for 6g mobile technologies. https://www.itu.int/en/mediacentre/Pages/PR-2023-12-01-IMT-2030-for-6G-mobile-technologies.aspx.

9 Cheng-Xiang Wang, Xiaohu You, Xiqi Gao, Xiuming Zhu, Zixin Li, Chuan Zhang, Haiming Wang, Yongming Huang, Yunfei Chen, Harald Haas, John S. Thompson, Erik G. Larsson, Marco Di Renzo, Wen Tong, Peiying Zhu, Xuemin Shen, H. Vincent Poor, and Lajos Hanzo. On the road to 6g: Visions, requirements, key technologies, and testbeds. *IEEE Communications Surveys & Tutorials*, 25(2): 905–974, 2023.

10 Vitaly Petrov, Alexander Pyattaev, Dmitri Moltchanov, and Yevgeni Koucheryavy. Terahertz band communications: Applications, research challenges, and standardization activities. In *2016 8th International Congress on Ultra Modern Telecommunications and Control Systems and Workshops (ICUMT)*, pages 183–190. IEEE, 2016.

11 Christina Chaccour, Mehdi Naderi Soorki, Walid Saad, Mehdi Bennis, Petar Popovski, and Mérouane Debbah. Seven defining features of terahertz (THz) wireless systems: A fellowship of communication and sensing. *IEEE Communications Surveys & Tutorials*, 24(2): 967–993, 2022.

12 Ahmet M. Elbir, Kumar Vijay Mishra, Symeon Chatzinotas, and Mehdi Bennis. Terahertz-band integrated sensing and communications: Challenges and opportunities. *IEEE Aerospace and Electronic Systems Magazine*, 2024.

13 Lu Lu, Geoffrey Ye Li, A Lee Swindlehurst, Alexei Ashikhmin, and Rui Zhang. An overview of massive MIMO: Benefits and challenges. *IEEE Journal of Selected Topics in Signal Processing*, 8(5): 742–758, 2014.

14 Zhe Wang, Jiayi Zhang, Hongyang Du, EI Wei, Bo Ai, Dusit Niyato, and Merouane Debbah. Extremely large-scale MIMO: Fundamentals, challenges, solutions, and future directions. *IEEE Wireless Communications*, 31(3):117–124, 2023.

15 Haiquan Lu, Yong Zeng, Changsheng You, Yu Han, Jiayi Zhang, Zhe Wang, Zhenjun Dong, Shi Jin, Cheng-Xiang Wang, Tao Jiang, Xiaohu You, and Rui Zhang A tutorial on near-field XL-MIMO communications towards 6G. *IEEE Communications Surveys & Tutorials*, 26(4):2213–2257, 2024.

16 Chongwen Huang, Sha Hu, George C Alexandropoulos, Alessio Zappone, Chau Yuen, Rui Zhang, Marco Di Renzo, and Merouane Debbah. Holographic MIMO surfaces for 6G wireless networks: Opportunities, challenges, and trends. *IEEE Wireless Communications*, 27(5): 118–125, 2020.

17 Zijian Zhang and Linglong Dai. Pattern-division multiplexing for multi-user continuous-aperture MIMO. *IEEE Journal on Selected Areas in Communications*, 41(8): 2350–2366, 2023.

18 Yu Lu and Linglong Dai. Near-field channel estimation in mixed LoS/NLoS environments for extremely large-scale MIMO systems. *IEEE Transactions on Communications*, 71(6): 3694–3707, 2023.

19 Marco Giordani and Michele Zorzi. Non-terrestrial networks in the 6G era: Challenges and opportunities. *IEEE Network*, 35(2): 244–251, 2020.

20 Nhu-Ngoc Dao, Quoc-Viet Pham, Ngo Hoang Tu, Tran Thien Thanh, Vo Nguyen Quoc Bao, Demeke Shumeye Lakew, and Sungrae Cho. Survey on aerial radio access networks: Toward a comprehensive 6G access infrastructure. *IEEE Communications Surveys & Tutorials*, 23(2): 1193–1225, 2021.

21 Shadab Mahboob and Lingjia Liu. Revolutionizing future connectivity: A contemporary survey on AI-empowered satellite-based non-terrestrial networks in 6G. *IEEE Communications Surveys & Tutorials*, 2024.

22 Federica Rinaldi, Helka-Liina Maattanen, Johan Torsner, Sara Pizzi, Sergey, Andreev, Antonio Iera, Yevgeni Koucheryavy, and Giuseppe Araniti. Non-terrestrial networks in 5G & beyond: A survey. *IEEE Access*, 8: 165178–165200, 2020.

23 Angela Sara Cacciapuoti, Marcello Caleffi, Francesco Tafuri, Francesco Saverio Cataliotti, Stefano Gherardini, and Giuseppe Bianchi. Quantum internet: Networking challenges in distributed quantum computing. *IEEE Network*, 34(1): 137–143, 2019.

24 Seon-Geun Jeong, Pham Dang Anh Duc, Quang Vinh Do, Dae-Il Noh, Nguyen Xuan Tung, Trinh Van Chien, Quoc-Viet Pham, Mikio Hasegawa, Hiroo Sekiya, and Won-Joo Hwang. Quantum annealing-based sum rate maximization for multi-UAV-aided wireless networks. *IEEE Internet of Things Journal*, 12(12):21225–21239, 2025.

25 Brad Lackey. Quantum networking: A roadmap to a quantum internet, 2023. https://azure.microsoft.com/en-us/blog/quantum/2023/11/01/quantum-networking-a-roadmap-to-a-quantum-internet/. [Accessed on 06-12-2024].

26 Amoldeep Singh, Kapal Dev, Harun Siljak, Hem Dutt Joshi, and Maurizio Magarini. Quantum internet—Applications, functionalities, enabling technologies, challenges, and research directions. *IEEE Communications Surveys & Tutorials*, 23(4): 2218–2247, 2021.

27 Rodney Van Meter, Ryosuke Satoh, Naphan Benchasattabuse, Kentaro Teramoto, Takaaki Matsuo, Michal Hajdušek, Takahiko Satoh, Shota Nagayama, and Shigeya Suzuki. A quantum internet architecture. In *2022 IEEE International Conference on Quantum Computing and Engineering (QCE)*, pages 341–352. IEEE, 2022.

28 Stephanie Wehner, David Elkouss, and Ronald Hanson. Quantum internet: A vision for the road ahead. *Science*, 362(6412): eaam9288, 2018.

29 Shimin Gong, Xiao Lu, Dinh Thai Hoang, Dusit Niyato, Lei Shu, Dong In Kim, and Ying-Chang Liang. Toward smart wireless communications via intelligent reflecting surfaces: A contemporary survey. *IEEE Communications Surveys & Tutorials*, 22(4): 2283–2314, 2020.

30 Huimei Han, Jun Zhao, Wenchao Zhai, Zehui Xiong, Dusit Niyato, Marco Di Renzo, Quoc-Viet Pham, Weidang Lu, and Kwok-Yan Lam. Reconfigurable intelligent surface aided power control for physical-layer broadcasting. *IEEE Transactions on Communications*, 69(11): 7821–7836, 2021.

31 Qingqing Wu, Shuowen Zhang, Beixiong Zheng, Changsheng You, and Rui Zhang. Intelligent reflecting surface-aided wireless communications: A tutorial. *IEEE Transactions on Communications*, 69(5): 3313–3351, 2021.

32 Wenxuan Long, Rui Chen, Marco Moretti, Wei Zhang, and Jiandong Li. A promising technology for 6G wireless networks: Intelligent reflecting surface. *Journal of Communications and Information Networks*, 6(1): 1–16, 2021.

33 Aly S. Abdalla, Pratheek S. Upadhyaya, Vijay K. Shah, and Vuk Marojevic. Toward next generation open radio access networks: What o-ran can and cannot do! *IEEE Network*, 36(6): 206–213, 2022.

34 Azadeh Arnaz, Justin Lipman, Mehran Abolhasan, and Matti Hiltunen. Toward integrating intelligence and programmability in open radio access networks: A comprehensive survey. *IEEE Access*, 10: 67747–67770, 2022.

35 Cong T. Nguyen, Yuris Mulya Saputra, Nguyen Van Huynh, Tan N. Nguyen, Dinh Thai Hoang, Diep N. Nguyen, Van-Quan Pham, Miroslav Voznak, Symeon Chatzinotas, and Dinh-Hieu Tran. Emerging technologies for 6G non-terrestrial-networks: From academia to industrial applications. *IEEE Open Journal of the Communications Society*, 5: 3852–3885, 2024.

36 Akram Shafie, Nan Yang, Sheeraz A. Alvi, Chong Han, Salman Durrani, and Josep Miquel Jornet. Spectrum allocation with adaptive sub-band bandwidth for terahertz communication systems. *IEEE Transactions on Communications*, 70(2): 1407–1422, 2021.

37 Akram Shafie, Nan Yang, Chong Han, Josep Miquel Jornet, Markku Juntti, and Thomas Kürner. Terahertz communications for 6G and beyond wireless networks: Challenges, key advancements, and opportunities. *IEEE Network*, 37(3): 162–169, 2022.

38 Li-Hsiang Shen, Kai-Ten Feng, and Lajos Hanzo. Five facets of 6G: Research challenges and opportunities. *ACM Computing Surveys*, 55(11): 1–39, 2023.

39 Ericsson. 6G network architecture – A proposal for early alignment, 2023. https://www.ericsson.com/en/reports-and-papers/ericsson-technology-review/articles/6g-network-architecture. [Accessed on 06-12-2024].

40 Yushan Siriwardhana, Pawani Porambage, Madhusanka Liyanage, and Mika Ylianttila. AI and 6G security: Opportunities and challenges. In *2021 Joint European Conference on Networks and Communications & 6G Summit (EuCNC/6G Summit)*, pages 616–621. IEEE, 2021.

41 Pawani Porambage, Gürkan Gür, Diana Pamela Moya Osorio, Madhusanka Liyanage, Andrei Gurtov, and Mika Ylianttila. The roadmap to 6G security and privacy. *IEEE Open Journal of the Communications Society*, 2: 1094–1122, 2021.

42 Van-Linh Nguyen, Po-Ching Lin, Bo-Chao Cheng, Ren-Hung Hwang, and Ying-Dar Lin. Security and privacy for 6G: A survey on prospective technologies and challenges. *IEEE Communications Surveys & Tutorials*, 23(4): 2384–2428, 2021.

43 Nokia Bell Labs. 6G technologies – Security, trust and privacy, 2023. https://www.bell-labs.com/research-innovation/what-is-6g/6g-technologies/security-and-trust. [Accessed on 06-12-2024].

44 Ning Hu, Zhihong Tian, Xiaojiang Du, and Mohsen Guizani. An energy-efficient in-network computing paradigm for 6G. *IEEE Transactions on Green Communications and Networking*, 5(4): 1722–1733, 2021.

45 Usman Mahmood Malik, Muhammad Awais Javed, Sherali Zeadally, and Saif ul Islam. Energy-efficient fog computing for 6G-enabled massive IoT: Recent trends and future opportunities. *IEEE Internet of Things Journal*, 9(16): 14572–14594, 2021.

46 Azadeh Pourkabirian, Mohammad Sadegh Kordafshari, Anish Jindal, and Mohammad Hossein Anisi. A vision of 6G URLLC: Physical-layer technologies and enablers. *IEEE Communications Standards Magazine*, 8(2): 20–27, 2024.

47 Feng Wang, Shengyu Zhang, Huiting Yang, and Tony QS Quek. Non-terrestrial networking for 6G: Evolution, opportunities, and future directions. *arXiv preprint arXiv:2412.00820*, 2024.

48 Nada Abdel Khalek, Deemah H. Tashman, and Walaa Hamouda. Advances in machine learning-driven cognitive radio for wireless networks: A survey. *IEEE Communications Surveys & Tutorials*, 26(2): 1201–1237, 2024.

49 Thien Huynh-The, Quoc-Viet Pham, Toan-Van Nguyen, Thanh Thi Nguyen, Rukhsana Ruby, Ming Zeng, and Dong-Seong Kim. Automatic modulation classification: A deep architecture survey. *IEEE Access*, 9: 142950–142971, 2021.

50 Abdullah Zayat, Mahmoud A. Hasabelnaby, Mohanad Obeed, and Anas Chaaban. Transformer masked autoencoders for next-generation wireless communications: Architecture and opportunities. *IEEE Communications Magazine*, 62(7): 88–94, 2024.

51 Yandong Shi, Lixiang Lian, Yuanming Shi, Zixin Wang, Yong Zhou, Liqun Fu, Lin Bai, Jun Zhang, and Wei Zhang. Machine learning for large-scale optimization in 6G wireless networks. *IEEE Communications Surveys & Tutorials*, 25(4): 2088–2132, 2023.

52 Ephrem Fola, Yang Luo, and Chunbo Luo. *AttenReEsNet*: Attention-aided residual learning for effective model-driven channel estimation. *IEEE Communications Letters*, 28(8): 1855–1859, 2024.

53 Zhen Gao, Minghui Wu, Chun Hu, Feifei Gao, Guanghui Wen, Dezhi Zheng, and Jun Zhang. Data-driven deep learning based hybrid beamforming for aerial massive MIMO-OFDM systems with implicit CSI. *IEEE Journal on Selected Areas in Communications*, 40(10): 2894–2913, 2022.

54 Thien Huynh-The, Toan-Van Nguyen, Quoc-Viet Pham, Daniel Benevides da Costa, and Dong-Seong Kim. MIMO-OFDM modulation classification using three-dimensional convolutional network. *IEEE Transactions on Vehicular Technology*, 71(6): 6738–6743, 2022.

55 Chao Xu, Thien Van Luong, Luping Xiang, Shinya Sugiura, Robert G. Maunder, Lie-Liang Yang, and Lajos Hanzo. Turbo detection aided autoencoder for multicarrier wireless systems: Integrating deep learning into channel coded systems. *IEEE Transactions on Cognitive Communications and Networking*, 8(2): 600–614, 2022.

56 Xiangyu Ren, Hamed Mosavat-Jahromi, Lin Cai, and David Kidston. Spatio-temporal spectrum load prediction using convolutional neural network and *ResNet*. *IEEE Transactions on Cognitive Communications and Networking*, 8(2): 502–513, 2022.

57 Shreshth Tuli, Shashikant Ilager, Kotagiri Ramamohanarao, and Rajkumar Buyya. Dynamic scheduling for stochastic edge-cloud computing environments using A3C learning and residual recurrent neural networks. *IEEE Transactions on Mobile Computing*, 21(3): 940–954, 2022.

58 Shu-Ping Yeh, Sonia Bhattacharya, Rashika Sharma, and Hassnaa Moustafa. Deep learning for intelligent and automated network slicing in 5G open RAN (ORAN) deployment. *IEEE Open Journal of the Communications Society*, 5: 64–70, 2024.

59 Zheheng Rao, Yanyan Xu, Shaoming Pan, Jiabao Guo, Yuejing Yan, and Zhiheng Wang. Cellular traffic prediction: A deep learning method considering dynamic nonlocal spatial correlation, self-attention, and correlation of spatiotemporal feature fusion. *IEEE Transactions on Network and Service Management*, 20(1): 426–440, 2023.

60 Tao Zhang, Kun Zhu, and Dusit Niyato. Detection of sleeping cells in self-organizing cellular networks: An adversarial auto-encoder method. *IEEE Transactions on Cognitive Communications and Networking*, 7(3): 739–751, 2021.

61 Jie Mei, Xianbin Wang, Kan Zheng, Gary Boudreau, Akram Bin Sediq, and Hatem Abou-Zeid. Intelligent radio access network slicing for service provisioning in 6G: A hierarchical deep reinforcement learning approach. *IEEE Transactions on Communications*, 69(9): 6063–6078, 2021.

62 Yalin E. Sagduyu, Tugba Erpek, Aylin Yener, and Sennur Ulukus. Will 6G be semantic communications? Opportunities and challenges from task oriented and secure communications to integrated sensing. *IEEE Network*, 38(6): 72–80, 2024.

63 Xiangnan Liu, Haijun Zhang, Chao Ren, Haojin Li, Chen Sun, and Victor C. M. Leung. Multi-task learning resource allocation in federated integrated sensing and communication networks. *IEEE Transactions on Wireless Communications*, 23(9): 11612–11623, 2024.

64 Amal Feriani and Ekram Hossain. Single and multi-agent deep reinforcement learning for *AI*-enabled wireless networks: A tutorial. *IEEE Communications Surveys & Tutorials*, 23(2): 1226–1252, 2021.

65 Hasan Albinsaid, Keshav Singh, Sudip Biswas, and Chih-Peng Li. Multi-agent reinforcement learning-based distributed dynamic spectrum access. *IEEE Transactions on Cognitive Communications and Networking*, 8(2): 1174–1185, 2022.

66 Yu Gou, Tong Zhang, Tingting Yang, Jun Liu, Shanshan Song, and Jun-Hong Cui. A deep *MARL*-based power-management strategy for improving the fair reuse of *UWSNs*. *IEEE Internet of Things Journal*, 10(7): 6507–6522, 2023.

67 Xuanheng Li, Yulong Zhang, Haichuan Ding, and Yuguang Fang. Intelligent spectrum sensing and access with partial observation based on hierarchical multi-agent deep reinforcement learning. *IEEE Transactions on Wireless Communications*, 23(4): 3131–3145, 2024.

68 Ali Majid Hasan Alibraheemi, Mhd Nour Hindia, Tengku Faiz Tengku Mohmed Noor Izam, and Kaharudin Dimyati. Spectrum efficient mode selection and resource allocation optimization for D2D communication in HetNet: A multi-agent Q-learning approach. *IEEE Access*, 12: 131217–131229, 2024.

69 Beibei Wang, Hong Zhu, Honghua Xu, Yuqing Bao, and Huifang Di. Distribution network reconfiguration based on NoisyNet deep Q-learning network. *IEEE Access*, 9: 90358–90365, 2021.

70 Daniela M. Casas-Velasco, Oscar Mauricio Caicedo Rendon, and Nelson L. S. da Fonseca. DRSIR: A deep reinforcement learning approach for routing in software-defined networking. *IEEE Transactions on Network and Service Management*, 19(4): 4807–4820, 2022.

71 Marios Avgeris, Aris Leivadeas, and Ioannis Lambadaris. Reinforcement learning-enabled auctions for self-healing in service function chaining. In *2023 IEEE International Conference on Communications Workshops (ICC Workshops)*, pages 776–781, Rome, Italy, 2023.

72 Quoc-Viet Pham, Fang Fang, Vu Nguyen Ha, Md Jalil Piran, Mai Le, Long Bao Le, Won-Joo Hwang, and Zhiguo Ding. A survey of multi-access edge computing in 5G and beyond: Fundamentals, technology integration, and state-of-the-art. *IEEE Access*, 8: 116974–117017, 2020.

73 Quang Vinh Do, Quoc-Viet Pham, Zhaohui Yang, and Won-Joo Hwang. Learning-based QoE optimization for green edge computing networks with aerial servers. *IEEE Transactions on Green Communications and Networking*, 9(3):1326–1339, 2024.

74 Frezer Guteta Wakgra, Widhi Yahya, Binayak Kar, Yuan-Cheng Lai, Ying-Dar Lin, and Seifu Birhanu Tadele. Ratio-based offloading optimization for edge and vehicular-fog federated systems: A multi-agent *td*3 approach. *IEEE Transactions on Vehicular Technology*, 73(11): 17684–17696, 2024.

75 Seung-Seob Lee and Sukyoung Lee. Resource allocation for vehicular fog computing using reinforcement learning combined with heuristic information. *IEEE Internet of Things Journal*, 7(10): 10450–10464, 2020.

76 Rohit Kumar and Neha Agrawal. EDMA-RM: An event-driven and mobility-aware resource management framework for green *IOT*-edge-fog-cloud networks. *IEEE Sensors Journal*, 24(14): 23004–23012, 2024.

77 Xiang Tan, Li Zhou, Haijun Wang, Yuli Sun, Haitao Zhao, Boon-Chong Seet, Jibo Wei, and Victor C. M. Leung. Cooperative multi-agent reinforcement-learning-based distributed dynamic spectrum access in cognitive radio networks. *IEEE Internet of Things Journal*, 9(19): 19477–19488, 2022.

78 Trong-Binh Nguyen, Minh-Duong Nguyen, Jinsun Park, Quoc-Viet Pham, and Won Joo Hwang. Federated domain generalization with data-free on-server matching gradient. In *The Thirteenth International Conference on Learning Representations (ICLR)*, 2025.

79 Moming Duan, Duo Liu, Xianzhang Chen, Renping Liu, Yujuan Tan, and Liang Liang. Self-balancing federated learning with global imbalanced data in mobile systems. *IEEE Transactions on Parallel and Distributed Systems*, 32(1): 59–71, 2021. ISSN 1558-2183.

80 Zonghang Li, Yihong He, Hongfang Yu, Jiawen Kang, Xiaoping Li, Zenglin Xu, and Dusit Niyato. Data heterogeneity-robust federated learning via group client selection in industrial IoT. *IEEE Internet of Things Journal*, 9(18): 17844–17857, 2022. ISSN 2327-4662. https://doi.org/10.1109/JIOT.2022.3161943.

81 Mai Le, Thien Huynh-The, Tan Do-Duy, Thai-Hoc Vu, Won-Joo Hwang, and Quoc-Viet Pham. Applications of distributed machine learning for the internet-of-things: A comprehensive survey. *IEEE Communications Surveys & Tutorials*, 27(2):1053–1100, 2024.

82 Brendan McMahan, Eider Moore, Daniel Ramage, Seth Hampson, and Blaise Aguera y. Arcas. Communication-efficient learning of deep networks from decentralized data. In *International Conference on Artificial Intelligence and Statistics (AISTATS)*, pages 1273–1282, Fort Lauderdale, Florida, USA, 2017.

83 Zhaohui Yang, Mingzhe Chen, Walid Saad, Choong Seon Hong, and Mohammad Shikh-Bahaei. Energy efficient federated learning over wireless communication networks. *IEEE Transactions on Wireless Communications*, 20(3): 1935–1949, 2021.

84 Guangjing Huang, Xu Chen, Tao Ouyang, Qian Ma, Lin Chen, and Junshan Zhang. Collaboration in participant-centric federated learning: A game-theoretical perspective. *IEEE Transactions on Mobile Computing*, 22(11): 6311–6326, 2023. ISSN 1558-0660. https://doi.org/10.1109/TMC.2022.3194198.

85 Dinh C. Nguyen, Quoc-Viet Pham, Pubudu N. Pathirana, Ming Ding, Aruna Seneviratne, Zihuai Lin, Octavia Dobre, and Won-Joo Hwang. Federated learning for smart healthcare: A survey. *ACM Computing Surveys (CSUR)*, 55(3): 1–37, 2022.

86 Rongqing Zhang, Jingxin Mao, Hanqiu Wang, Bing Li, Xiang Cheng, and Liuqing Yang. A survey on federated learning in intelligent transportation systems. *IEEE Transactions on Intelligent Vehicles*, 10(5):3043–3059, 2024.

87 Joohyung Lee, Faranaksadat Solat, Tae Yeon Kim, and H Vincent Poor. Federated learning-empowered mobile network management for 5G and beyond networks: From access to core. *IEEE Communications Surveys & Tutorials*, 2024.

88 Zachary Charles, Nicole Mitchell, Krishna Pillutla, Michael Reneer, and Zachary Garrett. Towards federated foundation models: Scalable dataset pipelines for group-structured learning. *Advances in Neural Information Processing Systems*, 36:32299–32327, 2024.

89 Number of internet of things (IoT) connections worldwide from 2022 to 2023, with forecasts from 2024 to 2033. 2015. https://www.statista.com/statistics/1183457/iot-connected-devices-worldwide/.

90 Xin Cheng, Chuan Ma, Jun Li, Haiwei Song, Feng Shu, and Jiangzhou Wang. Federated learning-based localization with heterogeneous fingerprint database. *IEEE Wireless Communications Letters*, 11(7): 1364–1368, 2022.

91 Feng Yin, Zhidi Lin, Qinglei Kong, Yue Xu, Deshi Li, Sergios Theodoridis, and Shuguang Robert Cui. FedLoc: Federated learning framework for data-driven cooperative localization and location data processing. *IEEE Open Journal of Signal Processing*, 1: 187–215, 2020.

92 Xiangjie Kong, Haoran Gao, Guojiang Shen, Gaohui Duan, and Sajal K Das. FedVCP: A federated-learning-based cooperative positioning scheme for social internet of vehicles. *Transactions on Computational Social Systems*, 9(1): 197–206, 2022.

93 Nguyen Van Huynh, Jiacheng Wang, Hongyang Du, Dinh Thai Hoang, Dusit Niyato, Diep N Nguyen, Dong In Kim, and Khaled B Letaief. Generative AI for physical layer communications: A survey. *IEEE Transactions on Cognitive Communications and Networking*, 2024.

94 Guangyuan Liu, Nguyen Van Huynh, Hongyang Du, Dinh Thai Hoang, Dusit Niyato, Kun Zhu, Jiawen Kang, Zehui Xiong, Abbas Jamalipour, and Dong In Kim. Generative AI for unmanned vehicle swarms: Challenges, applications and opportunities. *arXiv preprint arXiv:2402.18062*, 2024.

95 Thai-Hoc Vu, Senthil Kumar Jagatheesaperumal, Minh-Duong Nguyen, Nguyen Van Huynh, Sunghwan Kim, and Quoc-Viet Pham. Applications of generative AI (GAI) for mobile and wireless networking: A survey. *IEEE Internet of Things Journal*, 12(2):1266–1290, 2025. https://doi.org/10.1109/JIOT.2024.3487627.

2

Generative AI Background and Its Potentials for Future Communications Systems

Asmaa Abdallah[1,*], *Abdulkadir Celik*[2], *and Ahmed M. Eltawil*[1]

[1] *Computer, Electrical and Mathematical Sciences and Engineering (CEMSE) Division, King Abdullah University of Science and Technology (KAUST), Thuwal, Saudi Arabia*
[2] *School of Electronics and Computer Science, University of Southampton(UoS), Southampton, Hampshire, United Kingdom*

*Corresponding author: asmaa.abdallah@kaust.edu.sa

2.1 Introduction

As we transition from the era of 5G to the emerging horizon of 6G, the integration of novel communication paradigms and advanced technological trends becomes crucial. This chapter explores the pivotal role of generative artificial intelligence (GenAI) in addressing the complexities inherent in next-generation wireless networks. Unlike discriminative AI (DAI), which relies on extensive real-world datasets, GenAI excels in scenarios where data is scarce, incomplete, or challenging to model.

This chapter provides an in-depth exploration of GenAI models, organized to reflect their foundational concepts, prominent examples, and wireless communication applications. Section 2.2 introduces a taxonomy of generative models, distinguishing between explicit and implicit density modeling techniques and providing a framework for understanding their core principles. Section 2.3 focuses on key generative models, covering generative adversarial networks (GANs) and their variations, variational autoencoders (VAEs), flow-based models, diffusion-based models, generative transformers, autoregressive models, and large language models (LLMs). Each model is discussed in terms of its unique architecture and capabilities. Section 2.4 delves into the potential applications of GenAI in wireless

Generative AI for Communications Systems: Fundamentals, Applications, and Prospects,
First Edition. Edited by Diep N. Nguyen et al.

communications, including physical layer design, network optimization, network traffic analytics, cross-layer network security, and localization and positioning. In Section 2.5, we explore the emerging potential applications of GenAI in future networks, such as semantic communications, integrated sensing and communications, digital twins, AI-generated content services, mobile edge computing (MEC), edge AI (EAI), adversarial machine learning (AML), and trustworthy AI. Finally, Section 2.6 addresses the critical regulatory and policy considerations for integrating generative models into next-generation wireless networks.

By examining the challenges and potential strategies for leveraging GenAI, this chapter serves as a pivotal resource for researchers and professionals aiming to harness the power of AI in the development of future wireless technologies.

2.2 A Taxonomy of Generative Models

In accordance with the framework proposed by Goodfellow [1], and as illustrated in Figure 2.1, generative models (GMs) can be systematically classified based on their approach to modeling the probability distribution function (PDF) of data. This classification divides models into explicit and implicit density models.

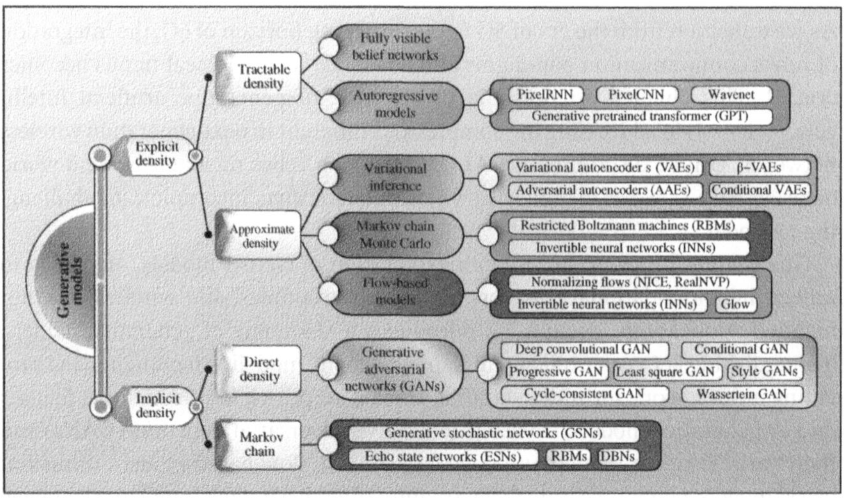

Figure 2.1 A taxonomy of GMs [2] / IEEE / CC BY 4.0.

2.2.1 Explicit Density Models

Explicit density models, sometimes called prescribed models, define the probability density function $p(x)$ explicitly, where x denotes a sample from the target distribution. This group includes two primary subtypes: tractable and approximate density models.

1. **Tractable Density Models:** These models feature well-defined mathematical formulations that enable efficient evaluation of $p(x)$ or straightforward data sampling. Their explicit structure supports practical likelihood computations.
2. **Approximate Density Models:** When direct calculation of $p(x)$ is infeasible, approximate models provide an alternative by leveraging sophisticated training techniques and architectures to capture complex data patterns.

Explicit density models have the distinct advantage of providing a clear mathematical form of $p(x)$, facilitating likelihood-driven training and enabling insightful analysis of data structures. However, their computational demands can be significant, and enforcing a specific form for the density function might limit the model's adaptability to highly intricate datasets.

2.2.2 Implicit Density Models

Unlike explicit models, implicit density models do not specify the data distribution mathematically. Instead, they learn to generate samples that reflect the target distribution, relying on stochastic processes to approximate the underlying PDF. These models fall into two categories: direct and Markov chain-based models.

1. **Direct Implicit Density Models:** These models create data samples directly from the target distribution without formulating an explicit density. GANs and their numerous variants are prominent examples of this class.
2. **Markov Chain-based Implicit Density Models:** These rely on Markov processes to iteratively generate samples. In this setup, the next sample depends only on the current state and transition probabilities, forming a stochastic sequence.

Implicit density models prioritize the generation of realistic samples over the explicit modeling of probability functions. While this makes them particularly suitable for applications emphasizing sample quality, it also poses challenges for model evaluation, stability during training, and interpretability, distinguishing them from explicit density models.

2.2.3 Ways GenAI Complements Discriminative AI

GenAI offers valuable enhancements to discriminative AI (DAI) by addressing several challenges in data availability, robustness, and flexibility. One significant contribution is in *data augmentation*, where generative models can create synthetic data to supplement real data, increasing dataset diversity without the need for costly and time-intensive data collection. This capability is particularly useful in complex environments where data acquisition is challenging. In scenarios with *missing data*, GenAI can perform imputation by generating plausible values to fill in gaps, thus improving the accuracy and robustness of DAI models working with incomplete datasets. Furthermore, GenAI serves as a form of *regularization*, generating varied examples that reduce overfitting and encourage DAI models to learn more generalized features. GenAI also supports *dimensionality reduction* by creating lower-dimensional representations of high-dimensional data, capturing essential structures and correlations that simplify complex data for DAI processing. In *semisupervised learning*, GenAI can generate labeled data from unlabeled samples, enhancing model performance even with limited labeled data. Finally, in *multitask learning*, GenAI can produce data relevant to multiple tasks, enabling DAI models to learn shared representations and improve performance across various applications. Together, these roles demonstrate how GenAI effectively complements DAI, boosting its versatility and efficacy in handling diverse and complex data requirements.

2.3 Prominent Generative Models

2.3.1 Generative Adversarial Networks

Generative adversarial networks (GANs), first proposed by Goodfellow et al. [3], are a powerful class of generative models designed to produce synthetic data that closely resembles real-world examples. At their core, GANs consist of two neural networks that engage in a competitive learning process, as illustrated in Figure 2.2:

- **The Generator ($G(\boldsymbol{z})$):** This network takes as input a random noise vector, denoted by $\boldsymbol{z}$, which is sampled from a simple probability distribution (e.g., uniform or Gaussian). The generator learns to transform this random noise into a data sample, $\boldsymbol{x}$, that mimics the real data distribution, denoted as $p(\boldsymbol{x})$. In simple terms, the generator tries to create fake samples that look like real data.
- **The Discriminator ($D(\boldsymbol{x})$):** This network receives a data sample $\boldsymbol{x}$ (which could be either a real sample from the dataset or a fake sample produced by the generator) and predicts whether the input is real (from the actual data distribution) or fake (from the generator). The discriminator outputs a probability score, indicating its confidence that $\boldsymbol{x}$ is real.

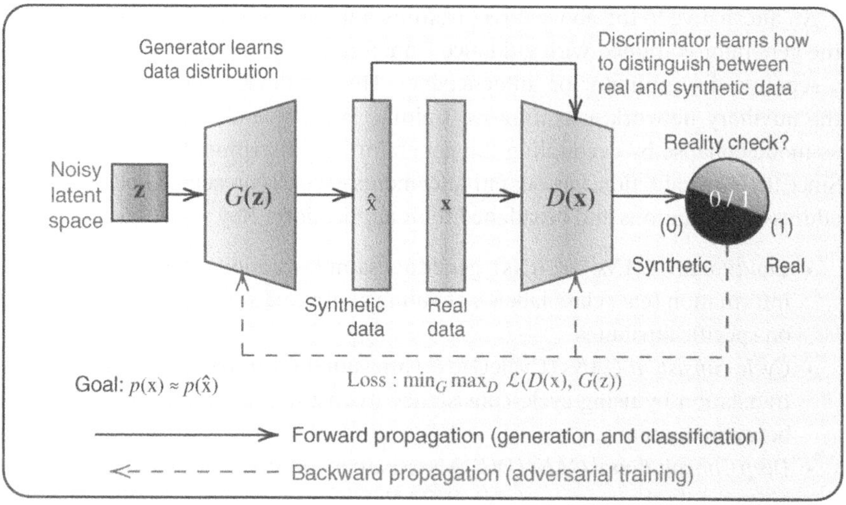

Figure 2.2 Schematic illustration of GAN architectures [2] / IEEE / CC BY 4.0.

The interaction between the generator and discriminator can be thought of as an adversarial game:

- The generator's goal is to produce samples that are so realistic that the discriminator cannot tell them apart from real data.
- The discriminator's goal is to correctly identify whether each sample it sees is real or generated.

This dynamic training process is formally captured by the following minimax objective function:

$$\min_{G(z)} \max_{D(x)} \mathcal{L}(D, G) = \mathbb{E}_x[\log D(\boldsymbol{x})] + \mathbb{E}_z[\log(1 - D(G(\boldsymbol{z})))], \tag{2.1}$$

where $\mathbb{E}_{x \sim p_{data}(x)}$ denotes the expectation over real data samples $\boldsymbol{x}$ drawn from the true data distribution $p_{data}(\boldsymbol{x})$, and $\mathbb{E}_{z \sim p_z(z)}$ denotes the expectation over noise vectors $\boldsymbol{z}$ sampled from the prior distribution $p_z(\boldsymbol{z})$ (typically uniform or Gaussian).

During training, the generator and discriminator weights are updated alternately, resulting in progressively better sample generation and discrimination capabilities. The generator tries to minimize the objective, making its fake samples increasingly realistic, while the discriminator tries to maximize the objective, improving its ability to correctly classify samples. This adversarial training gradually refines both networks, ideally leading to a generator that produces highly convincing synthetic data.

An alternative to the above direct training approach is indirect training, where the generator is trained with guidance from a pretrained auxiliary network (e.g., a separate discriminator or autoencoder). This method, which doesn't update the auxiliary network, can improve training stability and mitigate issues such as mode collapse by decoupling the generator and discriminator training [4, 5]. Since their introduction, various enhancements to GANs presented in Ref. [2] have addressed limitations and broadened their applications. Key variations include:

- *Conditional GANs* (cGANs) generate samples conditioned on additional information (e.g., class labels), enabling controlled sample generation based on specific attributes.
- *Cycle-consistent GANs* (CycleGANs) are suited for unpaired image-to-image translation by using cycle-consistency to ensure reversible transformations between domains.
- *Deep Convolutional GANs* (DCGANs) use convolutional layers to enhance stability, while *Wasserstein GANs* (WGANs) improve training stability by using the Wasserstein distance as a loss function.
- *Progressive GANs* (ProGANs) generate high-resolution images by progressively increasing resolution during training.
- *Bidirectional GANs* (BiGANs) add an encoder to learn mappings between data and latent space, aiding semisupervised and representation learning.
- *Information Maximizing GANs* (InfoGANs) enhance the disentangling of latent variables by maximizing mutual information, allowing for control over specific attributes in generated samples.
- *Categorical GANs* (CatGANs) improve semisupervised learning by incorporating a classifier in the discriminator, enhancing feature representation for classification.
- *Least Squares GANs* (LS-GANs) modify the loss function to use least squares, addressing stability and mode collapse issues.
- *Brainstorming GAN* (BGAN) enables distributed learning by allowing agents to exchange generated samples rather than sharing actual datasets, supporting collaborative learning across diverse architectures.

2.3.2 Variational Autoencoders

Conventional autoencoders (AEs) are neural network models designed to learn compact representations of data by encoding input samples into a lower-dimensional space and then reconstructing the original data from these compressed representations. As shown in Figure 2.3a, the encoder network, parameterized by θ, is denoted as $G_\theta(\boldsymbol{x})$. It transforms the input data $\boldsymbol{x}$ into a latent vector $\boldsymbol{z}$, capturing the essential features of the input. The decoder, parameterized

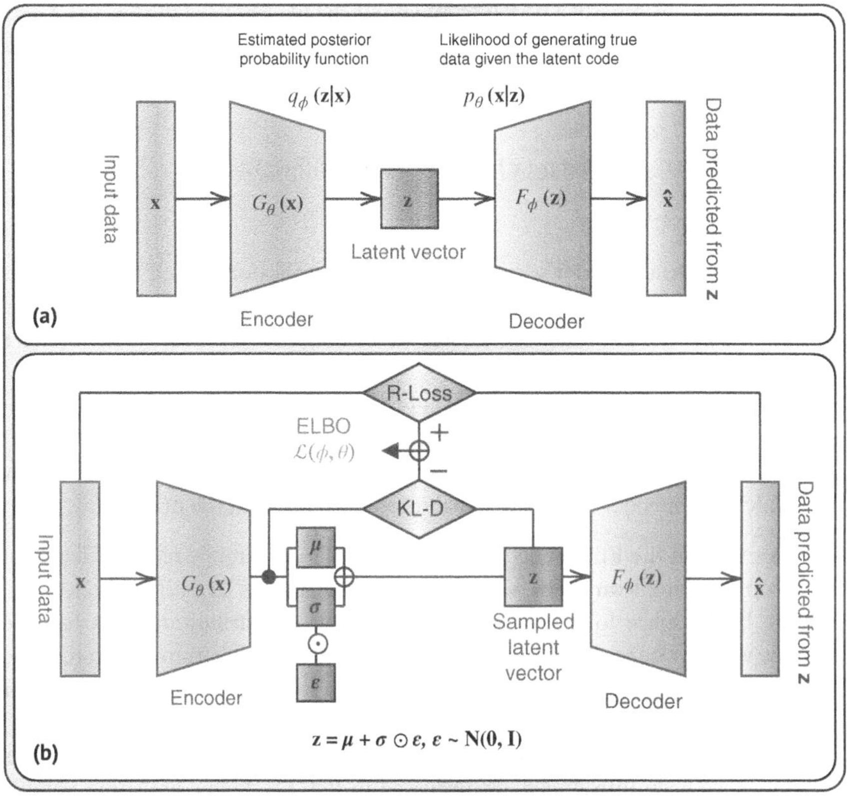

Figure 2.3 Schematic illustration of (a) AEs and (b) VAEs [2] / IEEE / CC BY 4.0.

by ϕ and denoted as $F_\phi(z)$, reconstructs the original data from this latent representation. While autoencoders excel at learning meaningful latent spaces for tasks such as data compression and reconstruction, they face challenges in generative modeling. Specifically, conventional AEs lack a probabilistic mechanism to ensure that new samples drawn from the latent space z correspond to valid data samples. This limitation hinders their ability to generate realistic new data instances.

To address this, variational autoencoders (VAEs) were introduced by Kingma and Welling [6], as depicted in Figure 2.3b. VAEs extend traditional autoencoders by framing the encoding-decoding process within a probabilistic model. Rather than mapping each input x to a single latent vector, VAEs learn a probabilistic distribution over latent variables z, capturing uncertainty in the data representation. This probabilistic approach allows for sampling new z values and generating realistic data samples through the decoder.

The training objective of a VAE is to maximize the likelihood of the observed data under the model. Since this likelihood is generally intractable to compute directly, VAEs optimize an approximation known as the evidence lower bound (ELBO), formulated as:

$$\mathcal{L}(\theta, \phi) = \mathbb{E}_{q_\phi(z|x)}[\log p_\theta(\boldsymbol{x} \mid \boldsymbol{z})] - \mathrm{KL}(q_\phi(\boldsymbol{z} \mid \boldsymbol{x}) || p(\boldsymbol{z})). \tag{2.2}$$

In this equation:

- $q_\phi(\boldsymbol{z} \mid \boldsymbol{x})$ is the encoder's learned posterior distribution, which approximates the true distribution of latent variables given input data.
- $p_\theta(\boldsymbol{x} \mid \boldsymbol{z})$ is the decoder's likelihood, describing how well the latent vector $\boldsymbol{z}$ can reconstruct the input $\boldsymbol{x}$.
- $p(\boldsymbol{z})$ is the prior distribution over latent variables, typically chosen as a standard multivariate Gaussian, $\mathcal{N}(\boldsymbol{z}; \boldsymbol{0}, \boldsymbol{I})$.
- $\mathrm{KL}(\cdot||\cdot)$ denotes the Kullback–Leibler (KL) divergence, a measure of how much the encoder's learned distribution deviates from the prior.

The first term in the ELBO encourages the model to accurately reconstruct input data from the latent variables, while the second term acts as a regularizer, pushing the latent space to align closely with the prior distribution. This balance enables VAEs to learn a smooth and continuous latent space from which realistic data samples can be generated by simply drawing random vectors from the prior distribution.

Following their emergence, several improvements and alternative formulations of VAEs have been introduced, as discussed in Ref. [2]. Below are some notable variants of VAEs:

- *β-VAE* incorporates a scaling factor β to adjust the balance between reconstruction fidelity and the regularization of the latent space, promoting disentangled and interpretable latent features for downstream tasks.
- *Wasserstein Autoencoders* (WAEs) substitute the KL divergence term with the Wasserstein distance, leading to more stable optimization and a better trade-off between data reconstruction and latent space regularization, enhancing the quality of generated samples.
- *VAE with Arbitrary Conditioning* (VAEAC) enables conditioning on any subset of observed variables, making it highly effective for scenarios involving incomplete data, such as data imputation tasks.
- *Vector Quantized VAEs* (VQ-VAEs) introduce vector quantization in the latent representation, resulting in a discrete latent space, which is particularly useful in symbolic data modeling tasks like text-to-speech synthesis and music composition.

- *Ladder-VAEs* organize the latent space in a hierarchical manner to capture features at multiple levels of abstraction, facilitating better interpretability and benefiting tasks that exploit hierarchical structures.
- *Conditional VAEs* (cVAEs) extend the standard VAE by incorporating conditional variables—such as labels or attributes—into the generation process, enabling more controlled and targeted sample generation.
- Although *normalizing flows* are not direct variants of VAEs, they can be combined with VAEs to model richer and more flexible posterior distributions, improving the model's ability to capture complex data structures (see Section 2.3.3 for details).
- *VAE-GAN* is a hybrid architecture that merges VAE and GAN methodologies, combining the interpretability of the VAE's latent space with the GAN's ability to generate sharp, realistic samples, while promoting more stable training dynamics.

These variants aim to improve VAEs in terms of sample quality and latent space disentanglement, broadening their usefulness across a variety of applications. Despite their stable training and interpretable latent representations, VAEs may still produce less visually sharp samples compared to GANs, which, although capable of high-fidelity outputs, suffer from complex training processes and challenges like mode collapse.

2.3.3 Flow-based Generative Models

Flow-based generative models (FGMs) are a class of models designed to map simple, well-understood probability distributions into complex, high-dimensional data distributions. Unlike other generative approaches that approximate or implicitly define the data distribution, FGMs leverage a sequence of invertible and differentiable transformations—known as flows—to perform this mapping exactly. This property allows FGMs to support precise likelihood computation, efficient sampling, and straightforward inference, as demonstrated by works such as Refs. [7] and [8].

As depicted in Figure 2.4, the process begins with a sample $\mathbf{z}$ drawn from a simple base distribution, typically a multivariate Gaussian distribution, denoted by $p(\mathbf{z})$. The model then learns a function f_θ, parameterized by θ, that transforms this simple sample into a data point $\mathbf{x}$ belonging to the target distribution $p(\mathbf{x})$. This transformation is invertible, meaning that one can recover $\mathbf{z}$ from $\mathbf{x}$ and vice versa:

- Inverse transformation: $\mathbf{x} = f_\theta^{-1}(\mathbf{z})$
- Forward transformation: $\mathbf{z} = f_\theta(\mathbf{x})$

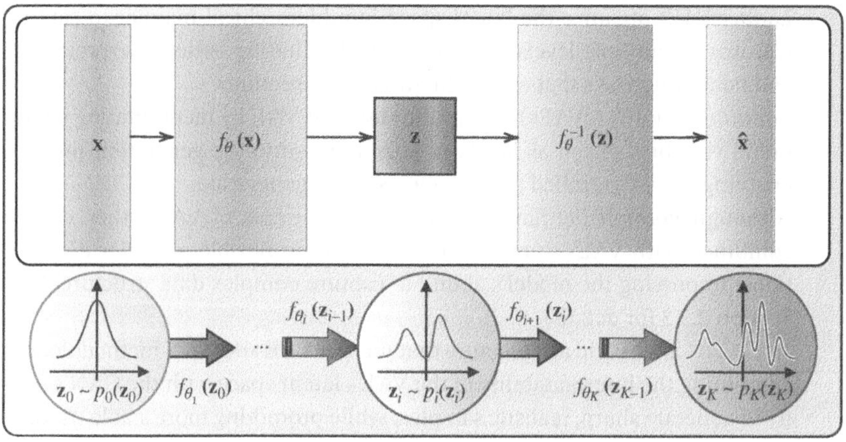

Figure 2.4 Schematic illustration of flow-based GMs [2] / IEEE / CC BY 4.0.

The probability density of a generated sample **x** can be computed exactly using the change of variables formula, which accounts for how the transformation stretches or compresses the space. This is mathematically expressed as:

$$p(\mathbf{x}) = p(\mathbf{z}) \left| \det \frac{df_\theta^{-1}(\mathbf{x})}{d\mathbf{x}} \right|, \tag{2.3}$$

where $\det \frac{df_\theta^{-1}(\mathbf{x})}{d\mathbf{x}}$ denotes the Jacobian determinant, describing how the transformation affects the volume in the data space. FGMs commonly use a sequence of invertible transformations, forming a normalizing flow:

$$p(\mathbf{x}) = p(\mathbf{z}) \left| \det \frac{df_\theta^{-1}(\mathbf{x})}{d\mathbf{x}} \right|, \tag{2.4}$$

and are trained by maximizing data likelihood. Ref. [2] presented prominent FGMs, which include:

- *Real-valued Non-volume Preserving* (RealNVP): Uses coupling layers to transform parts of the data with scale and translation functions, ensuring invertibility and efficient Jacobian computation, making it suitable for tasks such as image generation and representation learning.
- *Glow*: Builds on RealNVP with 1×1 invertible convolutions and actnorm layers, enhancing model expressiveness and training efficiency, and supports multiscale architecture for capturing both local and global data features, useful in image synthesis and translation.
- *Nonlinear Independent Components Estimation* (NICE): An early FGM using coupling layers without scaling functions, resulting in a simpler yet less expressive model, used for tasks such as image generation.

In addition, autoregressive flows combine the principles of autoregressive models (introduced in Section 2.3.6) with flow-based architectures. In this framework, each dimension of the data is modeled conditionally on the preceding ones, enabling the model to learn intricate dependencies within the data distribution and produce more realistic samples. This hybrid approach leverages the advantages of both autoregressive modeling and normalizing flows, achieving expressive modeling capabilities while preserving the exact likelihood computation and efficient sampling inherent to flow-based methods. Representative examples include the masked autoregressive flow (MAF) and the inverse autoregressive flow (IAF), which differ in the direction of their transformations: MAF applies an autoregressive structure in the forward pass, while IAF applies it during the inverse mapping.

2.3.4 Diffusion-based Generative Models

Diffusion-based generative models (DGMs) represent a novel approach to learning complex data distributions, drawing inspiration from the principles of nonequilibrium thermodynamics. The key idea behind DGMs is to gradually corrupt structured data by adding noise, transforming it into a simple, random distribution, and then training a model to reverse this corruption process, reconstructing the original data. This approach enables the generation of new, high-quality samples from random noise.

As illustrated in Figure 2.5, the forward process in DGMs is a continuous-time stochastic process, often discretized into T steps. At each step, a small amount

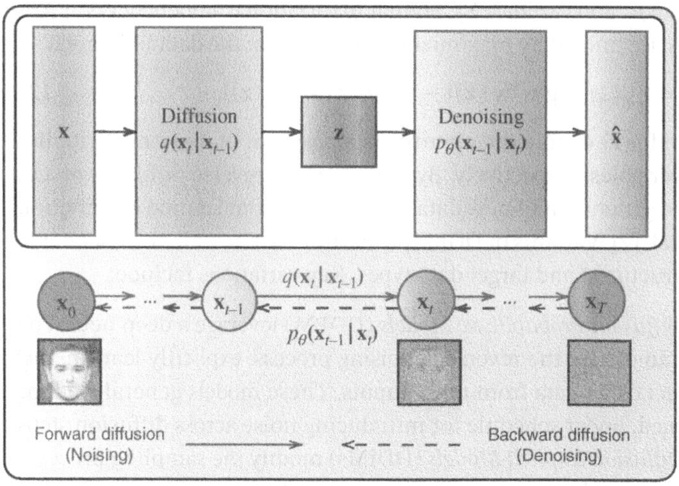

Figure 2.5 Schematic illustration of diffusion-based GMs [2] / IEEE / CC BY 4.0. The color version of this figure is available on Wiley.com.

of Gaussian noise is added to the data. This process, which can be modeled as a Markov chain, slowly transforms an initial data point $\mathbf{x}_0$, drawn from the true data distribution $q(\mathbf{x})$, into a highly noisy version $\mathbf{z} = \mathbf{x}_T$, which resembles an isotropic Gaussian distribution.

Mathematically, this progressive noising process is described by the following chain of conditional distributions:

$$q(\mathbf{x}_{1:T} \mid \mathbf{x}_0) = \prod_{t=1}^{T} q(\mathbf{x}_t \mid \mathbf{x}_{t-1}), \tag{2.5}$$

where each step $q(\mathbf{x}_t \mid \mathbf{x}_{t-1})$ is a Gaussian distribution:

$$q(\mathbf{x}_t \mid \mathbf{x}_{t-1}) \sim \mathcal{N}\left(\mathbf{x}_t; \mu_t = \sqrt{1-\beta_t}, \mathbf{x}_{t-1}, \boldsymbol{\Sigma}_t = \beta_t \mathbf{I}\right). \tag{2.6}$$

Here, β_t controls the amount of noise added at each step, and $\mathbf{I}$ denotes the identity matrix. Once the data has been fully diffused into noise ($\mathbf{z} = \mathbf{x}_T \sim \mathcal{N}(\mathbf{0}, \mathbf{I})$), the generative task involves reversing this process to recover a clean data sample. However, directly computing the exact reverse transitions $q(\mathbf{x}_{t-1} \mid \mathbf{x}_t)$ is infeasible. Instead, DGMs approximate these reverse steps using a parameterized model $p_\theta(\mathbf{x}_{t-1} \mid \mathbf{x}_t)$, defined as:

$$p_\theta(\mathbf{x}_{0:T}) = p(\mathbf{x}_T) \prod_{t=1}^{T} p_\theta(\mathbf{x}_{t-1} \mid \mathbf{x}_t), \tag{2.7}$$

where each reverse step is also modeled as a Gaussian:

$$p_\theta(\mathbf{x}_{t-1} \mid \mathbf{x}_t) \sim \mathcal{N}\left(\mathbf{x}_{t-1}; \boldsymbol{\mu}_\theta(\mathbf{x}_t, t), \boldsymbol{\Sigma}_\theta(\mathbf{x}_t, t)\right), \tag{2.8}$$

with learned mean $\boldsymbol{\mu}_\theta$ and variance $\boldsymbol{\Sigma}_\theta$. Similar to variational autoencoders (VAEs), diffusion models are trained by maximizing the ELBO on the data likelihood:

$$\mathcal{L}(\theta; \mathbf{x}, \mathbf{z}) = \mathbb{E}_{q(\mathbf{z}|\mathbf{x})}\left[\log p_\theta(\mathbf{x} \mid \mathbf{z})\right] - \mathrm{KL}\left(q(\mathbf{z} \mid \mathbf{x}) \parallel p(\mathbf{z})\right), \tag{2.9}$$

where $q(\mathbf{z} \mid \mathbf{x})$ and $p(\mathbf{z})$ denote the approximate posterior and prior distributions over the latent variables, respectively. By modeling the reverse diffusion process, DGMs enable the generation of new data by sampling from learned distributions, as described in Ref. [2]. Variants of DGMs primarily differ in their noise schedules, reverse model structures, and target data types. Key variations include:

- *Denoising Diffusion Probabilistic Models* (DDPM) leverage a deep neural network to parameterize the reverse denoising process, explicitly learning how to reconstruct clean data from noisy inputs. These models generally adopt a predetermined, linear schedule for introducing noise across diffusion steps.
- *Denoising Diffusion Implicit Models* (DDIMs) modify the sampling process to be deterministic rather than purely stochastic. DDIMs retain the same training objective as DDPMs but enable faster sample generation by reducing the number of required denoising steps through a non-Markovian formulation.

- *Score-based Models* estimate the data distribution's score function, the gradient of the log-density, guiding the reverse diffusion process. These models support various noise schedules and data types.
- *Denoising Score Matching* (DSM) trains energy-based models by aligning the model's score function with the data distribution. This approach enhances DGMs, enabling expressive generative modeling.

2.3.5 The Trilemma of GMs

GMs, such as (i) GANs, (ii) VAEs and normalizing flows, and (iii) diffusion models, are the three cornerstone of data generation tasks. However, these models are subject to the GenAI trilemma (c.f., Figure 2.6), which highlights three critical properties of generative models: fast sampling, mode coverage/diversity, and high sample quality. Each type of model excels in specific aspects while grappling with inherent limitations, as discussed below.

GANs are renowned for their ability to generate high-quality samples efficiently, making them particularly suitable for applications, including image synthesis, video generation, and real-time artistic style transfer. Their adversarial training framework allows GANs to produce outputs that are highly realistic. However, GANs often struggle with mode collapse, where the generator fails to capture the diversity of the underlying data distribution, resulting in repetitive or limited outputs. Additionally, their training process can be unstable and computationally intensive, requiring careful balancing between the generator and discriminator to avoid issues such as vanishing gradients or nonconvergence.

VAEs and normalizing flow-based models, on the other hand, prioritize mode coverage and diversity by explicitly learning the underlying data distribution

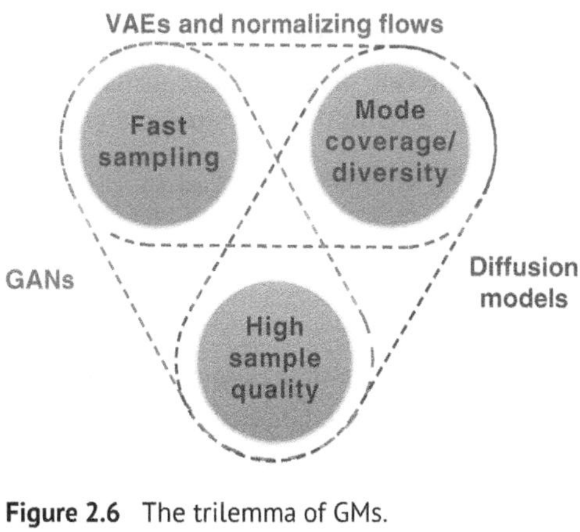

Figure 2.6 The trilemma of GMs.

through a latent variable representation. This makes them well-suited for applications including representation learning, anomaly detection, and clustering. Furthermore, VAEs offer efficient sampling due to their encoder-decoder architecture. However, their reliance on pixel-wise reconstruction loss often results in blurry or overly smooth outputs, which limits their sample quality compared to GANs. Additionally, the assumption of a predefined latent distribution may restrict their ability to generate highly detailed or complex data.

Diffusion models represent an emerging approach in generative modeling, excelling in both mode coverage and sample quality. By iteratively refining noisy data through a learned reverse diffusion process, these models are particularly robust in capturing complex data distributions. They have demonstrated exceptional performance in generating high-resolution images and videos, often surpassing GANs in certain domains. Despite these strengths, diffusion models face significant computational challenges due to their iterative sampling process, which is resource intensive. Moreover, their training demands large-scale datasets and extensive computational resources, making them less practical for real-time applications.

Exploring hybrid approaches or ensemble methods that combine the strengths of these aforementioned models may provide a path forward, offering solutions that balance the trade-offs highlighted by the trilemma.

2.3.6 Generative Autoregressive Models

Generative autoregressive models (GAMs) originate from traditional autoregressive models, where Yule and Box et al. [9, 10] initially developed for time series analysis. GAMs generate data by modeling the conditional probability distribution within a sequence, where each data point depends on preceding data points. For a data sequence $\mathbf{x} = [x_1, x_2, \dots, x_N]$ as illustrated in Figure 2.7, the joint probability is factorized as $p(\mathbf{x}) = \prod_{n=1}^{N} p(x_n \mid x_{<n})$, with each x_n conditioned on all prior data points $x_{<n}$. This structure allows GAMs to generate diverse and realistic data by sequentially sampling from these conditional distributions. Training is typically done via maximum likelihood estimation:

$$\mathcal{L}(\theta) = \sum_{n=1}^{N} \log p(x_n \mid \theta), \tag{2.10}$$

where θ are the model parameters.

Deep learning has expanded GAMs' capabilities, enabling them to capture complex dependencies, handle large and high-dimensional datasets, adapt flexibly to various data types, and capture long-range dependencies using techniques such

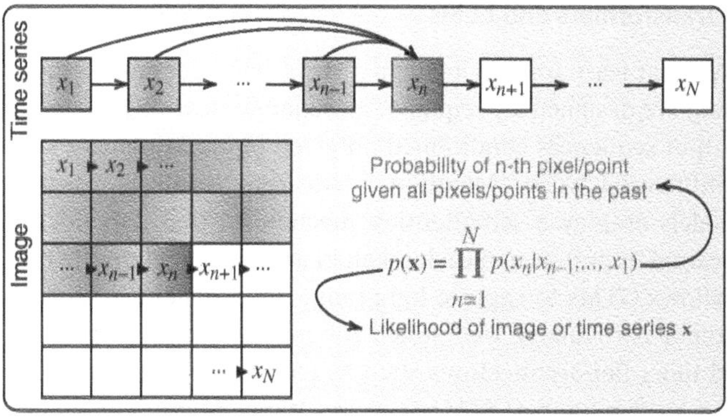

Figure 2.7 Schematic illustration of GAMs [2] / IEEE / CC BY 4.0.

as attention mechanisms and dilated convolutions. The most widely used deep GAMs, as summarized in Ref. [2]:

- *Fully Visible Sigmoid Belief Networks* (FVSBNs) utilize feedforward architectures with sigmoid activation functions to estimate the conditional probability of each variable based on all prior variables in a given order. These models consist of a single hidden layer and are typically trained using maximum likelihood optimization.
- *WaveNet* is specialized for raw audio generation, using dilated convolutional layers to effectively model long-term dependencies in audio sequences, enabling the generation of highly natural speech and music.
- *Neural Autoregressive Distribution Estimation* (NADE) builds on FVSBNs by employing a deep neural network with shared parameters across input dimensions, leading to improved training scalability and better generalization.
- *PixelRNN* and *PixelCNN* are autoregressive models developed for image synthesis. PixelRNN applies recurrent neural networks, while PixelCNN leverages convolutional layers to capture spatial dependencies and long-range correlations, both producing high-quality image outputs.
- *Masked Autoencoder for Distribution Estimation* (MADE) adopts an autoencoder framework where weight masking enforces autoregressive constraints. This design ensures that each output depends solely on preceding input dimensions, offering enhanced computational efficiency and improved performance over NADE.

2.3.7 Generative Transformers and LLMs

Unlike the above GMs that learn underlying data distributions, generative transformer models (GTMs) are designed for sequence-to-sequence learning, enabling the generation of output sequences conditioned on given input sequences. First introduced in the influential work "Attention is All You Need" by Vaswani et al. [11], these models employ a self-attention mechanism that dynamically determines the contextual importance of each token in a sequence relative to others. This approach allows GTMs to capture long-range dependencies with high efficiency, making them particularly well suited for natural language processing tasks. Compared to earlier architectures such as recurrent neural networks (RNNs) and long short-term memory (LSTM) networks, transformers demonstrate superior performance in modeling long and complex sequences.

2.3.7.1 GTM Architecture

A GTM is composed of two core modules: an encoder and a decoder, both built from several layers stacked sequentially, as depicted in Figure 2.8 (adapted from Ref. [11]). The encoder processes the input sequence and encodes it into a series of continuous latent vectors, which are then used by the decoder to generate the output sequence. Both encoder and decoder incorporate a self-attention mechanism, which enables the model to capture dependencies across the entire sequence. Specifically, the encoder can attend to all tokens in the input, while the decoder considers the full context of the previously generated tokens to predict the next one. In the sequel, we present a clear and intuitive description of the GTM's algorithmic structure and mathematical formulation.

- *Self-attention Mechanism:* In this process, each input token is projected into three distinct representations—query, key, and value—to compute attention scores that reflect how strongly each token relates to the others within the sequence. For an input sequence denoted as $X = [x_1, x_2, \dots, x_n]$, where each token x_i belongs to $\mathbb{R}^d$ (with d representing the feature dimension and n the total sequence length), the model applies learnable weight matrices W_Q, W_K, and W_V to obtain the respective query, key, and value vectors: $q_i = x_i W_Q$, $k_i = x_i W_K$, and $v_i = x_i W_V$.

  The attention score between token pairs (x_i, x_j) is computed by taking the dot product of their query and key vectors, scaled by the factor $1/\sqrt{d}$ to stabilize gradients, and normalized using the softmax function: $s_{ij} = \text{softmax}\left(q_i \cdot k_j^T / \sqrt{d}\right)$. Each token's contextualized representation z_i is then formed as a weighted sum of all value vectors v_j, where the attention scores s_{ij} serve as the weights: s_{ij}: $z_i = \sum_{j=1}^{n} s_{ij} v_j$. Finally, this result is transformed using an additional trainable matrix W_O to produce the final output of the

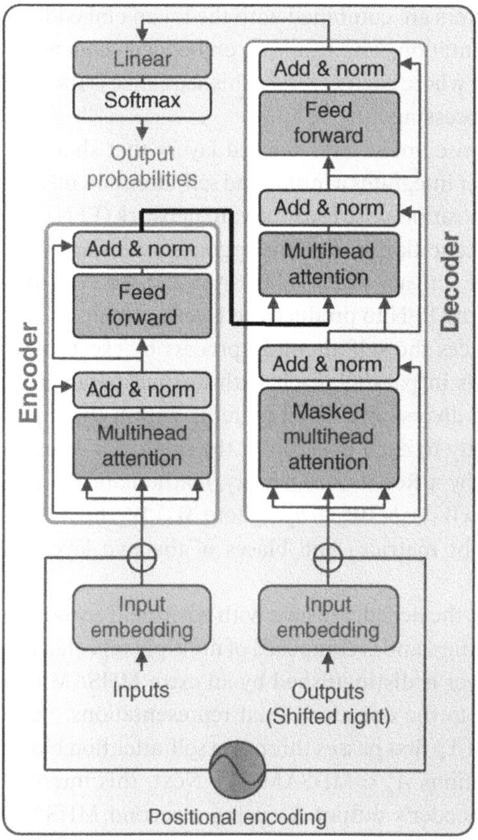

Figure 2.8 Schematic illustration of GTMs.

self-attention block for token x_i: $o_i = z_i W_O$. This mechanism enables the model to dynamically focus on the most relevant parts of the sequence when processing each token.

- *Input Embedding and Position Encoding:* The first step in the GTM pipeline involves converting discrete input tokens (such as words or subword units) into continuous vector representations. This is achieved using an embedding matrix, which is learned during training and serves to map each token to a dense vector space. Assuming a vocabulary of size V and an embedding dimension of d, the embedding matrix E is of size $V \times d$. Given an input sequence consisting of token indices $[i_1, i_2, \dots, i_n]$, the model looks up the corresponding embeddings for each token. Since embedding lookups alone do not convey the sequential order of tokens, the model integrates positional encodings to encode the relative or absolute position of each token within the sequence. These positional vectors are combined with the token embeddings to form the final input representation. The resulting embedded sequence is expressed as $X = [x_1, x_2, \dots, x_n]$, where each $x_i \in \mathbb{R}^d$. This sequence is then fed into the encoder for further processing.

- *Encoder:* The encoder is composed of several stacked layers that share the same structure, where each layer integrates a multihead self-attention mechanism (MHSAM) followed by a position-wise feed-forward network (FFN). For a given layer ℓ, the input representation X_ℓ first undergoes processing by the MHSAM, yielding intermediate activations $A_\ell = \text{MHSAM}(X_\ell)$. These activations are then passed through the FFN to produce the layer's output: $X_{\ell+1} = \text{FFN}(A_\ell)$. The MHSAM enhances the self-attention process by executing it across multiple attention heads in parallel, each with distinct parameters, enabling the model to capture diverse relational patterns within the input. The FFN, applied independently to each position in the sequence, consists of two linear layers separated by a ReLU nonlinearity, mathematically represented as: $\text{FFN}(x) = \max(0, xW_1 + b_1)W_2 + b_2$, where W_1, W_2, b_1, b_2 correspond to the trainable weight matrices and biases of the two layers of the FFN.
- *Decoder:* Similar to the encoder, the decoder begins with positional encodings combined with output embeddings, and is composed of multiple repeated layers. However, each decoder layer is distinguished by an extra MHSAM that enables the decoder to attend to the encoder's final representations. For a given decoder layer ℓ, the input Y_ℓ first passes through a self-attention block, producing intermediate activations $A_\ell^1 = \text{MHSAM}(Y_\ell)$. Next, this intermediate output attends to the encoder's output X_ℓ using a second MHSAM, resulting in $A_\ell^2 = \text{MHSAM}(A_\ell^1, X_\ell)$. The final step in the layer applies a feed-forward network to A_ℓ^2, producing the next layer's input: $Y_{\ell+1} = \text{FFN}(A_\ell^2)$. A key feature of the decoder is the masked multihead self-attention mechanism, which ensures that during training, each token only considers the tokens that come before it. This masking preserves the autoregressive nature of the model, preventing information leakage from future tokens and enforcing a strictly left-to-right generation process. Moreover, to stabilize training and improve convergence, both encoder and decoder layers integrate layer normalization and residual (skip) connections, allowing gradients to flow more effectively through the deep network.
- *Teacher Forcing Training:* Training in GTMs employs teacher forcing, where the correct previous output is fed into the decoder at each step during training. This method stabilizes the learning process by reducing error propagation across sequences. The training process involves embedding inputs, encoding, decoding with teacher forcing, computing the loss, and updating parameters through optimization, typically via stochastic gradient descent or its variants.

2.3.7.2 Large Language Models

GTMs have provided the architectural foundation for the emergence of large language models (LLMs), which have significantly advanced in natural language processing (NLP) applications such as machine translation, text summarization, and question answering. LLMs extend the core GTM design by stacking multiple transformer layers (denoted as L layers), each incorporating self-attention mechanisms, as illustrated in Figure 2.9, which outlines the generative pretrained transformer (GPT) framework. Unlike earlier models, LLMs are trained on vast text corpora—spanning hundreds of gigabytes—allowing them to capture complex linguistic structures, grammatical rules, and broader world knowledge. This extensive training enables them to produce fluent and contextually accurate language outputs.

LLMs are characterized by their enormous scale, typically involving billions to hundreds of billions of parameters. For instance, OpenAI's GPT-3 model has 175 billion parameters, while newer models such as GPT-4 and Google's PaLM exceed even this size, enhancing language comprehension and generation capabilities. This vast number of parameters and data allows LLMs to excel at a wide

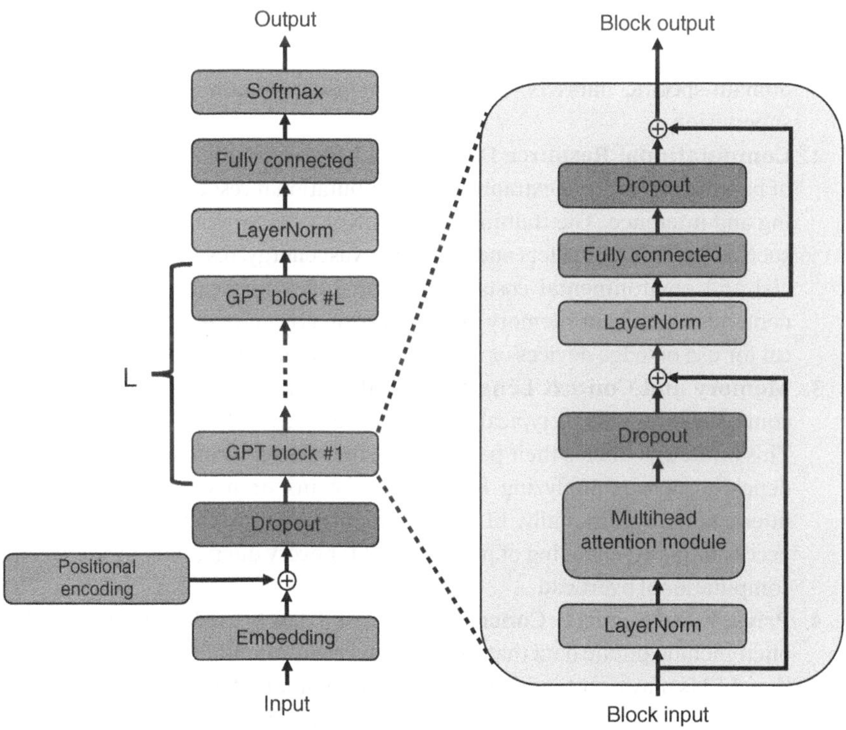

Figure 2.9 Conceptual architecture of a LLM GPT model.

range of NLP tasks, either through fine-tuning for specific applications or using zero-shot and few-shot prompting for flexibility across domains. Through prompt engineering or task-specific tuning, LLMs can generate accurate, contextually rich responses in applications ranging from summarization and question answering to conversational AI.

Although LLMs have become crucial tools in modern NLP, their large model size and data requirements pose challenges in terms of computational resources for both training and inference. Moreover, these models face significant limitations that constrain their efficiency, applicability, and scalability, especially for domain-specific challenges in fields such as wireless communications, economics, healthcare, and law. The sequel outlines these limitations and explores strategies to overcome them:

1. **Generalized Learning with Domain-specific Gaps:** LLMs are designed as general-purpose systems trained on diverse, large-scale datasets. While this enables versatility, it limits their effectiveness in domain-specific applications. For instance, in wireless communications, they may fail to account for nuances including propagation models or resource allocation protocols. Similarly, in healthcare or legal settings, LLMs might generate contextually inappropriate or factually incorrect responses, posing risks in high-stakes applications. Addressing these gaps often requires fine-tuning with curated, domain-specific datasets, a resource-intensive process demanding expert supervision.
2. **Computational Resource Demands:** LLMs, with billions or even trillions of parameters, require extraordinary computational resources for both training and inference. The training process involves large-scale hardware setups such as GPU/TPU clusters and consumes vast energy, resulting in high financial and environmental costs. Deploying LLMs for real-time applications demands significant memory and processing power, making them impractical for use on edge devices or in resource-constrained environments.
3. **Memory and Context Length Limitations:** LLMs are restricted by their context window size, typically ranging between 1,024 and 2,048 tokens. This limitation affects their performance in tasks requiring long-term dependencies, such as analyzing lengthy documents or maintaining persistent interactions. Additionally, LLMs lack memory persistence across sessions, necessitating re-encoding of prior context for every query, thereby increasing computational overhead.
4. **Privacy and Security Concerns:** The vast datasets used for training LLMs often include public data that might inadvertently contain sensitive information. LLMs may unintentionally reveal confidential information during inference. Ensuring strict isolation of sensitive data during training and inference remains challenging, particularly in regulated domains such as finance and healthcare.

5. **Hallucination and Reliability Issues:** LLMs often generate factually incorrect or fabricated outputs, a phenomenon known as "hallucination." These inaccuracies stem from their reliance on learned statistical patterns rather than grounded reasoning, limiting their reliability in tasks requiring precision.
6. **Static Knowledge and Adaptability:** LLMs are trained on static datasets and cannot dynamically update their knowledge. Consequently, they struggle to adapt to evolving information, such as new scientific discoveries or regulatory changes.

2.3.8 Strategies to Address LLM Limitations

Traditional fine-tuning of LLMs typically requires updating all model parameters, which can be both computationally intensive and costly. Fine-tuning a large model on a target dataset means adjusting billions of parameters, which consumes substantial GPU/TPU resources and extends training times. Additionally, deploying these models for inference requires considerable memory and energy, presenting practical challenges for real-time applications, especially on devices with limited resources. Moreover, these models are prone to overfitting when adapted to niche domains, which necessitates a method that can generalize well across various tasks without excessive retraining.

To make LLMs more adaptable to domain-specific needs without the extensive computational costs of full fine-tuning, parameter-efficient fine-tuning (PEFT) techniques have emerged. PEFT methods add small, task-focused layers or parameters to an already-trained LLM as presented in Ref. [12] and illustrated in Figure 2.10. By fine-tuning only these additional parameters rather than the entire model, PEFT makes the process more computationally feasible while enabling the LLM to learn domain-specific knowledge effectively. PEFT techniques

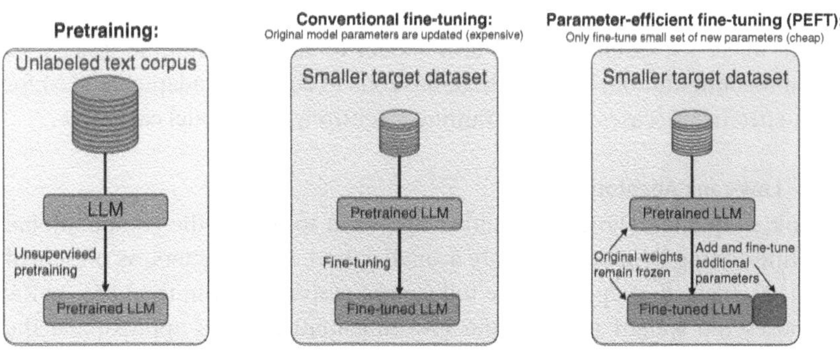

Figure 2.10 Fine-tuning techniques.

such as *adapters*, *low-rank adaptation*, and *quantized LoRA (QLoRA)* represent advancements in fine-tuning approaches, focusing on task-specific updates while keeping the base model largely intact. Another approach to overcoming LLM limitations is *retrieval-augmented generation (RAG)*. RAG enhances LLMs by integrating external domain-specific knowledge bases into the generative process. Queries are matched against an embedded knowledge base using semantic search, retrieving relevant context that is then incorporated into the model's response generation. This hybrid architecture allows LLMs to remain general-purpose while excelling in providing accurate, contextually informed answers for specialized tasks.

2.3.8.1 Adapters in LLM Fine-tuning

Adapters are additional layers inserted into each transformer block within the model. These adapter layers are lightweight neural networks trained specifically for the task at hand, while the primary model parameters remain frozen. For example, in each transformer block, adapters can be placed after key layers such as the self-attention mechanism and the feed-forward layer. When fine-tuning, only the adapter layers are modified, significantly reducing the number of parameters that require adjustment.

By freezing the main LLM parameters and fine-tuning only the adapters, we not only reduce memory and computational requirements but also create a modular system where different adapter layers can be swapped in and out for various tasks. This flexibility allows a single LLM to perform well across multiple domains by loading task-specific adapters when needed. Adapter tuning can be done using different strategies:

- *Additive adapters* add simple layers to capture domain-specific knowledge.
- *Reparametrization-based adapters* (such as LoRA) use structured decompositions to fine-tune efficiently.
- *Selective adapters* apply adapters only in certain layers, focusing computational power on the most relevant parts of the model.

Adapters make it possible for one base model to excel in multiple tasks with minimal reconfiguration, allowing organizations to deploy an adaptable LLM for domain-specific use cases without training an entirely new model each time.

2.3.8.2 Low-rank Adaptation

Low-rank adaptation (LoRA) is another approach to reduce the computational cost of fine-tuning by modifying only a small subset of parameters, as presented in Ref. [13]. Instead of updating the full weight matrices within the LLM, LoRA decomposes the weight updates into low-rank matrices that approximate the changes needed for a new task. Specifically, it introduces two smaller matrices, $\mathbf{A}$ and $\mathbf{B}$, that are used to approximate changes to the original weight matrix $\mathbf{W}$. The product **AB** serves as a low-rank update to **W**, capturing the task-specific adjustments with a fraction of the parameters required in traditional fine-tuning.

For instance, if **W** is a $d \times d$ matrix, LoRA learns two smaller matrices **A** (of size $d \times r$) and **B** (of size $r \times d$), where r is much smaller than d. During fine-tuning, only **A** and **B** are updated, while the original **W** remains unchanged, as depicted in Figure 2.11. The effective weight update then becomes $\mathbf{W} + \mathbf{AB}$, allowing the model to learn task-specific information without the need for full weight updates. This drastically reduces memory requirements, as the trainable parameters are now only a small fraction of the original parameters. This method allows LoRA to maintain high task performance with significantly reduced memory usage, as the storage and computational demands are limited to the smaller matrices **A** and **B**.

LoRA's approach of low-rank decomposition is especially beneficial for large models where full weight updates are impractical. Since only the matrices **A** and **B** are optimized, it allows for faster training times and more efficient deployment. Moreover, LoRA's flexibility makes it suitable for tasks that require nuanced adjustments, such as adapting LLMs for highly specialized language in technical fields.

2.3.8.3 Quantized Low-rank Adaptation

Building on the efficiency of LoRA, quantized low-rank adaptation (QLoRA) introduces an additional optimization step by reducing the precision of the low-rank matrices A and B through quantization, as introduced in Ref. [14]. As illustrated in Figure 2.12, QLoRA reduces the precision of weight parameters in pretrained LLMs to 4-bit. Conventionally, model parameters are stored in a standard 16- or 32-bit floating-point representations, but QLoRA compresses them to a quantized 4-bit representation. By representing matrix values with fewer bits, QLoRA significantly cuts down memory requirements without major sacrifices in model accuracy. This makes QLoRA particularly attractive for settings with

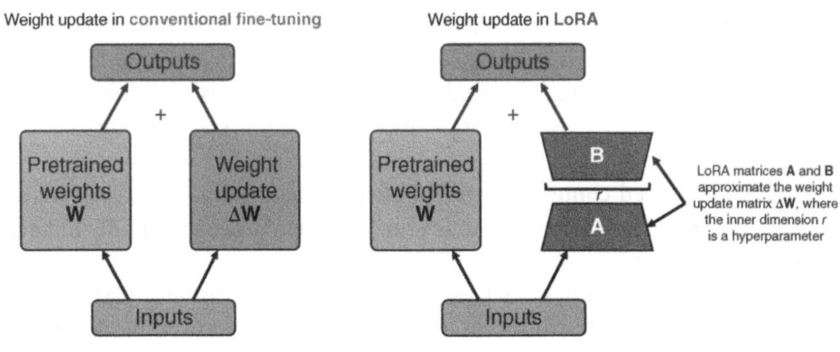

Figure 2.11 LoRA fine-tuning.

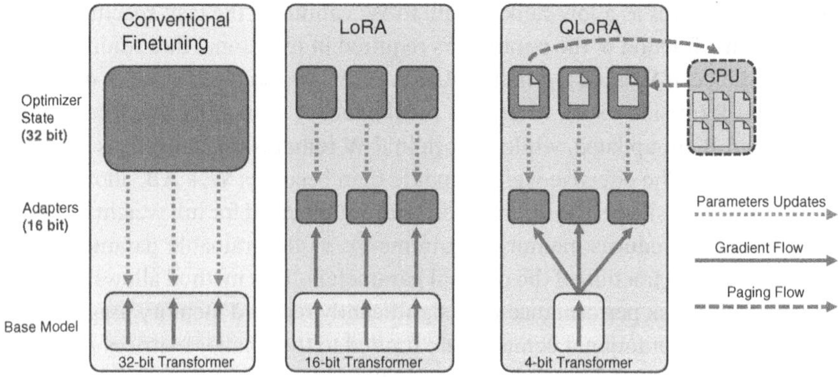

Figure 2.12 QLoRA fine-tuning.

limited hardware resources, as it allows fine-tuning with even lower computational overhead. QLoRA not only achieves comparable performance to full fine-tuning but also minimizes storage and computational costs. By quantizing the matrices involved in low-rank updates, QLoRA enables large models to be deployed on smaller, resource-constrained devices while retaining high accuracy on domain-specific tasks.

2.3.8.4 Retrieval-augmented Generation

RAG addresses key limitations of LLMs by integrating domain-specific knowledge directly into their generative processes. While LLMs excel at general language understanding and response generation, they often struggle with maintaining up-to-date information or providing highly accurate answers in specialized fields. RAG mitigates these limitations by combining the natural language generation capabilities of LLMs with external retrieval mechanisms to enhance their performance.

The RAG framework, as illustrated in Figure 2.13, begins with the construction of a curated, domain-relevant knowledge base. This knowledge base is embedded into a vector space using advanced embedding techniques, enabling efficient semantic search. When a user inputs a query, the query is also vectorized and matched against the knowledge base to identify and rank the most relevant information. This retrieved context is then supplied to the LLM as additional input, ensuring that the generated response is both informed and contextually aligned with the user's query.

In specialized applications such as wireless communications, RAG proves invaluable. It enables LLMs to access historical data, technical standards, or pretrained models tailored to specific protocols or technologies. By incorporating

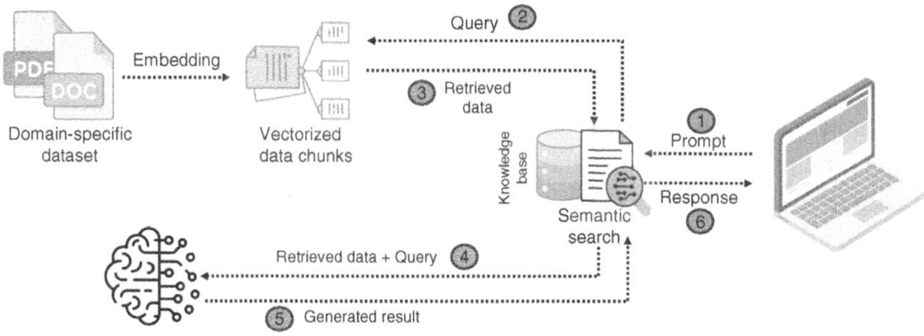

Figure 2.13 RAG framework illustration.

this domain-specific knowledge, RAG enhances the model's ability to generate accurate, precise, and contextually appropriate responses to complex technical inquiries. For example, an LLM integrated with RAG can effectively answer questions about wireless network standards, network optimization techniques, or troubleshooting scenarios by pulling information directly from authoritative sources in real-time.

Moreover, RAG's modular nature makes it highly adaptable to various domains. By updating the embedded knowledge base, RAG-equipped LLMs can remain current without requiring extensive retraining. This approach not only improves response accuracy but also reduces the computational overhead associated with training massive models on ever-expanding datasets. In essence, RAG serves as a bridge between the broad linguistic capabilities of LLMs and the precise demands of domain-specific applications, making it a critical innovation for enhancing the functionality and reliability of AI-driven systems.

2.4 GenAI Applications to Canonical Problems in Communications Systems

2.4.1 Physical Layer Design

GMs offer a robust approach to addressing the complex challenges associated with physical layer design in wireless communication systems. By learning underlying patterns and data structures, GMs support accurate channel modeling, estimation, and equalization. Specifically, they have proven effective for estimating nonlinear and time-varying channel characteristics, even under mathematically challenging conditions. For high-frequency domains, such as mmWave and THz, GMs such as cGANs and other GAN architectures assist in generating channel representations that adapt to high mobility and antenna diversity, improving reliability and data fidelity in mobile and multiantenna systems [15, 16, 17].

Moreover, GMs can approximate complex, stochastic channel behaviors, providing a data-driven alternative to traditional analytical methods for systems affected by various distortion sources [18, 19].

Beyond channel estimation, GMs contribute to blind equalization by compensating for channel-induced signal distortions. Approaches such as VQ-VAEs and maximum likelihood VAE models enhance the adaptability of equalizers, extending their applicability to nonlinear, high-order modulation systems and achieving error rates that surpass conventional algorithms [20, 21]. Additionally, GMs facilitate modulation recognition through the use of GANs, which accurately classify modulation schemes and enable spectrum-efficient communication by recognizing signal patterns and adjusting to varying conditions [22, 23]. These applications underscore the transformative potential of GMs in refining the physical layer, setting the stage for further innovations in high-frequency channel modeling and adaptive signal processing for future communication networks.

2.4.2 Network Resource Management

GMs have shown significant potential in enhancing network resource management by supporting efficient allocation, prediction, and optimization of resources across wireless networks. By generating synthetic data and improving the learning processes in reinforcement learning (RL) and deep reinforcement learning (DRL) frameworks, models such as GANs and VAEs have proven effective in addressing resource allocation challenges in nonstationary environments. For instance, Hua et al. introduced the GAN-based deep distributional Q network (GAN-DDQN) to enhance resource allocation in dynamic 5G network slicing by representing state-action values and mitigating issues linked to fluctuating network conditions [24]. The addition of a reward-clipping mechanism in GAN-DDQN and the separation of action and state-value distributions in dueling GAN-DDQN enable more efficient allocation across network slices, leading to improvements in spectrum efficiency and adherence to service level agreements.

In addition to resource allocation, GMs contribute to adaptive network slicing, which is essential for efficient management of network, computing, and storage resources. By predicting user demands, GMs allow network operators to allocate resources in a timely and flexible manner. Gu and Zhang [25] proposed GANSlicing, a GAN-based approach for IoT-driven dynamic network slicing, which enhances quality-of-experience (QoE) by efficiently allocating resources according to user requirements, outperforming traditional slicing approaches. Similarly, other studies have utilized cGANs to generate synthetic data that replicates real network conditions, thereby improving the flexibility and responsiveness of heterogeneous vehicular networks in dynamic environments [26]. These advancements underscore the value of generative models in improving resource management and adapting to changing network demands in real-time.

2.4.3 Network Traffic Analytics

GMs, particularly GANs, have shown exceptional potential in advancing network traffic analytics by addressing critical challenges such as synthetic traffic data generation, mitigating class imbalances, enabling dynamic traffic camouflaging, improving traffic super-resolution, and forecasting temporal and long-term traffic patterns. These capabilities have significantly enhanced network management through innovative approaches for data augmentation, anomaly detection, and predictive modeling. For instance, Aho et al. [27] demonstrated the utility of a GAN-based approach for generating realistic traffic across diverse network scenarios, significantly improving machine learning models for traffic classification. Similarly, Shahid et al. [28] combined AEs with GANs to create synthetic IoT network flows that mimic real traffic patterns, facilitating the evaluation of intrusion detection systems. Furthermore, FlowGAN, proposed in Ref. [29], was designed to address class imbalance in traffic classification by generating synthetic samples for underrepresented traffic classes, leading to notable improvements in metrics such as precision, recall, and F1-scores.

Beyond data generation, GMs have proven invaluable in predictive analytics and resource optimization. For instance, ZipNet-GAN, a GAN tailored for mobile traffic super-resolution presented in Ref. [30], can extract fine-grained mobile traffic patterns at up to 100x higher granularity compared to standard probing techniques, thereby optimizing resource allocation.

2.4.4 Cross-layer Network Security

GMs have demonstrated remarkable capabilities in enhancing cross-layer network security by addressing challenges such as physical layer security, jamming and spoofing recognition, anomaly detection, and intrusion detection. At the physical layer, RF fingerprinting has emerged as a robust method for device identification by leveraging unique hardware characteristics and channel features. Chen et al. [31] proposed a GAN-based method that achieves 95% accuracy in identifying rogue devices, showcasing its effectiveness in real-world environments. However, adversarial attacks continue to challenge these systems, as highlighted by studies revealing GAN-generated signals that mimic legitimate device characteristics, exposing vulnerabilities in RF fingerprinting mechanisms.

In jamming and spoofing scenarios, GMs have been employed for both attack and defense strategies. Tang et al. [32] introduced a jamming recognition model based on AC-VAEGAN, improving detection accuracy even with limited data. On the defensive side, adversarial ML strategies have shown promise in disrupting jamming attacks by employing selective spectrum access and data completion techniques. Similarly, GANs have been used to generate indistinguishable spoofing signals, posing significant challenges to existing anti-spoofing mechanisms. These dual roles of GMs—as tools for enhancing security and as enablers of sophisticated attacks—highlight their transformative yet complex role in network security.

2.4.5 Localization and Positioning

GMs have significantly contributed to solving challenges in localization and positioning, such as enhancing indoor accuracy, addressing issues in WiFi-based localization, automating radio map creation, and augmenting data for cost-efficient localization systems. A prominent example is the Tensor-GAN, developed by Liu and Wang [33], which enhances indoor localization by leveraging a three-player adversarial game involving a generator, discriminator, and regressor. This model not only improves localization accuracy but also reduces storage requirements, making it suitable for complex indoor environments.

In addition, Zhou et al. [34] introduced RaGAM, a generative model designed to automate the creation of radio maps for target intrusion sensing and localization. This model employs ray-tracing techniques to construct synthetic maps, significantly lowering computational costs while achieving high accuracy. Meanwhile, Njima et al. [35] showcased the application of GANs for data augmentation in indoor localization. By generating synthetic RSSI fingerprints from limited real data, their method improved localization accuracy by over 21%, demonstrating the potential of GAN-based techniques for addressing data scarcity. Furthermore, Suroso et al. [36] employed VAEs to enhance the density of offline fingerprinting databases, offering a cost-effective solution for expanding localization datasets while maintaining accuracy.

2.5 Future Communication Frontiers for GMs

GMs are set to play a transformative role in shaping the frontiers of future wireless networks. As next-generation wireless networks move beyond traditional connectivity to embrace intelligent, adaptive, and highly personalized services, GMs offer versatile tools to address the unique challenges of this new era. Key domains where GMs will leave their mark include semantic communications, integrated sensing and communication (ISAC), digital twins (DTs), AI-generated content services, MEC, EAI, and the broader context of AML and trustworthy AI.

2.5.1 Semantic Communications

Semantic communication represents a paradigm shift in wireless networks, prioritizing the transmission of meaningful information rather than raw bits. Unlike conventional communication systems that focus on minimizing distortion or loss, semantic communication incorporates context, prior knowledge, and intent to optimize data exchange. This transition is especially relevant for 6G networks, where vast and heterogeneous data streams must be intelligently processed. GenAI plays a pivotal role in enabling this vision by enhancing the efficiency, adaptability, and security of semantic communication systems.

LLMs contribute significantly to contextual analysis, semantic compression, and intent learning, enabling systems to interpret and prioritize relevant information before transmission. By analyzing message semantics, LLMs can extract essential details, reducing redundancy while preserving clarity. This is particularly beneficial for real-time applications where bandwidth efficiency is crucial. Additionally, LLMs facilitate the development of interactive AI agents that autonomously optimize communication strategies, evolving beyond traditional connected intelligence to a more self-governed, adaptive intelligence in telecom networks [37, 38].

Other generative models such as GANs, VAEs, and DGMs further advance semantic communication. GANs enable data synthesis, image translation, and adaptive modulation, ensuring that transmitted information is structured in an interpretable format. VAEs specialize in latent space compression, minimizing transmitted data size while preserving its essential meaning, which is crucial for resource-constrained networks. DGMs introduce new techniques for data enhancement and security, allowing for steganographic embedding of hidden messages within transmitted signals. FGMs further improve real-time adaptability and probabilistic reasoning, allowing systems to dynamically adjust encoding strategies based on changing conditions, ensuring that conveyed meaning remains accurate and relevant [39, 40].

By integrating GenAI into semantic communication, 6G networks can minimize transmission overhead, improve signal clarity, and enhance network intelligence through real-time learning and adaptation. These advancements position semantic communication as a cornerstone of next-generation wireless systems, fundamentally transforming how devices and users exchange information.

2.5.2 Integrated Sensing and Communications

ISAC is a fundamental component of future 6G networks, enabling systems to dynamically sense, interpret, and respond to environmental changes while optimizing communication performance. GMs play a pivotal role in ISAC by enhancing various aspects such as data augmentation, where GANs and VAEs generate synthetic yet realistic sensor data to enrich training datasets and improve model generalization [41]. Additionally, GMs contribute to anomaly detection, with normalizing flows from FGMs providing precise density estimations to detect outliers, while DGMs enhance detection robustness by incorporating uncertainty quantification [42]. In resource optimization, GANs simulate network conditions to improve decision-making processes, while FGMs help model complex network states, facilitating intelligent allocation of communication and sensing resources [43].

Beyond these applications, ISAC also benefits from time-series forecasting, where DGMs and LSTM models predict sensor readings based on historical data, enabling proactive decision-making [44]. Sensor fusion is another critical application, where VAEs and normalizing flows consolidate data from multiple sources into a unified representation, improving environmental perception and situational awareness [45]. Furthermore, decision-making is enhanced by transformer-based generative models (i.e., GTMs), which process sequential data to optimize communication strategies, while LSTMs contribute to efficient temporal reasoning [46]. Finally, uncertainty quantification, facilitated by VAEs and DGMs, ensures robust and reliable sensor-driven predictions by providing confidence intervals and probabilistic insights. By integrating these AI-driven techniques, ISAC can support highly adaptive and efficient 6G networks, paving the way for intelligent communication and sensing solutions.

2.5.3 Digital Twins

The integration of DTs into 6G networks is essential for creating dynamic, real-time virtual representations of network environments, enabling efficient monitoring, optimization, and predictive analytics. As 6G networks scale to accommodate large-scale IoT deployments and increasingly complex wireless environments, DTs facilitate the modeling of these intricate ecosystems. GMs play a crucial role in enhancing DTs by addressing key challenges such as high-resolution network simulation, uncertainty quantification, and continuous model refinement [47]. GANs are particularly effective in generating realistic network simulations, enabling the creation of high-fidelity digital replicas of network states based on sensory data [48]. VAEs, with their probabilistic nature, contribute to uncertainty quantification, ensuring that DT models accurately represent real-world conditions with confidence estimates, which is crucial for predictive maintenance and network reliability [49].

Beyond static modeling, GTMs provide a robust solution for temporal modeling and prediction, translating sequential network data into actionable forecasts that help operators anticipate performance fluctuations and optimize resources [50]. FGMs, particularly normalizing flows, enhance distribution modeling, capturing subtle variations in network behavior and ensuring DTs remain synchronized with real-world conditions [51]. Additionally, DGMs enable continuous DT refinement, iteratively updating digital models with new sensor data to improve accuracy and responsiveness in dynamic environments [52]. By integrating these AI-driven techniques, DTs in 6G networks can transition from static digital models to intelligent, self-evolving systems that enhance real-time decision-making, improve operational efficiency, and drive the next wave of intelligent connectivity.

2.5.4 AI-generated Content for 6G Networks

The emergence of 6G networks will fundamentally transform digital content creation, distribution, and consumption by integrating AI-generated content as a core capability. GMs will play a key role in enabling real-time, high-fidelity content generation tailored to user preferences and dynamic network conditions. LLMs and GPTs will power interactive AI services, such as conversational agents and personalized recommendations, while GANs and DGMs will generate and modify high-resolution images and videos, enabling seamless multimedia adaptation for augmented and virtual reality (AR/VR) applications [38]. Additionally, FGMs will support 3D object generation and transformations, further enhancing immersive digital experiences. These capabilities ensure that 6G networks provide real-time, context-aware content while reducing the dependency on centralized data storage and computational resources.

Beyond content creation, VAEs and FGMs will drive efficient content distribution by performing adaptive data compression and reconstruction, optimizing streaming performance across diverse network environments [49]. In smart cities and IoT ecosystems, GANs and DGMs will generate synthetic urban datasets, facilitating city planning and environmental monitoring. Meanwhile, EAI models, such as lightweight GANs, will enable localized content creation, allowing devices to generate and process content at the edge, minimizing latency and improving responsiveness for real-time applications like autonomous driving [52]. By integrating these generative AI technologies, 6G networks will shift from passive data transmission to intelligent, adaptive content generation, ultimately enhancing digital experiences while optimizing network efficiency.

2.5.5 MEC and EAI

The integration of MEC and EAI in 6G networks marks a paradigm shift from centralized cloud processing to decentralized, real-time computation at the network's edge. This transition is critical for latency-sensitive applications, remote deployments, and intelligent autonomous systems. GMs enhance MEC-EAI capabilities by enabling localized data augmentation, with GANs generating synthetic data directly at the edge to support real-time AI training and evaluation in data-scarce environments [53]. VAEs, with their probabilistic nature, facilitate on-device uncertainty modeling, helping edge devices assess confidence levels in predictions, which is crucial for safety-critical applications such as autonomous driving and remote healthcare [54]. Meanwhile, LSTMs and GTMs optimize real-time sequence processing, ensuring dynamic adjustments in AR/VR experiences and human-machine interactions.

Beyond data augmentation and processing, normalizing flows enable complex distribution modeling at the edge, allowing localized AI to interpret multi-modal sensor data more accurately. DGMs, known for their iterative learning capabilities, refine noisy or incomplete edge data over time, enhancing decision-making accuracy in unpredictable environments such as robotics and industrial automation. The fusion of MEC and EAI, supported by GMs, ensures that 6G networks can process, analyze, and act on data at the edge with minimal latency and maximal efficiency [55]. By decentralizing computation and leveraging adaptive generative models, 6G networks will be better equipped to support intelligent, real-time decision-making across diverse applications, ranging from smart cities to autonomous systems.

2.5.6 Adversarial Machine Learning and Trustworthy AI

As AI-driven 6G networks expand, they become increasingly vulnerable to AML attacks, where malicious actors manipulate inputs to mislead AI models. GMs play a dual role in AML: they can be leveraged to generate adversarial examples—for instance, GANs can craft deceptive samples that trick classifiers [56]—but also serve as a defense mechanism. By training a GAN discriminator to detect adversarial inputs, AI models can learn to distinguish between real and tampered data [57]. Additionally, VAEs can mitigate adversarial attacks by reconstructing inputs and filtering out perturbations before classification [58]. As 6G introduces MEC and EAI, real-time AI inferences at the edge become susceptible to adversarial manipulations, making GTMs and LSTMs critical for detecting anomalies in sequential data, such as compromised communication signals [59]. FGMs and DGMs further enhance security by modeling complex data distributions to flag discrepancies indicative of adversarial interference.

Beyond AML, trustworthy AI (TAI) ensures AI-driven 6G networks uphold ethical, transparent, and robust decision-making [60]. GMs contribute to key TAI pillars, such as privacy and data governance, where GANs generate synthetic datasets for training without exposing sensitive data, enabling federated learning at the edge. Transparency is enhanced by VAEs, which create interpretable latent space representations, helping users understand AI decisions. Diversity and fairness are supported by GANs, which test AI models across diverse data distributions to prevent bias. Additionally, GMs enable accountability by stress-testing AI models under various scenarios, allowing researchers to analyze and refine decision-making processes. As 6G networks become deeply integrated into society, the ability of GMs to fortify AI security, fairness, and reliability will be pivotal in establishing resilient and ethical AI systems for the next-generation intelligent infrastructure.

2.6 Regulation and Policy

As GMs are set to be integrated into next-generation wireless networks, regulation and policy frameworks will be essential to ensure their safe, ethical, and effective deployment. GMs, with their potential to revolutionize communication, analytics, and decision-making, also pose unique challenges, necessitating nuanced oversight to address privacy, security, and fairness concerns.

One critical frontier involves balancing innovation with data protection and ethical responsibility. GMs may inadvertently process sensitive data or replicate biases embedded in their training datasets. In the context of next-generation wireless networks, such biases could lead to discriminatory practices in service allocation or decision-making, raising ethical concerns and potential regulatory scrutiny. Ensuring compliance with frameworks such as the General Data Protection Regulation (GDPR) or future global data protection standards without stifling innovation will be paramount. The operational transparency of GMs, especially in high-stakes applications such as healthcare or national security, must also be maintained, fostering public trust in these technologies.

National security considerations add another dimension to regulatory challenges. The misuse of GMs for generating deepfakes, misinformation, or cyber-attacks could have far-reaching consequences. Governments may impose restrictions on certain types of GMs, regulating their development and deployment to mitigate such risks. These measures, while necessary, could impede the global adoption and interoperability of next-generation wireless networks.

Lastly, the global nature of next-generation wireless networks demands cross-border coordination and collaboration between public and private entities. Differing national regulations, varying levels of technological maturity, and diverse societal priorities make harmonization complex. Establishing international standards for GM behavior, data governance, and interoperability will be crucial. Additionally, fostering public-private partnerships can align commercial innovation with societal needs, ensuring shared accountability and mutual benefit.

As next-generation wireless networks evolve, addressing these regulatory and policy challenges will require a multifaceted approach that integrates technical, legal, and societal perspectives. The development of adaptive, transparent, and inclusive regulatory frameworks will be instrumental in realizing the full potential of GMs while safeguarding public interests.

2.7 Summary

This chapter has provided a comprehensive exploration of GenAI models and their transformative role in wireless communications and next-generation networks. Beginning with an overview of generative models, we established a taxonomy distinguishing explicit and implicit density models, detailing their mathematical foundations and practical implications. The discussion then covered key GenAI architectures, including GANs, VAEs, flow-based models, diffusion-based models, generative transformers, autoregressive models, and LLMs, explaining their structures, advantages, and limitations.

The latter sections explored how GenAI is revolutionizing multiple domains within wireless communications, including physical layer design, network optimization, network traffic analytics, security, localization, and positioning. Looking toward the future, we examined cutting-edge applications in 6G networks, such as semantic communications, integrated sensing and communications, digital twins, AI-generated content, MEC, EAI, AML, and trustworthy AI. Each of these frontiers benefits significantly from generative models by enabling more intelligent, efficient, and adaptive communication networks.

As generative models become integral to wireless systems, regulatory and ethical considerations must be addressed. Issues surrounding privacy, security, transparency, and fairness necessitate careful policy frameworks to balance innovation with responsible deployment. The continued evolution of GenAI in wireless networks holds immense potential, and this chapter provides a foundation for researchers and industry practitioners aiming to harness its capabilities for future communication systems.

References

1. Ian Goodfellow. Nips 2016 tutorial: Generative adversarial networks. *arXiv preprint arXiv:1701.00160*, 2016.
2. Abdulkadir Celik and Ahmed M. Eltawil. At the dawn of generative AI era: A tutorial-cum-survey on new frontiers in 6G wireless intelligence. *IEEE Open Journal of the Communications Society*, 5: 2433–2489, 2024. doi: 10.1109/OJCOMS.2024.3362271.
3. Ian Goodfellow, Jean Pouget-Abadie, Mehdi Mirza, Bing Xu, David Warde-Farley, Sherjil Ozair, Aaron Courville, and Yoshua Bengio. Generative adversarial nets. *Advances in Neural Information Processing Systems*, pages 2672–2680, 2014.
4. Jonathan Ho and Stefano Ermon. Generative adversarial imitation learning. *NeurIPS*, 2016.
5. Luke Metz, Ben Poole, David Pfau, and Jascha Sohl-Dickstein. Unrolled generative adversarial networks. *ICLR*, 2017.
6. Diederik P. Kingma and Max Welling. Auto-encoding variational bayes. *arXiv preprint arXiv:1312.6114*, 2013.
7. Danilo Rezende and Shakir Mohamed. Variational inference with normalizing flows. In *International Conference on Machine Learning*, pages 1530–1538. PMLR, 2015.

8 Durk P. Kingma and Prafulla Dhariwal. Glow: Generative flow with invertible 1x1 convolutions. *Advances in neural information processing systems*, 31, 2018.

9 Udny Yule. A method of investigating periodicities in disturbed series, with special reference to wolfer's sunspot numbers. *Philosophical Transactions of the Royal Society of London. Series A, Containing Papers of a Mathematical or Physical Character*, 226: 267–298, 1927.

10 George EP Box, Gwilym M. Jenkins, Gregory C. Reinsel, and Greta M. Ljung. *Time Series Analysis: Forecasting and Control.* John Wiley & Sons, 2015.

11 Ashish Vaswani, Noam Shazeer, Niki Parmar, Jakob Uszkoreit, Llion Jones, Aidan N Gomez, Lukasz Kaiser, and Illia Polosukhin. Attention is all you need. *Advances in Neural Information Processing Systems*, 30: 5998–6008, 2017.

12 Vladislav Lialin, Vijeta Deshpande, and Anna Rumshisky. Scaling down to scale up: A guide to parameter-efficient fine-tuning, arXiv preprint arXiv:2303.15647. 2023.

13 Edward J. Hu, Yelong Shen, Phillip Wallis, Zeyuan Allen-Zhu, Yuanzhi Li, Shean Wang, Lu Wang, and Weizhu Chen. LoRA: Low-rank adaptation of large language models. *arXiv preprint arXiv:2106.09685*, 2021. https://arxiv.org/abs/2106.09685.

14 Tim Dettmers, Artidoro Pagnoni, Ari Holtzman, and Luke Zettlemoyer. QLoRA: Efficient finetuning of quantized LLMs. *arXiv preprint arXiv:2305.14314*, 2023. https://arxiv.org/abs/2305.14314.

15 Xiaofeng Li, Ahmed Alkhateeb, and Cihan Tepedelenlioğlu. Generative adversarial estimation of channel covariance in vehicular millimeter wave systems. In *52nd Asilomar Conference on Signals, System and Computers*, pages 1572–1576, 2018. doi: 10.1109/ACSSC.2018.8645463.

16 Qianqian Zhang, Aidin Ferdowsi, and Walid Saad. Distributed generative adversarial networks for mmwave channel modeling in wireless UAV networks. In *IEEE International Conference on Communications (ICC)*, pages 1–6, 2021. doi: 10.1109/ICC42927.2021.9501056.

17 Eren Balevi and Jeffrey G. Andrews. Wideband channel estimation with a generative adversarial network. *IEEE Transaction Wireless Communications*, 20(5): 3049–3060, 2021. doi: 10.1109/TWC.2020.3047100.

18 Timothy J. O'Shea, Tamoghna Roy, and Nathan West. Approximating the void: Learning stochastic channel models from observation with variational generative adversarial networks. In *2019 International Conference on Computing, Networking and Communications (ICNC)*, pages 681–686, 2019. doi: 10.1109/ICCNC.2019.8685573.

19 Aaron Smith and Joseph Downey. A communication channel density estimating generative adversarial network. In *IEEE Cognitive Communications for Aerospace Applications Workshop (CCAAW)*, pages 1–7, 2019. doi: 10.1109/CCAAW.2019.8904907.

20 Jinxiang Song, Vincent Lauinger, Yibo Wu, Christian Häger, Jochen Schröder, Alexandre Graell i Amat, Laurent Schmalen, and Henk Wymeersch. Blind channel equalization using vector-quantized variational autoencoders, arXiv preprint arXiv:2302.11687. 2023.

21 Avi Caciularu and David Burshtein. Blind channel equalization using variational autoencoders. In *IEEE International Conference Communications Workshops (ICC Workshops)*, pages 1–6, 2018. doi: 10.1109/ICCW.2018.8403666.

22 Mansi Patel, Xuyu Wang, and Shiwen Mao. Data augmentation with conditional GAN for automatic modulation classification. In *Proceedings of the 2nd ACM Workshop on Wireless Security and Machine Learning*, WiseML '20, page 31–36, New York, NY, USA, 2020. Association for Computing Machinery. ISBN 9781450380072.

23 IV William H Clark, Steven Hauser, William C Headley, and Alan J. Michaels. Training data augmentation for deep learning radio frequency systems. *The Journal of Defense Modeling and Simulation*, 18(3): 217–237, 2021.

24 Yuxiu Hua, Rongpeng Li, Zhifeng Zhao, Xianfu Chen, and Honggang Zhang. GAN-powered deep distributional reinforcement learning for resource management in network slicing. *IEEE Journal on Selected Areas in Communications*, 38(2): 334–349, 2019.

25 Ruichun Gu and Junxing Zhang. Ganslicing: A GAN-based software defined mobile network slicing scheme for IoT applications. In *IEEE International Conference on Communications (ICC)*, pages 1–7. IEEE, 2019.

26 Farnoush Falahatraftar, Samuel Pierre, and Steven Chamberland. A conditional generative adversarial network based approach for network slicing in heterogeneous vehicular networks. In *Telecom*, volume 2, pages 141–154. MDPI, 2021.

27 Jon J. Aho, Alexander W. Witt, Carter B. F. Casey, Nirav Trivedi, and Venkatesh Ramaswamy. Generating realistic data for network analytics. In *MILCOM 2018 - IEEE Military Communications Conference (MILCOM)*, pages 401–406, 2018. doi: 10.1109/MILCOM.2018.8599782.

28 Mustafizur R. Shahid, Gregory Blanc, Houda Jmila, Zonghua Zhang, and Hervé Debar. Generative deep learning for internet of things network traffic generation. In *IEEE 25th Pacific Rim International Symposium on Dependable Computing (PRDC)*, pages 70–79, 2020. doi: 10.1109/PRDC50213.2020.00018.

29 ZiXuan Wang, Pan Wang, Xiaokang Zhou, ShuHang Li, and MoXuan Zhang. FlowGAN: Unbalanced network encrypted traffic identification method based on GAN. In *IEEE Intl Conf on Parallel & Distributed Processing with Applications, Big Data & Cloud Computing, Sustainable Computing & Communications, Social Computing & Networking (ISPA/BDCloud/SocialCom/SustainCom)*, pages 975–983, 2019. doi: 10.1109/ISPA-BDCloud-SustainCom-SocialCom48970.2019.00141.

30 Chaoyun Zhang, Xi Ouyang, and Paul Patras. Zipnet-GAN: Inferring fine-grained mobile traffic patterns via a generative adversarial neural network. In *Proceedings of the 13th International Conference on Emerging Networking Experiments and Technologies*, CoNEXT '17, page 363–375, New York, NY, USA, 2017. Association for Computing Machinery. ISBN 9781450354226.

31 Zekun Chen, Linning Peng, Aiqun Hu, and Hua Fu. Generative adversarial network-based rogue device identification using differential constellation trace figure. *EURASIP Journal on Wireless Communications and Net*working, 2021(1): 1–27, 2021.

32 Yan Tang, Zhijin Zhao, Xueyi Ye, Shilian Zheng, and Lijun Wang. Jamming recognition based on AC-VAEGAN. In *2020 15th IEEE International Conference on Signal Processing (ICSP)*, volume 1, pages 312–315, 2020. doi: 10.1109/ICSP48669.2020.9320987.

33 Xiao-Yang Liu and Xiaodong Wang. Real-time indoor localization for smartphones using tensor-generative adversarial nets. *IEEE Transactions on Neural Networks and Learning Systems*, 32(8): 3433–3443, 2021. doi: 10.1109/TNNLS.2020.3010724.

34 Mu Zhou, Yixin Lin, Nan Zhao, Qing Jiang, Xiaolong Yang, and Zengshan Tian. Indoor WLAN intelligent target intrusion sensing using ray-aided generative adversarial network. *IEEE Trans. Emerging Topics in Computational Intelligence*, 4(1): 61–73, 2020. doi: 10.1109/TETCI.2019.2892748.

35 Wafa Njima, Marwa Chafii, and Raed M. Shubair. GAN-based data augmentation for indoor localization using labeled and unlabeled data. In *2021 International Balkan Conference on Communications and Networking* (*BalkanCom*), pages 36–39, 2021. doi: 10.1109/BalkanCom53780.2021.9593240.

36 Dwi Joko Suroso, Panarat Cherntanomwong, and Pitikhate Sooraksa. Deep generative model-based RSSI synthesis for indoor localization. In *2022 19th International Conference on Electrical Engineering/Electronics, Computer, Telecommunications and Information Technology* (*ECTI-CON*), pages 1–5, 2022. doi: 10.1109/ECTI-CON54298.2022.9795409.

37 Lina Bariah, Hang Zou, Qiyang Zhao, Belkacem Mouhouche, Faouzi Bader, and Merouane Debbah. Understanding telecom language through large language models. *arXiv preprint arXiv:2306.07933*, 2023.

38 Hang Zou, Qiyang Zhao, Lina Bariah, Mehdi Bennis, and Merouane Debbah. Wireless multi-agent generative AI: From connected intelligence to collective intelligence. *arXiv preprint arXiv:2307.02757*, 2023.

39 Henry Gilbert, Michael Sandborn, Douglas C Schmidt, Jesse Spencer-Smith, and Jules White. Semantic compression with large language models. *arXiv preprint arXiv:2304.12512*, 2023.

40 Phillip Isola, Jun-Yan Zhu, Tinghui Zhou, and Alexei A Efros. Image-to-image translation with conditional adversarial networks. In *Proceedings of the IEEE Conference on Computer Vision and Pattern Recognition*, pages 1125–1134, 2017.

41 Ahmad El Sallab, Ibrahim Sobh, Mohamed Zahran, and Nader Essam. Lidar sensor modeling and data augmentation with GANs for autonomous driving. *arXiv preprint arXiv:1905.07290*, 2019.

42 Ling Yang, Zhilong Zhang, Yang Song, Shenda Hong, Runsheng Xu, Yue Zhao, Yingxia Shao, Wentao Zhang, Bin Cui, and Ming-Hsuan Yang. Diffusion models: A comprehensive survey of methods and applications. *arXiv preprint arXiv:2209.00796*, 2022.

43 Suihong Song, Tapan Mukerji, and Jiagen Hou. Gansim: Conditional facies simulation using an improved progressive growing of generative adversarial networks (gans). *Mathematical Geosciences*, pages 1–32, 2021.

44 Yan Li, Xinjiang Lu, Yaqing Wang, and Dejing Dou. Generative time series forecasting with diffusion, denoise, and disentanglement. *Advances in Neural Information Processing Systems*, 35: 23009–23022, 2022.

45 Fabian Duffhauss, Ngo Anh Vien, Hanna Ziesche, and Gerhard Neumann. Fusionvae: A deep hierarchical variational autoencoder for RGB image fusion. In *European Conference on Computer Vision*, pages 674–691. Springer, 2022.

46 Saeid Amiri, Mohammad Shokrolah Shirazi, and Shiqi Zhang. Learning and reasoning for robot sequential decision making under uncertainty. In *Proceedings of the AAAI conference on artificial intelligence*, volume 34, pages 2726–2733, 2020.

47 Lina Bariah and Mérouane Debbah. The interplay of AI and digital twin: Bridging the gap between data-driven and model-driven approaches. *arXiv preprint arXiv:2209.12423*, 2022.

48 Wafa Njima, Marwa Chafii, Arsenia Chorti, Raed M. Shubair, and H. Vincent Poor. Indoor localization using data augmentation via selective generative adversarial networks. *IEEE Access*, 9: 98337–98347, 2021. doi: 10.1109/ACCESS.2021.3095546.

49 Nuojin Cheng, Osman Asif Malik, Stephen Becker, and Alireza Doostan. Bi-fidelity variational auto-encoder for uncertainty quantification. *arXiv preprint arXiv:2305.16530*, 2023.

50 Elham Shabaninia, Hossein Nezamabadi-pour, and Fatemeh Shafizadegan. Transformers in action recognition: A review on temporal modeling. *arXiv preprint arXiv:2302.01921*, 2022.

51 Sam Bond-Taylor, Adam Leach, Yang Long, and Chris G. Willcocks. Deep generative modelling: A comparative review of VAEs, GANs, normalizing flows, energy-based and autoregressive models. *IEEE Transactions on Pattern Analysis and Machine Intelligence*, 44(11):7327–7347, 2021.

52 Hanqun Cao, Cheng Tan, Zhangyang Gao, Yilun Xu, Guangyong Chen, Pheng-Ann Heng, and Stan Z Li. A survey on generative diffusion model. *arXiv preprint arXiv:2209.02646*, 2022.

53 Yu Weng and Haiwen Zhou. Data augmentation computing model based on generative adversarial network. *IEEE Access*, 7: 64223–64233, 2019.

54 Arian Ahmadi, Omid Semiari, Mehdi Bennis, and Mérouane Debbah. Variational autoencoders for reliability optimization in multi-access edge computing networks. In *2022 IEEE Wireless Communications and Networking Conference (WCNC)*, pages 752–757. IEEE, 2022.

55 Yun-Cheng Wang, Jintang Xue, Chengwei Wei, and C. C. Jay Kuo. An overview on generative AI at scale with edge-cloud computing. *IEEE Open Journal of the Communications Society*, 4:2952–2971. 2023.

56 Jinyin Chen, Haibin Zheng, Hui Xiong, Shijing Shen, and Mengmeng Su. MAG-GAN: Massive attack generator via GAN. *Information Sciences*, 536: 67–90, 2020.

57 Yashar Deldjoo, Tommaso Di Noia, and Felice Antonio Merra. A survey on adversarial recommender systems: from attack/defense strategies to generative adversarial networks. *ACM Computing Surveys (CSUR)*, 54(2): 1–38, 2021.

58 Partha Ghosh, Arpan Losalka, and Michael J Black. Resisting adversarial attacks using gaussian mixture variational autoencoders. In *Proceedings of the AAAI Conference on Artificial Intelligence*, volume 33, pages 541–548, 2019.

59 Liyan Xu, Kang Xu, Yinchuan Qin, Yixuan Li, Xingting Huang, Zhicheng Lin, Ning Ye, and Xuechun Ji. TGAN-AD: Transformer-based GAN for anomaly detection of time series data. *Applied Sciences*, 12(16): 8085, 2022.

60 Davinder Kaur, Suleyman Uslu, Kaley J Rittichier, and Arjan Durresi. Trustworthy artificial intelligence: A review. *ACM Computing Surveys (CSUR)*, 55(2): 1–38, 2022.

3

Key Study Cases of Generative AI Applications to Communications Systems

Mehdi Letafati, Samad Ali, and Matti Latva-aho*

Center for Wireless Communications, University of Oulu, Finland

*Corresponding author: Mehdi.Letafati@oulu.fi

3.1 Overview on the Roles of Generative AI in Communication Systems

In this section, we review the potential study cases of key generative models in communication networks. The goal is to motivate readers about some of the potential applications of generative artificial intelligence (GenAI) in real-world wireless scenarios. Then, in Section 3.1.1, we focus on a specific use case based on denoising diffusion models.

Groundbreaking capabilities of GenAI in analyzing complicated data distributions implies its huge potential for wireless technologies [1]. As an example, diffusion models can learn complicated channel distributions and generate high quality synthetic channel realizations. Variational autoencoders (VAEs) can model the encoder-decoder functionalities in communication systems, and the generative-discriminative nature of generative adversarial networks (GANs) can be used for signaling classification tasks. In what follows, we provide an overview on some of these applications.

3.1.1 Use Cases of Generative Adversarial Networks in Communications

GANs are comprised of two main elements: the generator and the discriminator. These two models are employed to solve a minimax game, the former samples data, and the latter differentiates (classifies) the data as real or generated. The training

Generative AI for Communications Systems: Fundamentals, Applications, and Prospects, First Edition. Edited by Diep N. Nguyen et al.

process based on Nash equilibrium is aimed such that the discriminator cannot differentiate between the real or generated data.

GANs have been widely studied in the wireless AI literature, most typically in the context of modulation and signal classification. This can be explained due to the *generative-discriminative nature* of GAN's underlying mechanism. Traditional modulation classification algorithms require accurate knowledge of the underlying distribution of channels and the transmitted and received signals in order to achieve reasonable performance. Such approaches might not be effective in future wireless technologies due to the increased complexity of signals, diverse spectrum requirements, and high mobility of user equipment (UE). In addition, deep neural network (DNN)-based solutions typically rely on the availability of large datasets, working well only in specific environments that are "seen" during the training. In this regard, GANs have been studied to enrich the training dataset by generating synthetic data [2]. In addition, due to their *adversarial learning*-like characteristics, GANs can be leveraged to enhance the resilience of modulation and signaling classification schemes [3]. The generator-discriminator mechanism of GANs can also be leveraged for the use cases of signal anomaly detection of rogue or unexpected wireless signals, resulting in improved interference mitigation, wireless security, etc. For example, identifying unrecognized patterns within the wireless time-series data has been studied in Ref. [4], using the combination of GANs (utilizing its discriminative-generative capability) and long short-term memory recurrent neural network (LSTM), utilizing its sequence-type processing.

GANs' generative ability to learn complicated distributions also finds its applications in channel modeling as a typical basic application of generative models in wireless [5]. Although conventional DNNs perform well in discriminative tasks, they perform poorly in use cases that require modeling the complicated underlying distributions of wireless channels. This is one of the key use cases where GANs can be employed. In a more advanced setup, GAN's ability for *generating synthetic data* has been proposed for generating radio maps from environmental observations [6]. In particular, the generator creates radio map image masks, while the discriminator learns to distinguish the masks from the real dataset and the ones that are generated by the generator. GANs have shown some initial potential as an innovative building block for semantic communications as well. Thanks to the various perceptual quality metrics that can be employed within the framework of deep generative models, for example, pre-trained GANs, the decoder of a deep learning-based joint source channel coding (DeepJSCC) system can utilize GANs to enhance semantic image recovery [7].

To summarize, applications of GANs in communication systems mostly target modulation signal classification, anomaly detection, synthetic channel and radio map generation, and semantic communications. These applications leverage GANs due to the generative-discriminative nature of GANs, the data generation ability of generative models, and the ability to incorporate perceptual measures into GAN's framework.

3.1.2 Use cases of VAEs in Communications

VAEs can be seen as the extension of the traditional autoencoders by introducing a *probabilistic* framework for generating the latent representation of the input data. The encoder maps high-dimensional input data to a latent space, but not in the format of a single point as in autoencoders. Rather, the encoder generates a *probability distribution* over the latent space. The decoder then *samples* from this distribution and generates new data points. This probabilistic approach allows VAEs to learn in a structured manner, a continuous latent space representation, which will be exploited for generative modeling.

VAEs' use cases in communication systems can be considered in two main directions: (i) Channel equalization, estimation, and channel state information (CSI) compression; and (ii) end-to-end DeepJSCC schemes. Recently, DNN-based approaches have been widely adopted for channel modeling and equalization, relying heavily on a large volume of labeled dataset to learn sufficient characteristics of a specific channel environment. This makes it challenging for conventional DNNs to be used in dynamic environments with high levels of randomness, variability, and use case diversity. VAEs' characteristics in *variational learning and sampling* mechanism, generation of synthetic data under constraints, and *uncertainty estimation*, are particularly useful in the use cases of channel equalization and estimation in wireless systems. Accordingly, VAEs have been shown to be useful for channel equalization use cases. The encoder represents the channel model and noise, and the decoder counterpart is employed for approximating the (intractable) posterior distribution of transmitted symbols from the received signal [8]. Furthermore, VAEs can be employed for the applications of CSI compression at UEs under noisy channels extreme conditions [9]. Notably, conventional DNN-based CSI compression techniques, such as CSINet [10], are vulnerable to noisy feedback channels (which is the most common case in real-world scenarios). Since the VAE's encoder approximates distribution parameters instead of estimating a "point" for each data dimension, the compressed CSI appears to be robust against noise in the feedback channel, highlighting enhanced CSI compression when employing VAEs. Moreover, VAEs can also be employed for beamforming, due to their capability to efficiently approximate the probabilistic model of beam patterns. For example, deep recurrent VAEs are proposed by [11] to learn a probabilistic model of beam dynamics for predictive beam tracking. The VAE's loss function is modified so as to minimize the Kullback-Leibler (KL) divergence between the ground-truth (GT) and the learned model of beam dynamics within the training.

VAEs can be seen as a good candidate for end-to-end JSCC communication systems, by converting original data to a low-dimensional *variational latent space* and reconstructing the semantics at the decoder [12], while removing the need for exchanging pilot symbols. When having complicated high-dimensional sources, for example, large-scale image datasets, or when the channel bandwidth ratio decreases, the performance of conventional DNN-based JSCC schemes degrades significantly. This is because it cannot effectively learn the source distribution. In this scenario, the VAEs' generative capability in learning the latent space of high-dimensional signals at the encoder and regenerating the information signals at the decoder can be useful for the future generations of semantic communication-based wireless networks.

To summarize, *variational learning and sampling* mechanisms, as well as the *uncertainty estimation* capabilities of VAEs have made them a good choice for the applications of wireless channel estimation, CSI compression, and semantic JSCC in the realm of wireless GenAI.

3.1.3 Use-cases of Diffusion Models in Communications

Diffusion models have received significant attention as one of the recent breakthroughs in the realm of GenAI, having contributed to the popular solutions such as DALL.E 2[1] and ImageGen.[2] Existing diffusion model research has been mainly focused on computer vision-related tasks, as a direct application of diffusion generative models [13]. They have been applied to various fields, such as visual AI, natural language generation, waveform signal processing [14], and medical imaging as well (please see Ref. [15] and references therein). In wireless communications, as we will explore, the applications of diffusion models seem to have a great potential to be studied for solving some of the problems in this domain: [16, 17, 18, 19, 20, 21, 22]. Diffusion models have been applied to wireless network management for mobile AI-generated content services [16]. The *exploration capability* of diffusion models can enhance conventional reinforcement learning approaches for network management [23]. Diffusion models are utilized in [17] to augment wireless channel datasets via generating synthetic channel data generation. Notably, the diffusion-aided generated channels can then be used to tackle differentiable models within the back-propagation process of end-to-end training. The results in Ref. [17] highlight the performance of diffusion models compared to GANs. For the channel estimation use case in multi-input multi-output (MIMO) systems, diffusion models can be utilized to estimate the gradient of the log-prior of high-dimensional wireless channels [20]. Diffusion models are adapted to the

---

1 https://openai.com/dall-e-2
2 https://imagen.research.google/

applications of semantic communications [21, 22], being integrated with Deep-JSCC framework to enhance semantic data transmission. Such "hybrid" schemes can effectively enhance the perceptual quality of reconstruction. Nevertheless, methods like [21, 22] employ two different types of AI models, that is, both the discriminative models (autoencoder architecture for semantic transmission), and the generative models (denoising and reconstruction enhancement). Each of these models has their own distinct objective function, which can cause computational overhead and functionality inconsistencies in the training and inference pipelines.

In Section 3.2, we focus on diffusion models and delve into studying one of their use cases in communication system.

3.2 Case Study: Diffusion Models in Wireless Communications

In this section, we provide a closer look at denoising diffusion model and investigate a case study of diffusion models for communication systems. We also provide numerical results and comparative analysis.

We first provide a comprehensive overview of the theoretical foundations and technical details of denoising diffusion models to better understand the underlying concepts, from the perspective of mathematical backgrounds, as well as the practical implications.

The reasons of selecting diffusion generative models as case study can be outlined as follows:

- The idea behind diffusion models is to break down the data generation process into a sequence of "denoising" steps, gradually producing samples from pure noise. This characteristic naturally aligns with the fundamental goal of communication systems, which is to recover and reconstruct information signals from noisy observations. Our emphasis is on this powerful denoising capability.
- In contrast to VAEs and GANs, diffusion models achieve high sample quality and robust mode coverage through their uniquely detailed, stepwise denoising process. This quality makes them particularly well-suited for applications involving direct user interaction, where high-fidelity data reconstruction and generation are essential.

3.2.1 Working Mechanism of Diffusion Models

The core concept of diffusion modeling is that if we could develop an ML model that is capable of learning the *systematic decay of information* due to noise.

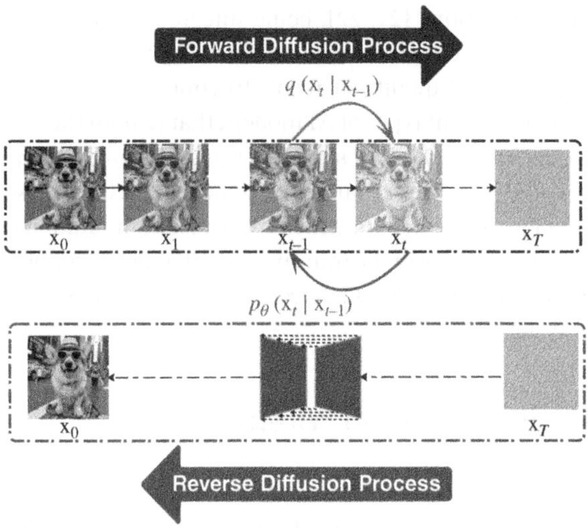

Figure 3.1 Diffusion models: a general overview [24]. The central concept is an AI model that can learn how information systematically degrades, while also being able to reverse this process to recover the original information from noise. The color version of this figure is available on Wiley.com.

It should therefore be possible to reverse the process and recover the information from noisy or corrupted data. This capability is fundamentally distinct from other generative models like GANs and VAEs. Our goal in this subsection is that by investigating the fundamentals of diffusion models, the close underlying relation between diffusion models' mechanism and the data reconstruction problem in communication systems becomes clarified.

Diffusion models represent a new class of state-of-the-art probabilistic generative models, inspired by principles of non-equilibrium thermodynamics [25]. These models consist of two main processes: a forward diffusion process and a parameterized reverse diffusion process, as shown in Figure 3.1. In the forward process, random noise is progressively added to the data until it becomes an isotropic Gaussian signal. In the reverse process, the model seeks to "decode" the data by denoising the corrupted samples in a hierarchical manner. Several diffusion-based generative models with similar core ideas have been explored in the machine learning literature, including diffusion probabilistic models [25], noise-conditioned score network (NCSN) [26], and denoising diffusion probabilistic models (DDPMs) [13]. In this book, we concentrate on DDPMs as a leading representative of this diffusion-model family.

3.2.1.1 Theoretical Foundations

In what follows, we delve into the fundamentals of DDPMs and their key probabilistic formulas of the forward and reverse process.

Forward Diffusion Process

Let $\boldsymbol{x}_0$ denote a data sample drawn from a distribution $q(\boldsymbol{x}_0)$, that is, $\boldsymbol{x}_0 \sim q(\boldsymbol{x}_0)$. For a finite number of time steps T, the forward diffusion process $q(\boldsymbol{x}_t|\boldsymbol{x}_{t-1})$ is defined by progressively adding Gaussian noise at each time step $t \in [T]$ according to a specified "variance schedule" $0 < \beta_1 < \beta_2 < \cdots < \beta_T < 1$. This process can be expressed as

$$q(\boldsymbol{x}_t|\boldsymbol{x}_{t-1}) \sim \mathcal{N}(\boldsymbol{x}_t; \sqrt{1-\beta_t}\boldsymbol{x}_{t-1}, \beta_t\boldsymbol{I}), \tag{3.1}$$

$$q(\boldsymbol{x}_{1:T}|\boldsymbol{x}_0) = \prod_{t=1}^{T} q(\boldsymbol{x}_t|\boldsymbol{x}_{t-1}). \tag{3.2}$$

Referring to (3.2), the data sample progressively loses its distinguishable characteristics as the time step advances, and as $T \to \infty$, $\boldsymbol{x}_T$ converges to an isotropic Gaussian distribution with covariance matrix $\Sigma = \sigma^2\boldsymbol{I}$ for some $\sigma > 0$ [13]. Invoking Eq. (3.1), each sample at time step t can be drawn from a conditional Gaussian distribution with mean vector $\boldsymbol{\mu}_t = \sqrt{1-\beta_t}\boldsymbol{x}_{t-1}$ and covariance matrix $\Sigma_t^2 = \beta_t\mathbf{I}$. Thus, the forward process can be expressed by sampling Gaussian noise $\epsilon_{t-1} \sim \mathcal{N}(\mathbf{0}, \mathbf{I})$ and computing

$$\boldsymbol{x}_t = \sqrt{1-\beta_t}\,\boldsymbol{x}_{t-1} + \sqrt{\beta_t}\epsilon_{t-1}. \tag{3.3}$$

An important property of the forward process in Eq. (3.3) is that $\boldsymbol{x}_t$ can be sampled directly at any arbitrary time step in closed form, by recursively applying the reparameterization trick from the machine learning literature [27], and using the additive properties of Gaussian random variables. This leads to the following formulation

$$\boldsymbol{x}_t = \sqrt{\bar{\alpha}_t}\boldsymbol{x}_0 + \sqrt{1-\bar{\alpha}_t}\epsilon_0, \tag{3.4}$$

$$q(\boldsymbol{x}_t|\boldsymbol{x}_0) \sim \mathcal{N}\left(\boldsymbol{x}_t; \sqrt{\bar{\alpha}_t}\boldsymbol{x}_0, (1-\bar{\alpha}_t)\boldsymbol{I}\right), \tag{3.5}$$

$$\bar{\alpha}_t = \prod_{i=1}^{t}(1-\alpha_i), \quad \alpha_t = 1-\beta_t. \tag{3.6}$$

Note that Eq. (3.4) indicates that we can sample $\boldsymbol{x}_t$ directly from x_0.

Reverse Diffusion Process

The next challenge is to invert the process described in Eq. (3.4) and sample from $q(\boldsymbol{x}_{t-1}|\boldsymbol{x}_t)$, in order to recover the true data samples starting from Gaussian noise $\boldsymbol{x}_T$. According to Sohl-Dickstein et al. [25], for sufficiently small β_t, the distribution $q(\boldsymbol{x}_{t-1}|\boldsymbol{x}_t)$, for all $t \in [T]$ also follows Gaussian distribution. However, directly estimating this distribution is impractical because it requires access to the distribution over all possible data samples (essentially, the full dataset) in order to compute the conditional probabilities. Therefore, to approximate these conditional probabilities and carry out the reverse diffusion process, a probabilistic model $p_\theta(\boldsymbol{x}_{t-1}|\boldsymbol{x}_t)$ parameterized by θ is learned. Accordingly, the following expressions can be established:

$$p_\theta(\boldsymbol{x}_{t-1}|\boldsymbol{x}_t) \sim \mathcal{N}(\boldsymbol{x}_{t-1}; \boldsymbol{\mu}_\theta(\boldsymbol{x}_t, t), \boldsymbol{\Sigma}_\theta(\boldsymbol{x}_t, t)), \tag{3.7}$$

$$p_\theta(\boldsymbol{x}_{0:T}) = p(\boldsymbol{x}_T)\prod_{t=1}^{T} p_\theta(\boldsymbol{x}_{t-1}|\boldsymbol{x}_t). \tag{3.8}$$

As shown in Eq. (3.7), both the mean and variance of the reverse process are conditioned on the time step t. Therefore, the problem reduces to learning the mean vector $\boldsymbol{\mu}_\theta(x_t, t)$ and the covariance matrix $\boldsymbol{\Sigma}_\theta(x_t, t)$. Here, a neural network can be trained to approximate the reverse process with θ representing its parameters.

Before diving into the specifics of learning the reverse diffusion process, it is worth noting that conditioning the reverse process on $\boldsymbol{x}_0$ makes the conditional probability tractable. In other words, if $\boldsymbol{x}_0$ is provided as a reference, one can take small backward steps from noise to generate data samples, and the reverse step can then be expressed as $q(\boldsymbol{x}_{t-1}|\boldsymbol{x}_t, \boldsymbol{x}_0)$. The intuition behind this could be explained by the following metaphor:

A painter (the generative model) uses a reference image $\boldsymbol{x}_0$ to gradually recreate (reverse diffuse via $q(\boldsymbol{x}_{t-1}|\boldsymbol{x}_t, \boldsymbol{x}_0)$) the painting.

Mathematically, by knowing the conditional probabilities $q(\boldsymbol{x}_t|\boldsymbol{x}_0)$ and $q(\boldsymbol{x}_{t-1}|\boldsymbol{x}_0)$, and applying Bayes' rule, one can obtain $q(\boldsymbol{x}_{t-1}|\boldsymbol{x}_t, \boldsymbol{x}_0)$ in a form similar to Eq. (3.7).

$$q(\boldsymbol{x}_{t-1}|\boldsymbol{x}_t, \boldsymbol{x}_0) \sim \mathcal{N}(\boldsymbol{x}_{t-1};\ \tilde{\mu}(\boldsymbol{x}_t, \boldsymbol{x}_0, t), \tilde{\beta}_t\boldsymbol{I}), \tag{3.9}$$

$$\tilde{\mu}(\boldsymbol{x}_t, \boldsymbol{x}_0, t) = \frac{\sqrt{\alpha_t}(1-\bar{\alpha}_{t-1})}{1-\bar{\alpha}_t}\boldsymbol{x}_t + \frac{\sqrt{\bar{\alpha}_{t-1}}\beta_t}{1-\bar{\alpha}_t}\boldsymbol{x}_0, \tag{3.10}$$

$$\tilde{\beta}_t = \frac{1-\bar{\alpha}_{t-1}}{1-\bar{\alpha}_t}\beta_t. \tag{3.11}$$

From this, one can observe that the covariance matrix in Eq. (3.9) does not involve any learnable parameters. Therefore, the only quantity that needs to be learned is the mean vector $\tilde{\mu}(\boldsymbol{x}_t, \boldsymbol{x}_0, t)$. In what follows, Eq. (3.10) is further simplified, and then the training framework is proposed.

Defining the Objective function

With the aid of the reparameterization trick, and using a similar approach as in Eq. (3.4), we can rewrite $\boldsymbol{x}_0$ as

$$\boldsymbol{x}_0 = \frac{1}{\sqrt{\bar{\alpha}_t}}(\boldsymbol{x}_t - \sqrt{1-\bar{\alpha}_t}\epsilon_t). \tag{3.12}$$

We substitute $\boldsymbol{x}_0$ in Eq. (3.10) by Eq. (3.12) and get

$$\tilde{\boldsymbol{\mu}}(\boldsymbol{x}_t, \boldsymbol{x}_0, t) = \frac{1}{\sqrt{\alpha_t}}\Big(\boldsymbol{x}_t - \frac{1-\alpha_t}{\sqrt{1-\bar{\alpha}_t}}\epsilon_t\Big). \tag{3.13}$$

We can now focus on learning the conditional probability distribution $p_\theta(\boldsymbol{x}_{t-1}|\boldsymbol{x}_t)$, realized via training a neural network as approximator for $\tilde{\boldsymbol{\mu}}(\boldsymbol{x}_t, \boldsymbol{x}_0, t)$. Thus, the approximated mean vector $\boldsymbol{\mu}_\theta(\boldsymbol{x}_t, t)$ is designed to match the form of the target mean $\tilde{\boldsymbol{\mu}}(\boldsymbol{x}_t, \boldsymbol{x}_0, t)$. Since $\boldsymbol{x}_t$ is known at time step t, the neural network can be reparameterized to predict the noise component ϵ_t from $\boldsymbol{x}_t$. Combining these insights lead to the following expression for $\boldsymbol{\mu}_\theta(\boldsymbol{x}_t, t)$

$$\boldsymbol{\mu}_\theta(\boldsymbol{x}_t, t) \;= \frac{1}{\sqrt{\alpha_t}}\Big(\boldsymbol{x}_t - \frac{1-\alpha_t}{\sqrt{1-\bar{\alpha}_t}}\epsilon_\theta(\boldsymbol{x}_t, t)\Big), \tag{3.14}$$

where $\epsilon_\theta(\boldsymbol{x}_t, t)$ denotes the model to predict ϵ_t.

Now the training loss function $\mathcal{L}_t$ can be defined, wherein the goal is to minimize the difference between $\boldsymbol{\mu}_\theta(\boldsymbol{x}_t, t)$ and $\tilde{\boldsymbol{\mu}}(\boldsymbol{x}_t, \boldsymbol{x}_0, t)$.

$$\begin{aligned}\mathcal{L}_t &= \mathbb{E}_{\substack{t\sim \mathsf{Unif}[T]\\ \boldsymbol{x}_0\sim q(\boldsymbol{x}_0)\\ \epsilon_0\sim\mathcal{N}(0,\boldsymbol{I})}}\left[\|\epsilon_t - \epsilon_\theta(\boldsymbol{x}_t, t)\|^2\right]\\ &= \mathbb{E}_{\substack{t\sim \mathsf{Unif}[T]\\ \boldsymbol{x}_0\sim q(\boldsymbol{x}_0)\\ \epsilon_0\sim\mathcal{N}(0,\boldsymbol{I})}}\left[\|\epsilon_t - \epsilon_\theta(\sqrt{\bar{\alpha}_t}\boldsymbol{x}_0 + \sqrt{1-\bar{\alpha}_t}\epsilon_t, t)\|^2\right].\end{aligned} \tag{3.15}$$

Invoking Eqs. (3.14) and (3.15), at each time step t, the DDPM takes $\boldsymbol{x}_t$ as input and returns the distortion components $\epsilon_\theta(\boldsymbol{x}_t, t)$, where ϵ_t denotes the diffused noise term at time step t.

To summarize, the concepts of the diffusion model training, forward diffusion process, and the reverse process are illustratively explained in Figure 3.2.

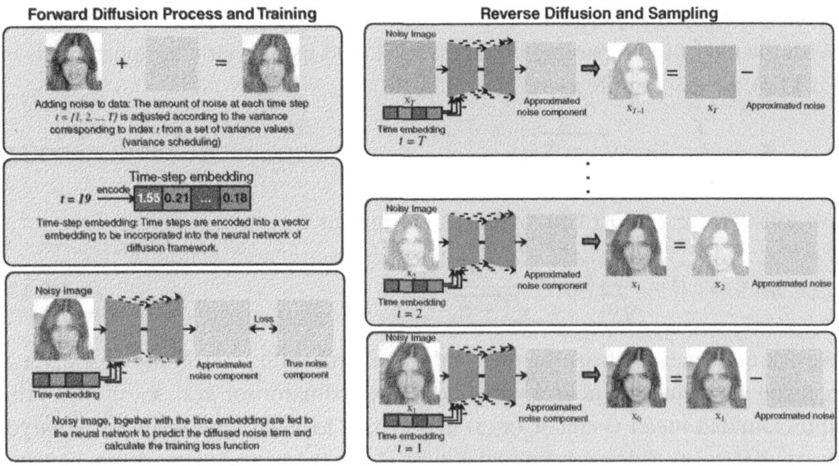

Figure 3.2 Details of forward diffusion process, training the diffusion model, and the reverse process [19]. The DDPM is trained using a loss function that measures the error between the true diffused noise and its approximated counterpart. During inference (sampling), the reverse process is applied to reconstruct the true data samples from noisy signals. Since the probabilistic model for the reverse process is difficult to estimate directly, a neural network is employed to approximate it. The color version of this figure is available on Wiley.com.

3.2.1.2 Experimental Methodologies

Here, we provide a brief overview of the general aspects related to the experimental and implementation of DDPMs.

Data Preprocessing

The data used for training a DDPM model is typically preprocessed and normalized. For image data, this often involves scaling pixel values to the range [0, 1] or [−1, 1] (other data types such as text or audio can also be preprocessed using similar normalization techniques). According to Ho et al. [13], the goal is to make sure the reverse process can function based on consistently scaled inputs, initially drawn from the standard normal prior $p(\boldsymbol{x}_T)$.

Noise Scheduling

A critical aspect of DDPM experimentation is the design of the noise schedule, that is, how the noise variance β_t evolves over time. The noise schedule can be linear, cosine, or learnable, affecting the quality and stability of the reverse process. Choosing a proper noise schedule can significantly improve model performance, especially for high-dimensional data representations.

Training Setup

The primary loss for DDPMs is based on the noise prediction error, as described earlier. The model is trained by optimizing this loss across all time steps using gradient-based optimization techniques such as Adam or Root Mean Square Propagation (RMSProp). Once trained, DDPMs generate samples by running the reverse diffusion process. One important challenge is that training DDPMs is computationally intensive, especially for high-dimensional data like images or videos. Graphical processing units (GPUs) or tensor processing units (TPU) are typically used to accelerate training, and models are trained for hundreds of thousands of iterations.

Evaluation Metrics

For image generation tasks, DDPMs are evaluated using metrics like Frechet inception distance (FID) [28] or inception score [29], which measure the quality and diversity of the generated samples. In addition, the model's ability to accurately model the data distribution can be assessed by estimating the log-likelihood of held-out data, for likelihood estimation. For the applications of image reconstruction, denoising or inpainting, reconstruction accuracy metrics such as peak signal-to-noise ratio (PSNR) can be used for evaluating the model's performance.

3.2.2 Case Study: Diffusion Models Applications for Data Reconstruction Enhancement in Communication Systems

We study how denoising diffusion models can be exploited in real-world wireless communication scenarios. We show that the *generative prior*, together with the denoising process of diffusion models helps improve the receiving performance under extreme conditions of low signal-to-noise ratio (SNR) regimes.

3.2.2.1 System Model

We study denoising diffusion models for removing the residual noise and distortions at the receiver of a practical communication system, aiming to enhance data reconstruction via diffusion models. Specifically, we exploit the *denoise-and-regenerate* capability of DDPMs for image transmission and reconstruction enhancement over wireless channels. Interested readers can apply a similar approach to other data modalities and wireless scenarios, using the general guidelines and experimental setups presented in this chapter. The investigated use case could be advantageous for communication scenarios where prior knowledge of the information content is available, such as *multimedia transmission*. In this context, rather than relying on complex channel codes that reduce the information rate—and may still provide insufficient error correction—diffusion priors can be leveraged for reliable data reconstruction, particularly under harsh channel conditions caused by low SNR or hardware impairments.

Assume a point-to-point communication system, where the transmitter aims to deliver to the receiver, the ground-truth images $x_0 \sim q(x_0)$. At the receiver side, a DDPM-assisted module is employed to reconstruct the original samples from the received batch of distorted signals. Let s denote the channel input signal, which is subject to the power constraint $\mathbb{E}[|s|^2] \leq P$. This information-bearing signal s is transmitted through the communication channel, and the receiver observes its degraded version y.

We consider the well-known communication model from Ref. [30]. The motivation behind this model can be explained as follows. In practice, communication systems are affected by hardware impairments due to the non-ideal transmitter and receiver hardware. Such effects are usually mitigated by signal processing compensation algorithms. However, these techniques cannot fully remove the impairments. Because there remains residual distortion noise due to the time-varying characteristics of RF hardware that cannot be parameterized and estimated via compensation algorithms [30]. This causes additional distortions in the received signals, making it challenging to reconstruct data symbols. Therefore, in our system model, we incorporate such residual RF impairments, and wish to show the performance of diffusion models for image reconstruction enhancement in such a practical setup.

Assume that the Rayleigh fading wireless channel coefficient is denoted by h. One can write

$$y = h(\sqrt{P}s + \tau^t) + \tau^r + n, \quad \tau^t \sim \mathcal{CN}(0, \kappa^t P), \quad \tau^r \sim \mathcal{CN}(0, \kappa^r P|h|^2), \tag{3.16}$$

where P represents the transmit power, τ^t models the distortion noise introduced by the transmitter hardware with κ^t, indicating its corresponding impairment level. Similarly, τ^r captures the hardware distortion at the receiver, with κ^r reflecting the impairment level of the receiver hardware. It is worth noting that τ^r is conditionally Gaussian, given the channel realization h. This is a well-established and experimentally validated model for hardware impairments, widely adopted in the wireless communications literature [30]. At the receiver, a decoding function attempts to recover the information signal, resulting in a noisy version of the ground-truth sample, denoted by $\hat{\boldsymbol{x}}$, such that $\hat{\boldsymbol{x}} \approx \boldsymbol{x}_0 + \boldsymbol{z}$, where $\boldsymbol{z}$ represents the unknown reconstruction error.

3.2.2.2 Diffusion Model-aided Solution

In the vanilla format of DDPMs, it is assumed that the denoising-generation process begins from isotropic Gaussian noise. However, in wireless systems, the signals received are instead a degraded version, $\hat{\boldsymbol{x}}$, of the ground-truth information signals, and are not necessarily distributed as isotropic Gaussian noise. Moreover, varying network conditions—such as different SNR levels and hardware impairments—further influence this behavior. Hence, we employ a conditional diffusion model, incorporating network conditions into account. For conditioning, we directly utilize the noisy degraded signals obtained at the receiver. Accordingly, the modified process of forward diffusion now models a weighted sum of the clean version of ground-truth samples $\boldsymbol{x}_0$ and the noisy samples $\widehat{\boldsymbol{x}}$ to be diffused at each time step, according to an interpolation hyperparameter λ_t. This can be rewritten as

$$q_{\text{cond}}(\boldsymbol{x}_t|\boldsymbol{x}_0, \widehat{\boldsymbol{x}}) \sim \mathcal{N}\left(\boldsymbol{x}_t; (1-\lambda_t)\sqrt{\bar{\alpha}_t}\boldsymbol{x}_0 + \lambda_t\sqrt{\bar{\alpha}_t}\widehat{\boldsymbol{x}}, \delta_t I\right). \tag{3.17}$$

For experiments, λ_t starts from 0 and gradually increases to $\lambda_T \approx 1$.

***Remark**: Intuitively speaking, this form of choosing λ_t "guides" the diffusion model to focus more on learning the mapping denoising from noisy samples (not from a pure Gaussian noise) to the original data points over the time.*

In the next step, the diffusion model should be able to remove the exact noise from $\widehat{\boldsymbol{x}}$ and refine the ground-truth in the conditional reverse. Accordingly, the reverse process should be modified as well. The modified conditional diffusion is supposed to predict $\boldsymbol{x}_{t-1}$ based on $\boldsymbol{x}_t$ and $\widehat{\boldsymbol{x}}$. This can be formulated as

$$p_{\text{cond}}(\boldsymbol{x}_{t-1}|\boldsymbol{x}_t, \widehat{\boldsymbol{x}}) \sim \mathcal{N}(\boldsymbol{x}_{t-1}; \boldsymbol{\mu}_\theta(\boldsymbol{x}_t, \widehat{\boldsymbol{x}}, t), \delta_t \boldsymbol{I}). \tag{3.18}$$

Note that the above conditional distribution is obtained with a similar approach to Eq. (3.7). The mean vector $\boldsymbol{\mu}_\theta(\boldsymbol{x}_t, \widehat{\boldsymbol{x}}, t)$ of the reverse transition probability should express a weighted sum of the denoised signals at step t, $\boldsymbol{x}_t$, the conditional information $\widehat{\boldsymbol{x}}$, and the estimated noise $\epsilon_\theta(\cdot, t)$. Hence, with a similar approach to Eq. (3.14), this can be reformulated as

$$\boldsymbol{\mu}_\theta(\boldsymbol{x}_t, \widehat{\boldsymbol{x}}, t) = \psi_x \boldsymbol{x}_t + \psi_{\hat{x}}\widehat{\boldsymbol{x}} - \psi_\epsilon \epsilon_\theta(\boldsymbol{x}_t, \widehat{\boldsymbol{x}}, t). \tag{3.19}$$

Finally, the loss function is also modified as

$$\mathcal{L}_t^{\text{cond}} = \mathbb{E}_{\boldsymbol{x}_0, \epsilon, \widehat{\boldsymbol{x}}} \,||\, \left(\frac{\lambda_t\sqrt{\bar{\alpha}_t}}{\sqrt{1-\bar{\alpha}_t}}(\widehat{\boldsymbol{x}} - \boldsymbol{x}_0) + \frac{\sqrt{\delta_t}}{\sqrt{1-\bar{\alpha}_t}}\epsilon\right) - \epsilon_\theta(\boldsymbol{x}_t, \widehat{\boldsymbol{x}}, t) \,||_2^2, \tag{3.20}$$

Notably, the loss function is modified in a way that the conditional diffusion learns to estimate both the Gaussian noise and the residual errors $\widehat{\boldsymbol{x}}$, that is, $\widehat{\boldsymbol{x}} - \boldsymbol{x}_0$ for decoding the information. We provide a schematic of the conditional DDPM as illustrated in Figure 3.3. The figure exhibits a visual overview of the main formulas discussed above. Below, we also define the coefficients of the modified forward

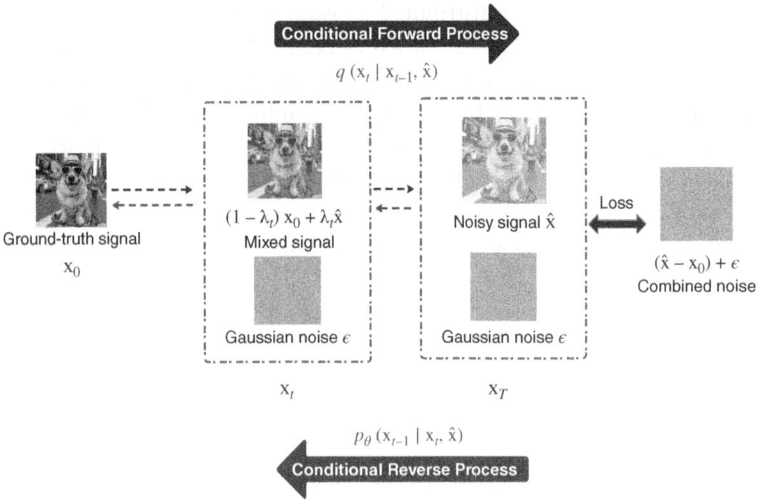

Figure 3.3 The framework of conditional DDPM [24]. The color version of this figure is available on Wiley.com.

and reverse diffusion process, which can be directly obtained from subsection 3.2.1 with a similar approach.

$$\psi_x = \frac{\delta_{t-1}(1-\lambda_t)}{\delta_t(1-\lambda_{t-1})}\sqrt{\alpha_t} + (1-\lambda_{t-1})\frac{\delta_{t|t-1}}{\delta_t\sqrt{\alpha_t}}, \tag{3.21}$$

$$\psi_{\hat{x}} = (\lambda_{t-1}\delta_t - \frac{\lambda_t(1-\lambda_t)}{1-\lambda_{t-1}}\alpha_t\delta_{t-1})\frac{\sqrt{\bar{\alpha}_{t-1}}}{\delta_t}, \tag{3.22}$$

$$\psi_\epsilon = (1-\lambda_{t-1})\frac{\delta_{t|t-1}\sqrt{1-\bar{\alpha}_t}}{\delta_t\sqrt{\alpha_t}}, \tag{3.23}$$

$$\delta_{t|t-1} = \delta_t - \left(\frac{1-\lambda_t}{1-\lambda_{t-1}}\right)^2 \alpha_t\delta_{t-1}, \delta_t := (1-\bar{\alpha}_t) - \lambda_t^2\bar{\alpha}_t. \tag{3.24}$$

Readers who are interested in even further mathematics are referred to Ref. [31]. In this chapter, we aim to keep a balance between theory and experiments, and hence we do not go through further details of the mathematical expressions. Rather, we focus on different experiments in the following.

3.2.2.3 Experimental Results

Communication Setup

For the experiments, we use NVIDIA Sionna system level simulator [32] for transmitting MNIST images[3] over wireless channel. The conditional DDPM has been

3 http://yann.lecun.com/exdb/mnist

developed based upon Hugging Face annotated diffusion,[4] and is integrated at the end of the receiver block within the Sionna simulator.[5] Its purpose is to improve the reconstruction of images transmitted through Sionna's communication pipeline. For the modulation and coding scheme, we employ 64-QAM modulation symbols with 5G low-density parity check codes (5G-LDPC). The transmit SNR is defined as $\Gamma = 10\log_{10}(\frac{P}{\sigma^2})$ dB. In all experiments, the average transmit power is fixed at $P = 1$, while the SNR is varied by adjusting the noise standard deviation σ [33]. Furthermore, impairment levels are considered to be in the range of $[0, 0.15^2]$ [18, 34], where smaller values indicate less severe transceiver hardware impairments.

Dataset Creation

The tuple of samples $(\boldsymbol{x}_0, \hat{\boldsymbol{x}})$ required for dataset of the conditional DDPM $\mathcal{D}$ is obtained by transmitting the samples of MNIST (split into 60,000 training samples and 10,000 test samples) using Sionna link-level simulator, where the decoded signals are obtained at the receiver output, after which each ground-truth sample is paired with its corresponding degraded version to construct the dataset $(\boldsymbol{x}_0, \hat{\boldsymbol{x}}) \in \mathcal{D}$.

Neural Network Architecture

The neural network used in the reverse process is designed to take the distorted image at each time step and predict the noise, producing an output with the same dimensions as the input image. The U-Net architecture is well-suited for this task. It features a contracting path and an expansive path, which respectively capture contextual information while reducing the spatial resolution of the input, and then upsample the feature maps while performing convolutional operations [35]. Additionally, it introduced residual connections between the encoder and decoder, which greatly improves the gradient flow. Hence, inspired by Ho et al. [13], we adopt the U-Net architecture [35] with depth of 4 for the neural network functionality $\epsilon_\theta(\boldsymbol{x}_t, \hat{\boldsymbol{x}}, t)$ for our reverse diffusion framework.

Training Phase Design

Training is performed over 10 epochs each for $T = 200$ denoising steps, using adaptive moment estimation (Adam) optimizer with learning rate $\lambda = 10^{-3}$. To stabilize the training process, we suggest employing exponential moving average (EMA) for model update during training [36]. This idea, which can be found in most of the diffusion model implementations, allows maintaining a form of "model

---

4 https://huggingface.co/blog/annotated-diffusion

5 Note that data type conversion would be required to ensure interoperability between Sionna's Tensorflow native codes and PyTorch-based diffusion algorithm.

momentum." To elaborate, we do not directly update the weights of the neural model. Instead, a copy of the previous weights are kept, and a weighted sum of the previous weights and the new weights are calculated for the model update. For the conditioning hyperparameter, λ_t, we consider a linear scheduling $0 < \lambda_t < 1$ over the time steps. For the variance scheduling β_t, linear scheduling starting from $\beta_0 = 0.0001$ to $\beta_T = 0.0095$ is considered. As an experimental note, we remark that the values for the variance scheduling are typically obtained via trial-and-error over a wide range of α_t values, searching over different scheduling functions, such as linear scheduling, quadrature scheduling, and sigmoid scheduling. Inspired by Ho et al. [13], time steps $t \in [T]$ are encoded and fed as an embedded input to each hidden layer of the neural network, viewed as positional embedding. The output of the hidden layers are multiplied by the time embeddings [13, 15, 20]. Intuitively speaking, this helps the model know at which particular time step it is operating for every sample in the batch, making it possible to use a single neural network across all denoising time steps rather than training T separate models for each step $t \in [T]$.

Figure 3.4 illustrates the performance of our conditional DDPM-aided scheme. In this example, we consider a challenging scenario of the extreme channel condition with 0 dB SNR, that is, $\Gamma = 0$ with the transmitter's impairment level $\kappa^t = 0.1$ and the receiver's impairment level of $\kappa^r = 0.15$. Under these harsh conditions, a conventional communication system struggles to effectively mitigate noise and distortions (see Figure 3.4b). In contrast, our proposed diffusion-aided approach

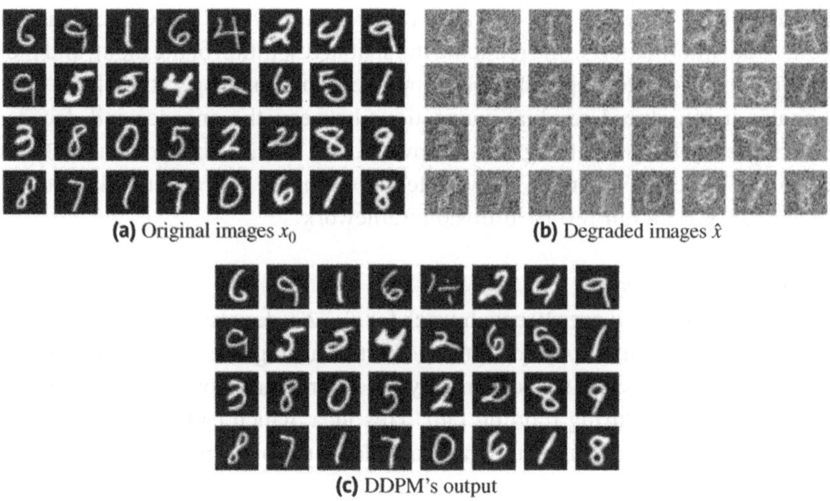

(a) Original images x_0

(b) Degraded images $\hat{x}$

(c) DDPM's output

Figure 3.4 Data visualization for DDPM performance for image reconstruction enhancement over wireless [24]. The color version of this figure is available on Wiley.com.

significantly enhances image reconstruction by removing noise and distortions conditioned on the degraded signals. This leads to a notable improvement in decoding performance, achieved by leveraging the prior knowledge learned by the conditional DDPM.

Figure 3.5 demonstrates the reconstruction performance in terms of PSNR metric. In this test, VAEs, which are also famous for image reconstruction, are compared with our DDPM approach. Notably, our approach considers the "contents" of the transmitted signal during denoising and reconstruction (a.k.a. *generative prior*). Therefore, to have a fair comparison from the perspective of "prior knowledge" utilization, we consider this baseline, which also considers the prior knowledge about the underlying distribution of the ground-truth when reconstructing.

The VAE baseline architecture

For this baseline, in the encoder part four 2D convolutional layers (`Conv2D`) are implemented with 32, 64, 64, and 64 filters, respectively, and kernel size of 3×3 and rectified linear unit (ReLU) activation functions. This is followed by a flattening layer and a dense layer of size 32 (to combine the down-sampled learned

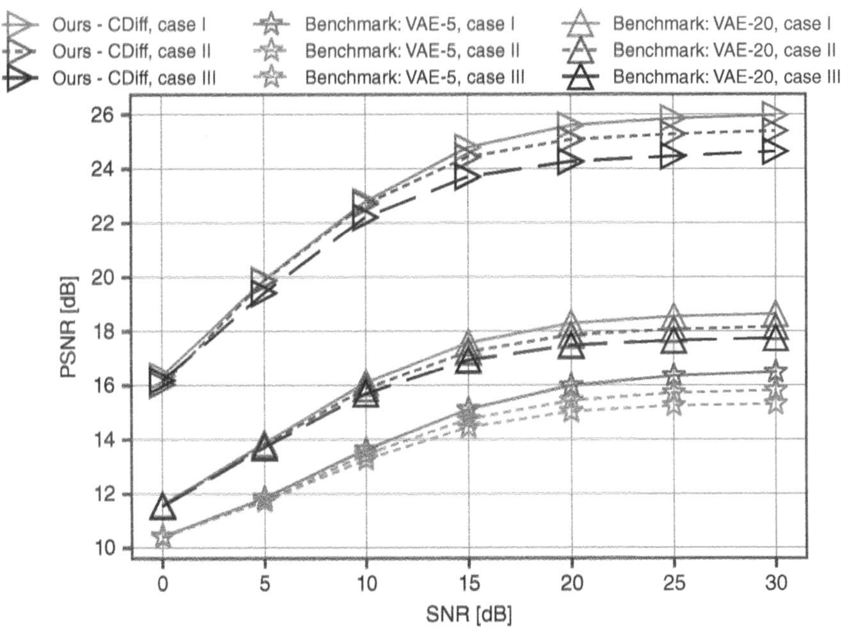

Figure 3.5 Reconstruction performance in terms of PSNR for DDPM versus VAE [24]. For case I to III, we consider $(\kappa^t, \kappa^r) = (0.01, 0.01)$, $(\kappa^t, \kappa^r) = (0.05, 0.1)$, and $(\kappa^t, \kappa^r) = (0.05, 0.15)$, respectively.

features before entering the latent space). Two dense layers further generate the parameters of the Gaussian distribution in the latent space (a.k.a. the mean vector μ and log variance vector σ, which represent the compressed latent features of the input), with latent space dimension `Dim` set to 5 and 20 for different scenarios (see Figure 3.5). The decoder reconstructs the images from the latent representation, by exactly mirroring the encoder in reverse, by progressively up-scaling the compressed data back to its original dimension. Accordingly, the decoder architecture is implemented by a dense layer and a reshape layer to up-scale to a higher dimension, and then followed by de-convolutional layers (`Conv2DTranspose`), to further upscale and refine the data to reconstruct the original input. In the figure, VAE-`Dim`, shows the latent space dimension of the implemented VAE, where higher latent dimensions can offer a slightly better reconstruction quality. As can be seen from the figure, our proposed DDPM outperforms the VAE baseline with more than 40% improvement in PSNR metric. This is due to the high-quality sampling characteristic of diffusion models, while VAEs lack this important generative feature. In contrast to VAEs, diffusion models maintain high sample quality and strong mode coverage, because of their unique feature for *fine-grained denoising*, which is realized via the reverse diffusion process.

Figure 3.6 presents the reconstruction performance of our proposed method when combined with Sionna's neural receiver.[6] We compare this to the baseline approach that uses Sionna's neural receiver without any diffusion-based enhancement.

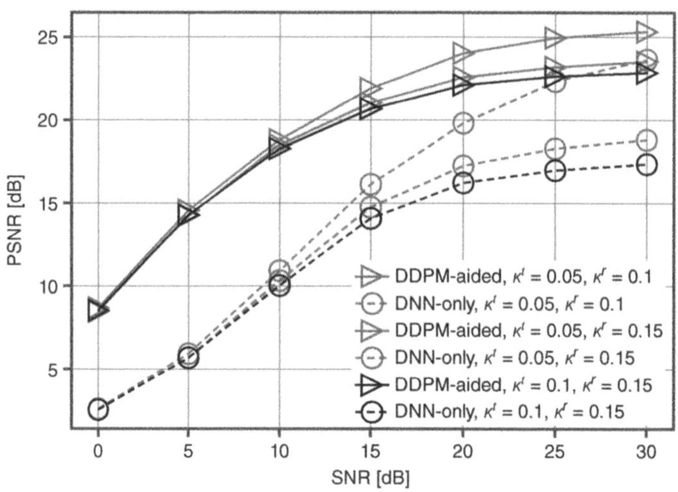

Figure 3.6 Comparing generative (DDPM-based) learning approach and discriminative (DNN-based) learning approach [24]. Reconstruction performance is evaluated in terms of PSNR.

6 https://nvlabs.github.io/sionna/examples/Sionna_tutorial_part4.html

The results demonstrate approximately a 10 dB gain in low-SNR scenarios and over 5 dB improvement at higher SNRs when employing our generative model. This performance boost arises because our DDPM-based approach considers the full underlying distribution of the ground-truth signals transmitted over the wireless channel. In contrast, conventional DNNs overlook the semantic "content" of the information, focusing instead on "bit-wise" transmission and relying on discriminative loss functions like binary cross-entropy, rather than adopting a content-aware approach.

Figure 3.7 illustrates the image reconstruction error measured by mean square error (MSE) as a function of SNR for impairment levels $\kappa^t = \kappa^r = 0.25$. The results indicate that under challenging conditions—specifically low- to mid-SNR regimes and highly impaired communications—a naive approach without our proposed diffusion model struggles to achieve good image reconstruction performance, even when the error correction code rate is reduced from 0.5 to 0.2. When the conditional DDPM is applied, reconstruction quality improves significantly, with gains exceeding 10 dB. This improvement stems from leveraging the denoising strengths of diffusion models, combined with conditioning on the degraded outputs from the naive communication system, which effectively guides the diffusion model to enhance data decoding and reconstruction, as demonstrated in the figure. The figure suggests that instead of lowering code rates to improve decoding quality, our proposed method can be used to maintain data rates while achieving better performance. Specifically, the DDPM-based scheme attains more than a 5 dB performance gain and approximately 2.5 times higher data rates under low SNR conditions, compared to reducing the code rate in a naive scheme from 0.5 to 0.2.

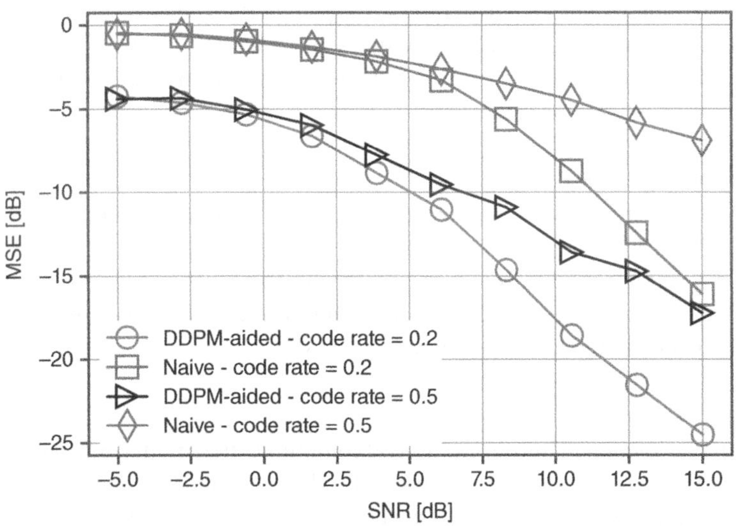

Figure 3.7 Reconstruction performance in terms of MSE [24].

This makes it suitable for scenarios where maintaining both high data rates and reliable communications is crucial. Additionally, although the performance gap between the naive low-code rate scheme (blue square) and the DDPM-aided scheme with a code rate of 0.5 (black triangle) narrows as SNR increases, the figure clearly shows that the DDPM approach can achieve comparable performance without sacrificing the data rate.

3.3 Future Implications and Potential Impacts on Communication Systems

In this chapter, we showed that DDPMs are inherently connected to the task of *denoising*, making them highly relevant to communication systems, where the unwanted noise is a significant challenge. In what follows, we provide a summary of some potential implications of DDPMs in communication systems.

Channel Noise Mitigation

One of the core tasks in communication systems is to mitigate the effect of noise introduced during transmission. DDPMs, being a powerful denoising model, can be applied to clean noisy signals received over communication channels. The reverse diffusion process could reconstruct original signals from noisy transmissions, improving signal quality [37].

Data Encoding and Decoding

In communication systems, data encoding and decoding are essential for transmitting information efficiently. DDPMs have the potential for advanced encoding schemes, where data is diffused into noisy states and then decoded using reverse processes. Such techniques could lead to improved error-correction capabilities or novel methods of embedding information [17].

Error Correction and Data Recovery

DDPMs could be extended to error-correction codes by treating communication errors as noise in a diffusion process. By learning how to reverse this noise, DDPMs may offer innovative approaches to correct errors in transmitted data [38]. This could significantly enhance the reliability of communication systems, especially in environments with high levels of interference [18].

End-to-End Learning in Communication Systems

DDPMs could be integrated into an end-to-end communication framework, where both transmitter and receiver are learned jointly. The transmitter could use a forward diffusion process to generate noisy encoding of the transmitted signal, while the receiver could use the reverse process to decode the original data, optimizing the overall communication pipeline for robustness and efficiency [17, 37].

In conclusion, the theoretical robustness and practical applications of DDPMs, as indicated above, highlight that they have the potential to play important roles in future communication systems. This can include, but is not limited to *resilient communications*, making communication systems more resilient to noise, interference, and distortions, particularly in extreme environments, such as space (non-terrestrial networks), underwater, or urban areas with heavy interference. High-fidelity signal reconstruction is another impact of employing DDPMs for communication networks, which can potentially result in clearer voice calls, sharper video transmissions, and more accurate data recovery in practical communication applications.

3.4 Chapter Summary

In this chapter, we first reviewed the main use cases of key GenAI models in communication systems. Then, we focused on DDPMs to showcase how diffusion generative models can enhance the performance of communication systems under extreme channel conditions. We provided a comprehensive overview of the theoretical foundations of DDPM to better understand the underlying concepts. Details of the experiments and the numerical results, as well as comparisons with baselines were provided to guide readers to do further experiments in the realm of GenAI. Suggested future directions were also provided to indicate potential implications of the use of DDPMs for communication systems.

References

1 Nguyen Van Huynh, Jiacheng Wang, Hongyang Du, Dinh Thai Hoang, Dusit Niyato, Diep N. Nguyen, Dong In Kim, and Khaled B. Letaief. Generative AI for physical layer communications: A survey. *IEEE Transactions on Cognitive Communications and Networking*, 10(3): 706–728, 2024. https://doi.org/10.1109/TCCN.2024.3384500.

2 Subin Lee, Young-Il Yoon, and Yong Ju Jung. Generative adversarial network-based signal inpainting for automatic modulation classification. *IEEE Access*, 11: 50431–50446, 2023. https://doi.org/10.1109/ACCESS.2023.3279022.

3 Eyad Shtaiwi, Ahmed El Ouadrhiri, Majid Moradikia, Salma Sultana, Ahmed Abdelhadi, and Zhu Han. Mixture GAN for modulation classification resiliency against adversarial attacks. In *GLOBECOM 2022 - 2022 IEEE Global Communications Conference*, pages 1472–1477, 2022. https://doi.org/10.1109/GLOBECOM48099.2022.10001138.

4 Blake Barnes-Cook and Timothy O'Shea. Scalable wireless anomaly detection with generative-lstms on rf post-detection metadata. In *2022 IEEE Wireless Communications and Networking Conference (WCNC)*, pages 483–488, 2022. https://doi.org/10.1109/WCNC51071.2022.9771754.

5 Yaqi Hu, Mingsheng Yin, William Xia, Sundeep Rangan, and Marco Mezzavilla. Multi-frequency channel modeling for millimeter wave and thz wireless communication via generative adversarial networks. In *2022 56th Asilomar Conference on Signals, Systems, and Computers*, pages 670–676, 2022. https://doi.org/10.1109/IEEECONF56349.2022.10052063.

6 Satya Kumar Vankayala, Swaraj Kumar, Ishaan Roy, D Thirumulanathan, Seungil Yoon, and Ignatius Samuel Kanakaraj. Radio map estimation using a generative adversarial network and related business aspects. In *2021 24th International Symposium on Wireless Personal Multimedia Communications (WPMC)*, pages 1–6, 2021. https://doi.org/10.1109/WPMC52694.2021.9700474.

7 Ecenaz Erdemir, Tze-Yang Tung, Pier Luigi Dragotti, and Deniz Gündüz. Generative joint source-channel coding for semantic image transmission. *IEEE Journal on Selected Areas in Communications*, 41(8): 2645–2657, 2023. https://doi.org/10.1109/JSAC.2023.3288243.

8 Avi Caciularu and David Burshtein. Unsupervised linear and nonlinear channel equalization and decoding using variational autoencoders. *IEEE Transactions on Cognitive Communications and Networking*, 6(3): 1003–1018, 2020. https://doi.org/10.1109/TCCN.2020.2990773.

9 Mostafa Hussien, Kim Khoa Nguyen, and Mohamed Cheriet. Prvnet: A novel partially-regularized variational autoencoders for massive MIMO CSI feedback. In *2022 IEEE Wireless Communications and Networking Conference (WCNC)*, pages 2286–2291, 2022. https://doi.org/10.1109/WCNC51071.2022.9771642.

10 Chao-Kai Wen, Wan-Ting Shih, and Shi Jin. Deep learning for massive MIMO CSI feedback. *IEEE Wireless Communications Letters*, 7(5): 748–751, 2018. https://doi.org/10.1109/LWC.2018.2818160.

11 Muddassar Hussain and Nicolò Michelusi. Learning and adaptation for millimeter-wave beam tracking and training: A dual timescale variational framework. *IEEE Journal on Selected Areas in Communications*, 40(1): 37–53, 2022. https://doi.org/10.1109/JSAC.2021.3126086.

12 Qiyu Hu, Guangyi Zhang, Zhijin Qin, Yunlong Cai, Guanding Yu, and Geoffrey Ye Li. Robust semantic communications with masked VQ-VAE enabled codebook. *IEEE Transactions on Wireless Communications*, 22(12): 8707–8722, 2023. https://doi.org/10.1109/TWC.2023.3265201.

13 Jonathan Ho, Ajay Jain, and Pieter Abbeel. Denoising diffusion probabilistic models. In H. Larochelle, M. Ranzato, R. Hadsell, M.F. Balcan, and H. Lin, editors, *Advances in Neural Information Processing Systems*, volume 33, pages 6840–6851. Curran Associates, Inc., 2020. https://proceedings.neurips.cc/paper_files/paper/2020/file/4c5bcfec8584af0d967f1ab10179ca4b-Paper.pdf.

14 Zhifeng Kong, Wei Ping, Jiaji Huang, Kexin Zhao, and Bryan Catanzaro. Diffwave: A versatile diffusion model for audio synthesis. In *International Conference on Learning Representations*, 2021. https://openreview.net/forum?id=a-xFK8Ymz5J.

15 Ling Yang, Zhilong Zhang, Yang Song, Shenda Hong, Runsheng Xu, Yue Zhao, Wentao Zhang, Bin Cui, and Ming-Hsuan Yang. Diffusion models: A comprehensive survey of methods and applications. *ACM Computing Surveys*, 56(4):1–39, 2023. ISSN 0360-0300. https://doi.org/10.1145/3626235.

16 Yinqiu Liu, Hongyang Du, Dusit Niyato, Jiawen Kang, Zehui Xiong, Dong In Kim, and Abbas Jamalipour. Deep generative model and its applications in efficient wireless network management: A tutorial and case study. *IEEE Wireless Communications*, 31(4): 199–207, 2024. https://doi.org/10.1109/MWC.009.2300165.

17 Muah Kim, Rick Fritschek, and Rafael F. Schaefer. Learning end-to-end channel coding with diffusion models. In *WSA & SCC 2023; 26th International ITG Workshop on Smart Antennas and 13th Conference on Systems, Communications, and Coding*, pages 1–6, 2023.

18 Mehdi Letafati, Samad Ali, and Matti Latva-Aho. Diffusion model-aided data reconstruction in cell-free massive MIMO downlink: A computation-aware approach. *IEEE Wireless Communications Letters*, 13(11): 3162–3166, 2024. https://doi.org/10.1109/LWC.2024.3457008.

19 Mehdi Letafati, Samad Ali, and Matti Latva-aho. Diffusion models for wireless communications. *arXiv preprint arXiv:2310.07312*, 2023. https://arxiv.org/abs/2310.07312.

20 Nicolas Zilberstein, Chris Dick, Rahman Doost-Mohammady, Ashutosh Sabharwal, and Santiago Segarra. Annealed Langevin dynamics for massive MIMO detection. *IEEE Transactions on Wireless Communications*, 22(6): 3762–3776, 2023. https://doi.org/10.1109/TWC.2022.3221057.

21 Xueyan Niu, Xu Wang, Deniz Gündüz, Bo Bai, Weichao Chen, and Guohua Zhou. A hybrid wireless image transmission scheme with diffusion. In *2023 IEEE 24th International Workshop on Signal Processing Advances in Wireless Communications (SPAWC)*, pages 86–90, 2023. https://doi.org/10.1109/SPAWC53906.2023.10304477.

22 Jiakang Chen, Di You, Deniz Gündüz, and Pier Luigi Dragotti. Commin: Semantic image communications as an inverse problem with INN-guided diffusion models. In *ICASSP 2024 - 2024 IEEE International Conference on Acoustics, Speech and Signal Processing (ICASSP)*, pages 6675–6679, 2024. https://doi.org/10.1109/ICASSP48485.2024.10448462.

23 Hongyang Du, Ruichen Zhang, Yinqiu Liu, Jiacheng Wang, Yijing Lin, Zonghang Li, Dusit Niyato, Jiawen Kang, Zehui Xiong, Shuguang Cui, Bo Ai, Haibo Zhou, and Dong In Kim. Enhancing deep reinforcement learning: A tutorial on generative diffusion models in network optimization. *IEEE Communications*

Surveys & Tutorials, 26(4): 2611–2646, 2024. https://doi.org/10.1109/COMST.2024.3400011.

24 Mehdi Letafati, Samad Ali, and Matti Latva-Aho. Conditional denoising diffusion probabilistic models for data reconstruction enhancement in wireless communications. *arXiv preprint arXiv:2310.19460*, 2024.

25 Jascha Sohl-Dickstein, Eric Weiss, Niru Maheswaranathan, and Surya Ganguli. Deep unsupervised learning using nonequilibrium thermodynamics. In *International Conference on Machine Learning*, pages 2256–2265. PMLR, 2015.

26 Yang Song and Stefano Ermon. Generative modeling by estimating gradients of the data distribution. *Advances in Neural Information Processing Systems*, 32, 2019.

27 Durk P. Kingma, Tim Salimans, and Max Welling. Variational dropout and the local reparameterization trick. *Advances in Neural Information Processing Systems*, 28, 2015.

28 Martin Heusel, Hubert Ramsauer, Thomas Unterthiner, Bernhard Nessler, and Sepp Hochreiter. GANs trained by a two time-scale update rule converge to a local Nash equilibrium. In *Advances in Neural Information Processing Systems*, pages 6626–6637, 2017.

29 Tim Salimans, Ian Goodfellow, Wojciech Zaremba, Vicki Cheung, Alec Radford, and Xi Chen. Improved techniques for training GANs. In *Advances in Neural Information Processing Systems*, pages 2234–2242, 2016.

30 Emil Björnson, Jakob Hoydis, Marios Kountouris, and Mérouane Debbah. Massive MIMO systems with non-ideal hardware: Energy efficiency, estimation, and capacity limits. *IEEE Transactions on Information Theory*, 60(11): 7112–7139, 2014. https://doi.org/10.1109/TIT.2014.2354403.

31 Yen-Ju Lu, Zhong-Qiu Wang, Shinji Watanabe, Alexander Richard, Cheng Yu, and Yu Tsao. Conditional diffusion probabilistic model for speech enhancement. In *2022 IEEE International Conference on Acoustics, Speech and Signal Processing (ICASSP)*, pages 7402–7406. IEEE, 2022.

32 Jakob Hoydis, Sebastian Cammerer, Fayçal Ait Aoudia, Avinash Vem, Nikolaus Binder, Guillermo Marcus, and Alexander Keller. Sionna: An open-source library for next-generation physical layer research, 2023. https://arxiv.org/abs/2203.11854.

33 Seyyed Amirhossein Ameli Kalkhoran, Mehdi Letafati, Ecenaz Erdemir, Babak Hossein Khalaj, Hamid Behroozi, and Deniz Gündüz. Secure deep-JSCC against multiple eavesdroppers. In *2023 IEEE Global Communications Conference*, pages 3433–3438, 2023. https://doi.org/10.1109/GLOBECOM54140.2023.10436928.

34 Mehdi Letafati, Samad Ali, and Matti Latva-aho. Denoising diffusion probabilistic models for hardware-impaired communications. In *2024 IEEE Wireless Communications and Networking Conference (WCNC)*, pages 1–6, 2024. https://doi.org/10.1109/WCNC57260.2024.10570533.

35 Olaf Ronneberger, Philipp Fischer, and Thomas Brox. U-net: Convolutional networks for biomedical image segmentation. In *International Conference on Medical Image Computing and Computer-assisted Intervention*, pages 234–241. Springer, 2015.

36 Yang Song and Stefano Ermon. Improved techniques for training score-based generative models. *Advances in Neural Information Processing Systems*, 33: 12438–12448, 2020.

37 Tong Wu, Zhiyong Chen, Dazhi He, Liang Qian, Yin Xu, Meixia Tao, and Wenjun Zhang. CDDM: Channel denoising diffusion models for wireless communications. In *2023 IEEE Global Communications Conference*, pages 7429–7434, 2023. https://doi.org/10.1109/GLOBECOM54140.2023.10436728.

38 Yoni Choukroun and Lior Wolf. Denoising diffusion error correction codes. *arXiv preprint arXiv:2209.13533*, 2022.

4

Generative AI at PHY Layer: Native AI or Trainable Radios

Eren Balevi

New York, Brooklyn, USA

*Corresponding author: erenbalevi@gmail.com

4.1 Wireless Communications Empowered with Generative Models

Generative AI holds great potential to transform state-of-the-art physical (PHY) layer communication algorithms into more innovative methodologies associated with the capabilities of generative models to analyze distribution characteristics. To give more insights, this part introduces the motivations and unique aspects of generative AI for communication systems and briefly explains representative applications that make use of generative AI at the PHY layer.

4.1.1 Motivations of GenAI at the PHY

A generative model extracts features within a dataset and learns how to combine these features to generate new data samples. This mechanism makes generative AI introduce more powerful models than traditional discriminative AI, which mainly excels at regression and classification tasks. There has recently been unprecedented interest in academia and industry for developing generative AI-based communication techniques, and the expectations are pretty high with respect to the previous technology enablers. One reason for that is the enormous success of the

Generative AI for Communications Systems: Fundamentals, Applications, and Prospects,
First Edition. Edited by Diep N. Nguyen et al.

rich set of generative models, for example, GANs, VAEs, normalizing flows, diffusion models, or autoregressive models, including transformer decoders on different research domains. Another motivation is the fast-evolving nature of generative architectures that promise to surpass existing benchmarks.

One of the distinguishing properties of utilizing AI in the physical layer is the extreme dynamic conditions the learned model can be subject to due to changing environments, user mobility, and different system configurations. Classical deep learning models are optimized according to the training dataset, which can represent only a tiny part of the underlying distribution. Hence, their generalization capability is limited. On the other hand, generative AI can learn the entire distribution and is expected to generalize better, making it favorable for dynamic communication conditions. Additionally, deep learning models for communication systems can require a large volume of channel measurements for each site. Therefore, it is obligatory to generate synthetic data and/or replace them with less data-hungry architectures. All these factors, summarized in Figure 4.1, make adapting generative models to data-driven wireless communication algorithms/methods intriguing.

4.1.2 Applications of GenAI at the PHY

Listing the use cases that can merit most from generative AI and perceiving what features/properties of generative AI can be utilized to enhance the currently used algorithms in physical communication are good starting points, but unfortunately, these are not yet clear and by no means trivial. To take steps in the exciting research direction of employing generative AI at the PHY, the applications that directly exploit the characteristics and modeling of wireless channels are first given. The other use cases then follow.

4.1.2.1 Downstream Tasks with Channel Modeling

A simple illustration of adapting generative AI models for wireless channels with some related downstream tasks is given in Figure 4.2. One of the first use cases of generative AI targets channel modeling. It is attractive to attain channel samples

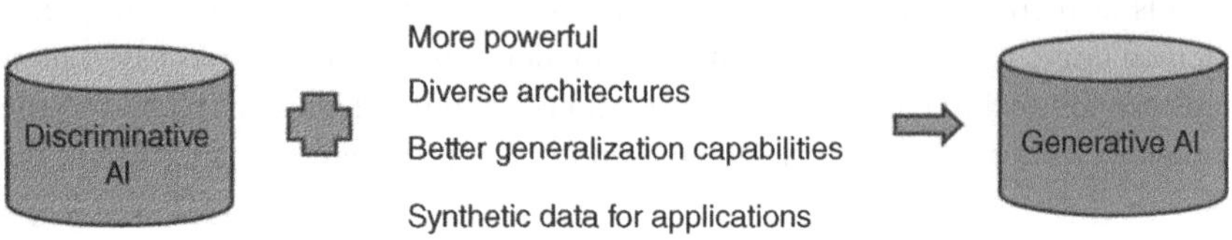

Figure 4.1 Motivations and unique aspects of generative AI over traditional discriminative AI. *Source:* Eren Balevi.

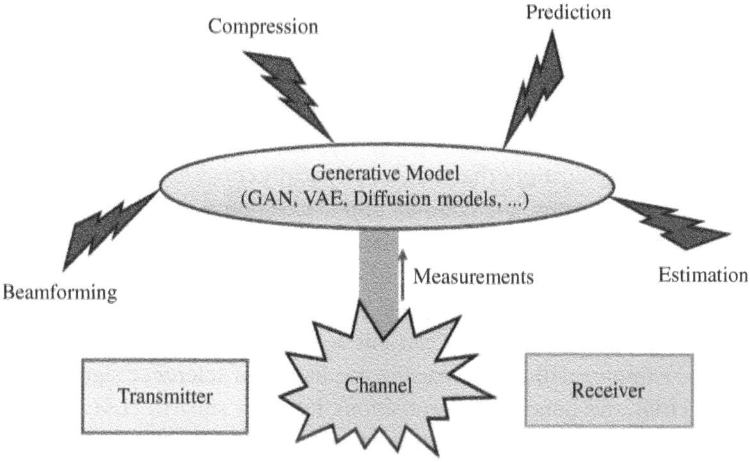

Figure 4.2 Generative models incorporated into a point-to-point communication with some example use cases. *Source:* Eren Balevi.

from a generative channel model similar to the measurements of a given environment since this can be utilized for multiple scopes. The initial studies leverage the generative AI as a black-box optimization by directly taking the channel samples as inputs and do not use any domain knowledge, for example, GAN framework for AWGN channels [1], clustered delay line (CDL) models [2], and tapped delay line (TDL) models [3]. Showing the capability of a generative model to produce realistic samples from the practical 3GPP-defined statistical models such as CDL and TDL motivates reducing the cost of channel measurements. Specifically, to have a sufficient dataset for channel characterization, considerable effort must be put into holding extensive measurement campaigns. The generative channel modeling approach can reduce this by generating synthetic measurements from the limited real channel measurements.

Exploiting the wireless domain knowledge for channel modeling yields a parametric model. However, heterogeneity and the resulting complexity of channel modeling due to new frequency bands, diverse environments, and antenna configurations make figuring out the complex relationships among various parameters challenging. Generative AI excels in learning and capturing probabilistic relations among parameters. Motivated by this feature, generative models can be incorporated into channel modeling to obtain the joint distributions of the relevant parameters as an alternative to black-box optimization. For instance, a generative model can generate random instances of small-scale parameters for a given channel model structure, for example, CDL models for MIMO channels. Then, these parameters can be integrated into the overall channel model.

Unlike the widely used generic stochastic channel models, generative channels can produce site-specific environments with limited channel measurements. Site-specific modeling can be helpful for network capacity and coverage optimization, as well as configuration optimization, for example, parameters of a reconfigurable intelligent surface (RIS). Additionally, site-specific generative models can be utilized for future applications that look for physically consistent environments between the physical location and channel impulse response, for example, positioning and integrated sensing and communication [4].

Fueled with channel information, exploiting a generative model for channel estimation is another research topic for the physical layer. A naive use case to improve channel estimation accuracy can mimic the received sequence to generate additional samples without increasing the length of reference signals and combine the actual and generated observations for least squares (LS) channel estimation [5]. However, more sophisticated approaches must be thoroughly explored to incorporate channel knowledge into channel estimation. One promising method that suggests digesting channel information within the trained generative model iteratively has been applied to narrowband channel estimation [2] and wideband channel estimation [3]. The additional (prior) knowledge that comes from the trained generative AI well serves to optimize the number of reference signals (or pilots) and the superiority of the iterative framework over fundamental compressed sensing algorithms is clearly illustrated by Balevi et al. and Balevi and Andrews [2, 3]. In those papers, a trained GAN was used to test the performance of the proposed framework without any loss of generality, that is, replacing the GAN with any other generator model is straightforward.

Following channel estimation, the compression of a channel, particularly for massive MIMO systems is an exciting research direction to which generative AI can contribute.[1] Traditional deep learning architectures, such as CNNs, LSTMs, and transformers, have been intensely investigated in terms of performance and complexity for spatial, frequency, and time domain [6]. It is motivating to investigate this topic further regarding what generative AI can bring to the table in addition to the traditional models. Such an extension is highly needed to enhance the reconstruction quality while reducing the overhead despite the additional cost of machine learning models due to increased computational complexity. Some earlier ideas have been centered around using the latent vector of a generator as the compressed channel [7]. Semantic compression incorporated into joint source-channel coding (JSCC) for semantic communication is also a related field with generative compression [8]. However, understanding the role of generative AI for

1 This problem can be interpreted as precoder compression as well if the receiver, for example, a user device, reports the best precoder to the transmitter, for example, a base station instead of raw channel measurements.

compression is still in its infancy. Channel prediction is another application that can exploit the merits of generative AI. To illustrate, channel prediction that uses the attention mechanism of a transformer can map the past channel samples to future channels [9].

Beamforming design to produce destructive and constructive interference at some angles for directional communication (beam synthesis) and the alignment of the beams of a transmitter and a receiver are crucial for the deployment of millimeter wave (mmWave) spectrum within 5G New Radio (NR) to secure a safe link budget for both the initial access and data transmission phase since mmWave signals can undergo high attenuation. Various estimation and search algorithms can be used for beam synthesis and beam sweeping (e.g., a predefined exhaustive search at some angles in the angle domain). With the rise of generative AI, the beamforming problem has been studied in the low-dimensional channel manifold that holds channel characteristics, resulting in feedback overhead reduction [7]. The highly changing conditions depending on environmental factors and the complexity of the problem make generative AI appealing because generative models can capture the statistical characterization of the beam dynamics and inject this information into the relevant beamforming algorithms.

4.1.2.2 Other Use Cases

Data-driven methods that avoid the need for an accurate channel model and perfect CSI knowledge are intriguing research directions for many transceiver tasks, for example, detection, signal classification, and modulation recognition. However, the expense of collecting data, in addition to the computational complexities of training, challenges this approach. More importantly, the limited generalization capability of data-driven techniques narrows the impact on the environment where the training dataset is collected. That is, a non-negligible performance drop is widely expected for non-similar environments.

At this point, generative AI that enables learning the distribution of data is promising instead of fitting the likelihood of data in accordance with the underlying actual distribution as classical deep learning models do. To this end, GAN architectures have been inspected in many areas. Incorporating signal detection as a complicated classification problem with unknown channel models, a GAN learns the channel transition probability for Viterbi detection [10]. Alternatively, the signal detection problem in the context of MIMO systems with unknown statistical knowledge of noise is approached by a normalizing flow model, including an unsqueeze layer, K flow steps, and a squeeze layer [11]. GANs are also utilized to create a synthetic dataset to enhance the performance for modulation classification [12].

VAEs can be adopted for the general problems that require estimating intractable posterior distributions. More precisely, VAEs can model the posterior distribution of the transmitted signals to equalize the distorted received signal.

In this regard, VAEs are employed for blind channel equalization without channel estimation [13]. Despite the promise of initial generative AI studies, the dynamics of wireless environments and the impact of configuration at the base station or user device incognizant of the other side need to be thoroughly investigated to have the whole picture.

Foundation models, which pretrain a large-scale transformer model with a massive dataset and adapt to the downstream task with minimum effort, for example, prompting and fine-tuning, have recently become quite popular in natural language processing (NLP) with many emerging large language models (LLMs) such as GPT, BERT, and Llama. Applying the same idea of foundation models in the physical layer has been researched to eliminate deploying a separate AI model for each task [14]. One exciting idea is to pretrain a powerful channel transformer model trained with extensive channel measurements and utilize it for multiple physical layer functions. However, this idea is still nascent, and its feasibility needs to be discovered.

New tools for physical layer research can be needed to embrace generative AI and enable trainable radios. One prominent effort for link-level simulations is Sionna, which implements the basic building blocks in a differentiable fashion to backpropagate gradients [4]. Such tools facilitate and enable end-to-end learning and AI native air interfaces. Security and privacy are other aspects, and many tutorials nicely cover these topics, for example, Ref. [15].

4.2 Channel Modeling

Modeling propagation environment with channel impulse response or transfer function is central to developing novel communication technologies and evaluating the performance of wireless systems. Channel modeling aims to characterize LOS probability and received signal strengths (or path loss) with other large-scale parameters, for example, shadow fading, Rician K-factor, delay spread, Doppler spread, and angle spread in addition to small-scale parameters, for example, delays, angle of arrivals and departures, and cluster powers. It covers frequencies from ultra-high frequency (UHF) and super-high frequency (SHF) to mmWave and beyond, including indoor environments, outdoor macrocells, and microcells.

Propagation details can be fully acquired by solving Maxwell's equation with radar cross-section (RCS) of obstructing objects and the related wireless system parameters. This, however, is a computationally heavy and complicated task since it is hard to know every single detail/parameter of the propagation environment. Ray tracing models have emerged to approximate electromagnetic waves as rays in geometric optics and eliminate the complexity of Maxwell equations as a canonical example of deterministic channel modeling. Combining the 3D view of the environment with reflected, diffracted, and scattered rays, ray tracing models can also be computationally and memory inefficient. Empirical models that derive channel parameters from raw channel measurements are another method. They are computationally simpler and can reflect the propagation environments more accurately. On the other hand, data collection with channel measurements is too costly.

Stochastic channel modeling is favorable over deterministic models if there are many multipath components with non-deterministic time variations and unknown parameters, such as geometry and dielectric properties of materials in the propagation environment. One popular approach is geometry-based statistical models that separate antenna configuration and channel parameters and extract the parameters from the statistical distribution of channel measurements. The main drawback of stochastic models is the accuracy of a model in specific environments since they are developed for generic scenarios, for example, indoor hotspots, and dense urban or rural areas.

4.2.1 Generative Channel Modeling

Increased bandwidth, for example, from 15–20 MHz for 3G systems to 100–400 MHz (with aggregated bandwidth) for 5G NR, and multiple antenna elements at one or both ends of the communication link, creating new parameters such as angle of arrival and angle of departure make channel modeling tricky. Furthermore, operating over multiple frequency bands as the combination of sub-6 GHz and mmWave bands significantly increases channel model parameters. In addition to handling the difficulties of all these issues, there are inherent deficiencies in deterministic and stochastic approaches, for example, deterministic models may not be accurate due to the unknown material properties and approximations, and stochastic models become more precise for regular geometries and uniform dielectric properties. All these factors motivate novel, innovative methodologies, and generative AI can become a promising tool. Generative AI offers a new approach for channel modeling by learning distributions over channel samples for multiple parameters. It makes no (or minimum) prior assumptions since assumptions may lead to inaccuracies for some environments.

Channel distribution can be acquired fundamentally differently with modern generative architectures. Specifically, for the classical parameterized distribution estimation, a parameterized distribution is first defined, and then the Kullback-Leibler (KL) divergence between the target (real) and the parameterized (model) distribution is minimized. However, explicitly obtaining/learning such a parameterized distribution in closed form can be intractable. Modern generative models take an innovative approach by defining a random variable with a distribution and passing it through a parametric function, for example, a neural network, to generate samples that follow the target distribution. Optimizing the parameters of that function changes the distribution, and the objective is to make the learned distribution closer to the actual distribution as much as possible for a given distance measure.

The ultimate goal of generative channel modeling is to generate synthetic channel samples using the available channel measurements from an unknown distribution, such that the generated sample is similar to the real measurements but not the same. This is accomplished by capturing dependencies in the dataset without explicitly obtaining the channel distribution. Generative channel modeling is promising in multiple directions. One is related to scalability to the extension of frequency bands and the number of antenna elements. Extending the state-of-the-art channel models for new bands with increased dimensions is tricky. In contrast, generative AI with expanded architecture dimensions can tackle this with minimum effort. Generative models can enable combining channel measurements with other sensor outputs, for example, an image captured by cameras or a radar trace, for more accurate modeling since they can support processing multiple modalities. Another one is the convenience of integrating generative channel models with stochastic or deterministic channel models as a complementary tool to enhance them.

The quality and quantity of channel measurements are crucial in generative modeling. There are other practical considerations as well stemming from the cost of holding channel measurement campaigns and data provisioning, the limitations of the model to the dataset, that is, generalization capability for different environments than the one the dataset is collected, and a large number of parameters, bringing additional computational complexity. However, before addressing these concerns, it is imperative to figure out the performance of the generative AI-based channel modeling.

The initial efforts have been led by GANs, for example, see Ref. [1] for AWGN channel modeling, Ref. [2] for CDL modeling, and Ref. [3] for TDL modeling. The GAN framework adapted to generative channel modeling is depicted in Figure 4.3. GAN comprises a pair of neural networks labeled "generator" and "discriminator." While random noise samples feed the generator, the discriminator takes channel measurements as inputs.[2] Raw channel measurements can be preprocessed to simplify the model architecture and/or optimize performance. The preprocessing step can contain multiple transformations among domains, for example, from spatial-frequency-time to angle-delay-Doppler domain, and can be essential to exploit sparsity. Note that preprocessing may not be needed if black-box optimization is done for channel modeling. To convert the standard GAN architecture to a conditional GAN, additional input, such as the distance between a transmitter and a receiver, can be provided to the generator and the discriminator.

2 If channel measurements are not available, the channel samples can be obtained via a simulator-based tool, for example, a ray tracing or sampling a statistical model.

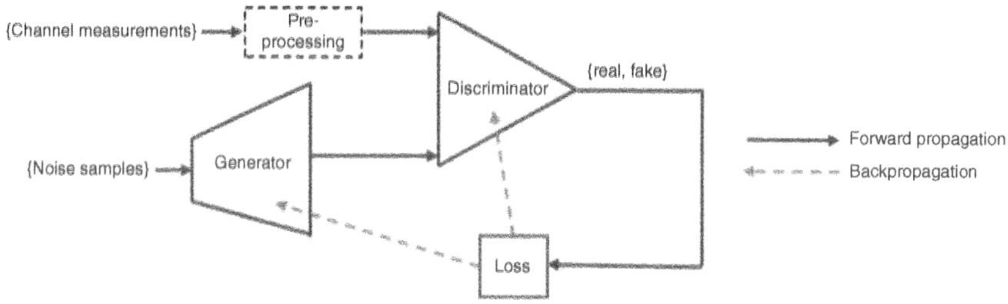

Figure 4.3 A GAN framework for channel modeling with noise samples, channel measurements, and a loss function, for example, Wasserstein loss with gradient penalty. *Source:* Eren Balevi.

In the channel GAN framework, the goal of the generator is to produce channel samples similar to the measurements so that the discriminator cannot distinguish the real and fake (or synthetic) channel samples. On the other hand, the discriminator aims to clearly separate real and fake channels by maximizing the probability of assigning correct labels to real and fake samples. After epochs of training, the training stops when the discriminator cannot distinguish the real and fake samples, that is when it assigns a 0.5 probability for both of them. This is an ideal condition since the generator and discriminator are trained to reach Nash equilibrium within a minimax game optimization framework. More research is needed to understand how closely the Nash equilibrium of a non-convex minimax optimization with gradient descent can be reached. It is also expected that the generator fully learns the underlying distribution behind the channel measurements to be capable of producing samples from that distribution. However, the mode collapse problem can tackle learning the entire distribution, leading to generating less diverse samples from only a part of the distribution.

The experiments done for GAN-based channel modeling show the capability of a generator to produce synthetic channel measurements. It is, however, challenging to fully capture the correlations in multiple domains and generate high-dimensional channel realizations with entirely accurate statistical properties. As a stepping stone and to facilitate the problem a little bit, generative models can be employed to learn the joint distribution of the relevant parameters of a parameterized channel model instead of directly outputting raw channels. The parameters can be extracted via the preprocessing block shown in Figure 4.3. For instance, the joint distribution of the paths between a transmitter and receiver is learned by a GAN to generate the cluster parameters [16]. Then, the generated parameters are combined with antenna array configurations to obtain a MIMO channel.

VAE is another generative model that can be utilized for channel modeling as illustrated in Figure 4.4. VAE architectures comprise an encoder and a decoder inherited by the traditional autoencoder structures. The decoder part can be used

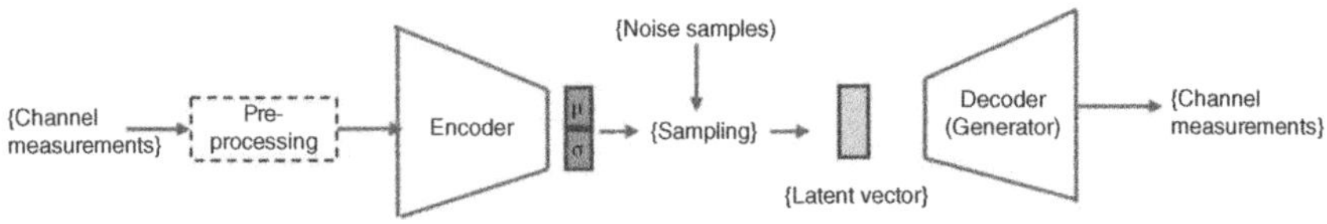

Figure 4.4 A VAE framework for channel modeling with noise samples, sampling, and channel measurements. *Source:* Eren Balevi.

as the generator network. A random sample taken from a prior distribution with learned statistics, for example, with mean μ and variance σ^2, can generate a channel realization through the parameterized and trained decoder such that the generated sample is close to the measurements in the training dataset. The underlying goal is the same as that of GAN, and what differs is the optimization framework and the architectures. VAEs are optimized by approximating log-likelihood or evidence lower bound (ELBO) instead of the adversarial loss function in GANs. The generator-discriminator network pair is also replaced with the encoder-decoder.

For channel modeling, VAEs can be leveraged to produce a channel sample from the target distribution after the learning process or to approximate the joint distribution of the channel parameters. For the latter, the path losses, delays, and angle of arrivals and departures are generated by a trained VAE for each link conditioned on the link state such as LOS, NLOS, or outage for mmWave unmanned aerial vehicle (UAV) communication [17]. The approach of generating model parameters via the training dataset can be regarded as the automation of deriving the parameters with stochastic channel modeling.

After their success in synthesizing images, diffusion models, for example, denoising diffusion probabilistic models (DDPMs), score-based generative models (SGMs), and stochastic differential equations (SDEs) attract attention for generative channel modeling. The optimization framework of diffusion channel modeling is composed of a forward and a reverse diffusion process. The forward path samples data from the target (real) distribution and gradually adds slight Gaussian noise with a scheduled variance in multiple steps until it turns into an isotropic Gaussian distribution. This happens because the data loses the defining features with the steps. The reverse process approximates conditional distribution using a variational lower bound.

Training a diffusion model with a 3GPP-defined stochastic channel model, Sengupta et al. [18] illustrates the spatial and frequency domain performance and compares the results with a Wasserstein GAN. According to their results, although Wasserstein GAN is as good as diffusion models for the fidelity of the generated channel, it produces less diverse channel samples. Likewise, user-specific high-dimensional MIMO channels are generated via diffusion models conditioned on the user's location [19]. The proposed conditional denoising diffusion implicit model was trained with a synthetic dataset, and its efficacy has been shown with downstream tasks, namely channel compression and beam alignment.

In contrast to some initial observations that argue the superiority of diffusion models over GANs for channel modeling, GANs can be better than diffusion models in some cases/metrics. Thus, more work is needed to compare different architectures with each other. Unlike visual inspection in the image domain, which also uses established metrics such as maximum mean discrepancy, no explicit measures exist to assess the quality of generated channel samples. Spectral efficiency analysis and codebook-based approach are discussed to compare the generative channel models since they capture the channel power and directionality [20].

4.2.2 Site-specific Generative Models

In the preliminary evaluations of generative AI-based channel modeling, synthetic datasets derived from generic ray tracing tools or samples of statistical channel models have been typically used. These experiments could be repeated with the real samples acquired by site-specific measurement campaigns to see the full potential of generative AI. Real channel measurements are more meaningful because they contain all the correlations and structures of a propagation environment, which may not be captured by ray tracing or statistical models. However, data acquisition from natural system setups is always costly. Therefore, utilizing a mixture of datasets from real measurements and simulator-based models can be a good compromise. For instance, a model trained with synthetic samples can be fine-tuned with site-specific data. Training the model with all the data without fine-tuning is also possible.

Collecting channel data from one specific environment and training the model accordingly yields a site-specific channel model. Compared to generic models, such a specialized channel model can promise performance improvements for various channel-related downstream tasks, for example, estimation, compression, and prediction, due to better leveraging of correlations. On the other hand, site-specific modeling can negatively impact the generalization capability, leading to training multiple models for different cases. This brings additional network complexities to maintain multiple models with related signaling and lifecycle management procedures. Thorough and systematic evaluations should follow the initial studies to investigate how a trained generative model with a dataset collected from one environment could perform for others.

To make site-specific generative channel modeling practically viable, the challenges related to the choice of architecture must be handled. To illustrate, the mode collapse problem in GANs that refers to learning a few samples to fool the discriminator could significantly degrade the modeling quality since this leads to capturing only a tiny portion of the distribution. Many different architectural solutions have been studied extensively in the literature on this problem, including loss functions. VAEs or any models trained based on likelihood measures, for example, diffusion models, can considerably decrease performance in practical implementations depending on the approximation of the likelihood function or the usage of a surrogate loss function with the need for specialized architectures [21]. Additionally, the computational cost of a generative model should be deeply analyzed, and the necessary optimizations should be made to limit the compute and memory complexity.

4.3 Generative Channel Estimation

Channel estimation is classically performed by transmitting reference signals (or pilot tones) known by the receiver. The problem can be formulated in terms of the received signal **Y**, given by

$$\mathbf{Y} = \sqrt{\rho}\mathbf{H}\mathbf{P} + \mathbf{N}, \tag{4.1}$$

where ρ is the signal-to-noise ratio (SNR), **H** denotes the channel matrix, **P** is the unitary pilot matrix, and **N** represents the additive zero-mean Gaussian noise matrix. Specifically, the task is to estimate the channel **H** from the signal **Y** for a known pilot matrix **P** and ρ at the receiver.[3] Note that the dimension of the matrices of Eq. (4.1), as well as the structures, depend on the channel characteristics, and beamforming, that is, **H** can be a beamformed (or effective) channel. To illustrate, for a narrowband MIMO channel with N_t transmit antennas and N_r receive antennas, the complex-valued matrices are $\mathbf{Y} \in \mathcal{C}^{N_r \times N_p}$, $\mathbf{H} \in \mathcal{C}^{N_r \times N_t}$, $\mathbf{P} \in \mathcal{C}^{N_t \times N_p}$, and $\mathbf{N} \in \mathcal{C}^{N_r \times N_p}$ for N_p number of pilots.

Vectorizing the matrices in Eq. (4.1) produces

$$\mathbf{y} = \sqrt{\rho}\mathbf{A}\mathbf{h} + \mathbf{n}, \tag{4.2}$$

where $\mathbf{A} = \mathbf{P}^T \otimes \mathbf{I}_{N_r}$ such that $(.)^T$ denotes the transpose operator, $\mathbf{I}_{N_r}$ is an $N_r \times N_r$ identity matrix, $\otimes$ is the Kronecker product, and thus $\mathbf{A} \in \mathcal{C}^{N_r N_p \times N_r N_t}$. Additionally, **y** is an $N_r N_p$ dimensional vector, since $\mathbf{y} = \text{vec}(\mathbf{Y})$. Likewise, $\mathbf{h} = \text{vec}(\mathbf{H})$ and $\mathbf{n} = \text{vec}(\mathbf{N})$ are $N_r N_t$ and $N_r N_p$ dimensional vectors, respectively. Note that the extension of narrowband MIMO channels to OFDM is straightforward and the details can be found in Ref. [3].

3 Within the scope of problem formulation, it is equivalent to consider **H** either as the raw channel between the antenna elements or the channel with analog transmit and/or receive beamforming. Likewise, **P** can refer to either reference signals or precoded reference signals.

Reducing the number of pilots to save resources and enhance spectral efficiency can lead to an underdetermined system. To be more precise, if $N_p < N_t$, the number of equations ($N_r N_p$) becomes fewer than the number of unknowns ($N_r N_t$) in Eq. (4.2). In this case, the estimation becomes an ill-posed inverse problem and can be solved as a regularized optimization problem of

$$\hat{\mathbf{h}} = \arg\min_{\mathbf{h}} ||\mathbf{y} - \sqrt{\rho}\mathbf{A}\mathbf{h}||_2^2 + \lambda r(\mathbf{h}), \tag{4.3}$$

where $||.||_2^2$ denotes the square of the Euclidean ($\mathcal{L}_2$) norm, λ is a hyperparameter, and $r(\mathbf{h})$ is a regularization term. The optimization problem given in Eq. (4.3) can be reformulated to enforce sparsity-based priors (channel sparsity) and enable compressed sensing algorithms, that is,

$$\begin{aligned} \hat{\mathbf{h}} = \arg\min_{\mathbf{h}} \quad & ||\mathbf{h}||_1 \\ \text{subject to} \quad & ||\mathbf{y} - \sqrt{\rho}\mathbf{A}\mathbf{h}||_2^2 < \epsilon, \end{aligned} \tag{4.4}$$

where $||.||_1$ denotes the $\mathcal{L}_1$ norm, and ϵ is a threshold. Generative AI can uniquely transform sparsity-based compressed sensing solutions to learnable generative priors [2].

The fundamental challenge in utilizing generative AI for channel estimation is the lack of frameworks since, as opposed to traditional discriminative AI models that map one set of inputs to another set of outputs, it is not trivial how to use generative models. This issue can be addressed with an iterative framework empowered by a pretrained generator, as illustrated in Figure 4.5. The iterative framework starts with a random noise vector $\mathbf{z}$ labeled as a "latent vector" in the generative AI terminology. Specifically, a pretrained generator produces a channel sample $\hat{\mathbf{h}}$ in response to random input. However, this is not the channel estimate that is seeking. This generated channel sample is passed through the measurement matrix,

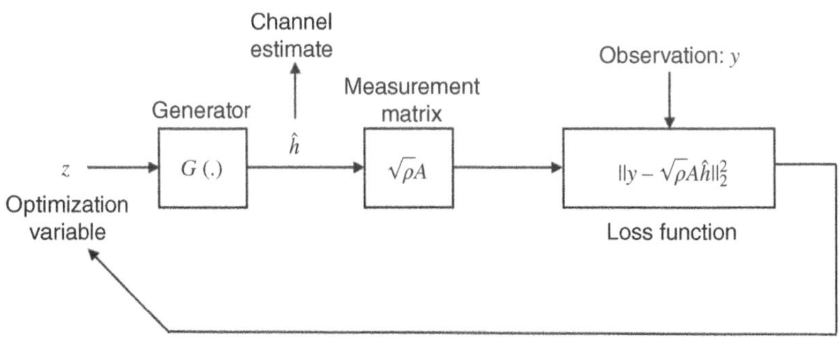

Figure 4.5 An iterative framework to adapt generative models to channel estimation. *Source:* Eren Balevi.

that is, multiplied with $\sqrt{\rho}\mathbf{A}$ and then this vector is optimized toward the observation $\mathbf{y}$ in an iterative fashion to minimize the loss function $||\mathbf{y} - \sqrt{\rho}\mathbf{A}\widehat{\mathbf{h}}||_2^2$. A regularization term can be added to this loss function. This implies that the generator produces the best channel estimate to minimize the loss function by optimizing the latent vector $\mathbf{z}$.

The underlying mathematical formulation for the iterative framework is as follows. First, a generative model, which can be the generator of a GAN, the decoder of a VAE, or any other generative network, is pretrained according to channel measurements, as explained in channel modeling. Then, the input of the generator network is periodically optimized with a period of channel coherence time according to the received signal in Eq. (4.2) to optimize

$$\mathbf{z}^* = \arg\min_{\mathbf{z}} ||\mathbf{y} - \sqrt{\rho}\mathbf{A}G(\mathbf{z})||_2^2. \tag{4.5}$$

The typical choice for initializing $\mathbf{z}$ is to sample it from an independent identically distributed (i.i.d.) zero-mean unit variance Gaussian random variables. The result of Eq. (4.5) provides the generative channel estimate (GCE), specified by

$$\widehat{\mathbf{h}}_{\text{GCE}} = G(\mathbf{z}^*). \tag{4.6}$$

These steps are summarized in Algorithm 1.

Algorithm 1: Generative channel estimation

Input: Gaussian i.i.d. noise vector $\mathbf{z}$
Output: $\widehat{\mathbf{h}}_{\text{GCE}}$

Offline Phase:
Train a generative model
Extract the generator part G
Online Phase:
for each coherence time interval **do**
  Sample $\mathbf{z}$ from i.i.d. Gaussian random variables
  Given the noisy received signal $\mathbf{y}$, SNR estimate $\sqrt{\rho}$ and $\mathbf{A}$, solve Eq. (4.5)
  Obtain the channel estimate in Eq. (4.6)
end for

Discriminative AI with deep learning architectures has been heavily investigated for many sorts of channel estimation [22]. One of the considerations is whether generative AI is superior to channel estimation. The channel estimation with discriminative deep learning models is mainly based on CNNs such as ResNet or neural networks with fully connected layers. These traditional models are typically trained with supervised learning according to a labeled dataset composed of pairs of processed or unprocessed noisy channel measurements and ideal (noiseless) channel samples that serve as labels. It could be interesting to compare the iterative framework explained in Algorithm 1 for generative channel estimation with a model trained according to supervised learning. The same neural network architectures are used for generative and supervised learning to make a fair comparison. The performance metric is normalized mean square error (NMSE), which is defined as

$$\text{NMSE} = \mathbb{E}\left[\frac{||\mathbf{h} - \hat{\mathbf{h}}||_2^2}{||\mathbf{h}||_2^2}\right]. \tag{4.7}$$

As shown in Figure 4.6, the generative channel estimation outperforms the traditional AI/ML-based channel estimator with supervised learning by a large margin when there are insufficient pilots. The gap becomes much more prominent for large SNRs.

High dimensionality due to larger bandwidth and more antenna elements at both ends of the communication link can bring significant user-perceived or network throughput gain. On the other hand, increased dimensionality leads to transmitting more reference signals, wasting additional resource elements. Minimizing the reference signals for channel estimation is thus appealing for high-dimensional channels. Compressed sensing algorithms that leverage channel

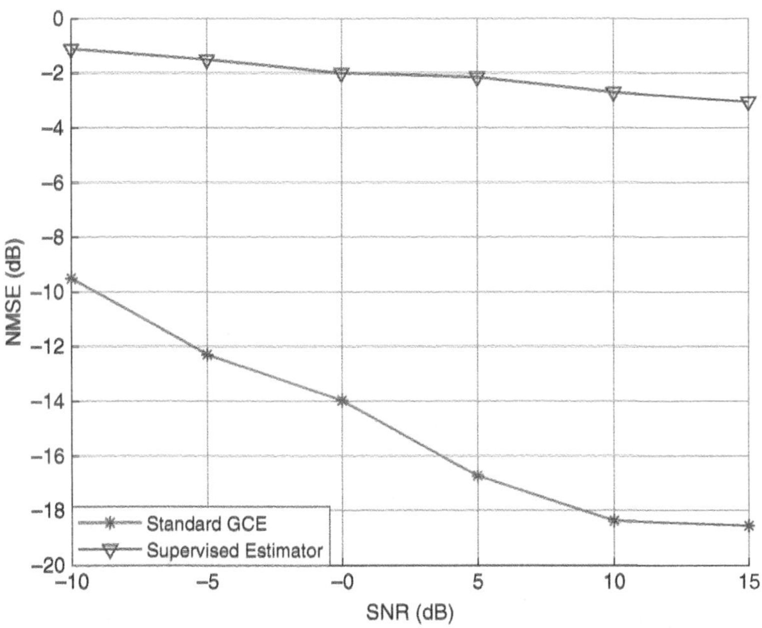

Figure 4.6 Comparison of the generative estimation with the traditional supervised learning for the same neural network architecture. *Source:* Eren Balevi.

sparsity are one of the main techniques to save reference signals if there are insufficient measurements. However, the level of sparsity may not be enough for the desired performance, and figuring out other priors related to channel distribution is a challenge. The generative channel estimator tackles this by learning a pretrained generator that can be used as the prior information, as explained in Figure 4.5. Next, the generative approach is experimented against the state-of-the-art compressed sensing techniques for narrowband and wideband channels.

4.3.1 Narrowband Channel Estimation with Reduced Pilots

The generative channel estimation can be regarded as the generalization of compressed sensing algorithms that replaces and generalizes the prior based on sparsity with a learned prior. This implies that if the sparsity is the optimum structure for a problem, the learned prior becomes sparse, that is, equivalent to the sparsity prior. Thus, it is compared with two fundamental compressed sensing algorithms, namely Orthogonal Matching Pursuit (OMP) and Expectation-Maximization Gaussian-Mixture Approximate Message Passing (EM-GM-AMP) [23] when there is a lack of pilot signals for MIMO channels. Without any loss of generality, a synthetic dataset from the 3GPP-defined CDL-D channel model samples is acquired with a delay spread of 5 ms and a Doppler shift of 5 Hz. A Wasserstein GAN is selected and trained for the generative model to address the training issues of a GAN [24]. Additionally, the channel samples are normalized in the preprocessing step for training stability.

The details of the generator and discriminator architectures are given in Tables 4.1 and 4.2 for $N_t = 64$ and $N_r = 16$. The number of pilots is set $N_p = 16$ so that $N_p < N_t$. The generator architecture involves a fully connected layer followed by several upsampling, convolutional operations, and batch normalization with the ReLU activation function. The discriminator architecture mainly comprises convolutional layers with leaky ReLU activation function, batch normalization, zero padding, and a fully connected layer at the end. Note that these are just example architectures used to obtain simulation results, that is, any architecture can be utilized for the generative estimation. The results for all the generative and compressed sensing-based channel estimators are obtained by running each algorithm until they converge.

As shown in Figure 4.7, the channel estimation with generative AI outperforms the state-of-the-art compressed sensing baselines, whose parameters are also optimized over the SNR ranges from −10 dB to 15 dB. To illustrate, for a fixed SNR of 0 dB, the NMSE of the GCE becomes −12 dB, whereas this is −8 dB for the OMP and the EM-GM-AMP algorithm. Likewise, the GCE provides nearly 9 dB SNR gain for a fixed NMSE of −10 dB. It is worth noting that the EM-GM-AMP gives an NMSE slightly above 0 dB at −10 dB SNR. However, this is just an imperfection in the simulation since any reasonable estimator should yield values below 0 dB NMSE.

Table 4.1 Generator architecture.

Layer	Input shape	Output shape
Fully Connected	1 x 35	1 x 8192
ReLU	1 x 8192	1 x 8192
Reshape	1 x 8192	128 x 16 x 4
Upsample	128 x 16 x 4	128 x 32 x 8
Conv2D	128 x 32 x 8	128 x 32 x 8
Batch Normalization	128 x 32 x 8	128 x 32 x 8
ReLU	128 x 32 x 8	128 x 32 x 8
Upsample	128 x 32 x 8	128 x 64 x 16
Conv2D	128 x 64 x 16	128 x 64 x 16
Batch Normalization	128 x 64 x 16	128 x 64 x 16
ReLU	128 x 64 x 16	128 x 64 x 16
Conv2D	128 x 64 x 16	2 x 64 x 16
Batch Normalization	2 x 64 x 16	2 x 64 x 16

Table 4.2 Discriminator architecture.

Layer	Input shape	Output shape
Conv2D	2 x 64 x 16	16 x 32 x 8
Leaky ReLU	16 x 32 x 8	16 x 32 x 8
Conv2D	16 x 32 x 8	32 x 16 x 4
Batch Normalization	32 x 16 x 4	32 x 16 x 4
Zero Padding	32 x 16 x 4	32 x 17 x 5
Leaky ReLU	32 x 17 x 5	32 x 17 x 5
Conv2D	32 x 17 x 5	64 x 9 x 3
Batch Normalization	64 x 9 x 3	64 x 9 x 3
Leaky ReLU	64 x 9 x 3	64 x 9 x 3
Conv2D	64 x 9 x 3	128 x 9 x 3
Batch Normalization	128 x 9 x 3	128 x 9 x 3
Leaky ReLU	128 x 9 x 3	128 x 9 x 3
Flatten	128 x 9 x 3	3456
Fully Connected	3456	1

The observed benefit of the generative approach with the initial evaluations for the narrowband channels motivates more extensive experimental works, including various generative architectures such as VAEs, diffusion models, or autoregressive models with different frameworks and loss functions.

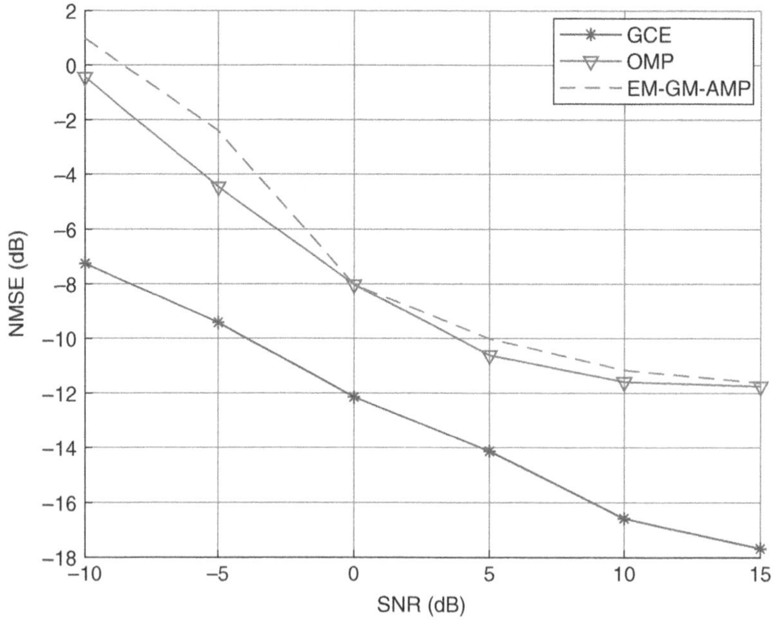

Figure 4.7 MIMO channel: The comparison of the GCE with the baseline compressed sensing algorithms for $N_t = 64$, $N_r = 16$, and $N_p = 16$. *Source:* Eren Balevi.

4.3.2 Wideband Channel Estimation with Reduced Pilots

With wideband MIMO-OFDM channels, the problem dimension further increases linearly in the frequency domain over a MIMO channel. The granularity of the frequency domain depends on the channel coherence bandwidth. That is, if the coherence bandwidth is in the order of a subcarrier, then the dimension scales up with the number of subcarriers. Otherwise, the granularity can become a group of subcarriers at the subband level. In this case, the number of subcarriers or subbands is denoted by N_f, and the generative AI-based channel estimation is compared with the EM-GM-AMP algorithm in terms of NMSE performance when there is a limited number of pilots or reference signals to perceive the impact of generative models for wideband channel estimation.

The same Wasserstein GAN architecture is given by Tables 4.1 and 4.2 are trained similarly to the narrowband channels, and the extracted G is used as the generative channel estimation. The channel parameters are set as $N_t = 64$, $N_r = 4$, and $N_f = 64$. A 64×64 channel matrix is measured at each receive antenna in the spatial-frequency domain. This implies that four different generative networks G are acquired at the end of the training, that is, one for each receive antenna to estimate the channel via the generative approach. Hence, the performance of the

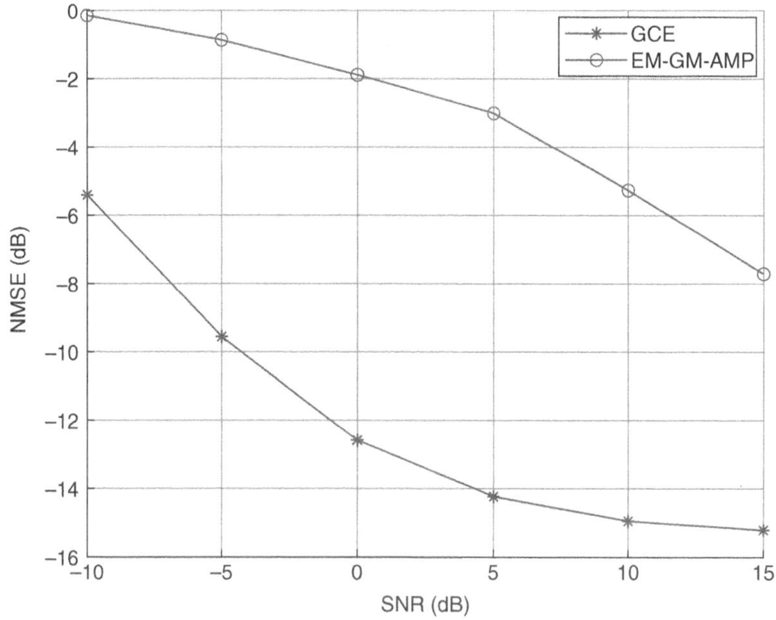

Figure 4.8 MIMO-OFDM channel: The comparison of the GCE with the EM-GM-AMP algorithm for $N_t = 64$, $N_f = 64$, $N_r = 4$, and $N_p = 16$. *Source:* Eren Balevi.

GCE is presented as the average of these 4 estimations over the received antennas. Figure 4.8 indicates the NMSE enhancement of the GCE method over the baseline compressed sensing or the EM-GM-AMP technique when the number of pilots is 16, that is, $N_p = 16$. To illustrate, at 15 dB SNR, the NMSE becomes nearly −15 dB for the GCE, whereas this is −8 dB for the EM-GM-AMP.

4.4 Channel Compression

Massive MIMO is one of the key technology enablers for 5G and beyond. Having equipped several antennas enables base stations to mitigate multiuser interference by precoding the multiple streams, resulting in enhanced cell throughput at the expense of the need for accurate channel state information (CSI). In time-division-duplex (TDD) systems, the estimated uplink channel at the base station can be utilized for the downlink precoder optimization due to reciprocity. However, there can be multiple factors that disturb reciprocity such as RF calibration differences between transmitters and receivers. On the other hand, frequency-division-duplex (FDD) systems necessitate estimating the downlink channel at the user device and sharing it with a base station. This, however, can lead to a large communication overhead associated with many antenna elements and increasingly wide bands. Alternatively, the precoders can be optimized at the user devices after channel estimation and transmitted to the base station. However, this approach also has a non-negligible overhead with a non-trivial codebook design for precoder; thus, incorporating efficient compression methods seems inevitable.

Channel compression and feedback are fundamental problems between base stations and user devices. The main goal of compression is to minimize the distortion in the reconstructed channel for a target rate to prevent performance drop. Many studies have been conducted on compression, including vector quantization, codebook design, and compressed sensing algorithms. Deep learning algorithms have recently been applied to the problem. For instance, autoencoder-based architectures have been studied to compress the channel or precoder estimate more efficiently before transmitting it to the base station for FDD systems in different channel domains, for example, angular-delay domain compression [25]. In this framework, there is an encoder at the user device for compression and a decoder at the base station for reconstruction. Despite some promising results, the performance gain is limited, and there is additional room for optimization to outweigh the additional computational complexity due to deep learning models.

Leveraging generative AI for CSI compression can further enhance compression performance regarding the rate-distortion trade-off. The initial results of a paper show that the generator network input of a trained Wasserstein GAN can be utilized for compressed channel representation [7], for example, a channel sample with 2048 complex symbols can be reduced to 15 real values. This concept can be easily generalized to share the latent vector of a generative channel model for compression. Another method is to use the generator of a GAN as the decoder for a CNN-based encoder to recover the compressed CSI with a loss function composed of the weighted sum of the adversarial loss and the reconstruction loss [26]. VAE has also been studied for CSI compression [27]. The idea is to add the weighted KL divergence between the encoder distribution and the prior distribution as a regularization term to the reconstruction loss. In this setting, the relative importance of the KL divergence was investigated, and the training was completed using the reparameterization trick with the selected hyperparameters. The superiority of the VAEs over traditional autoencoders for channel compression is illustrated in the paper as well.

Semantic compression is tightly coupled with generative AI-based compression. The driving idea for semantic compression is to focus on a source's conceptual features instead of perceptual features and obtain perceptually equivalent compressed signals despite much more aggressive compression ratios. The high feature distilling capability of transformer-based LLMs can be exploited to perform semantic compression. Jointly optimizing compression and transmission yields JSCC. Integrating semantic compression into JSCC for CSI transmission is also possible, which can bring benefits when transmitting under noisy dynamic conditions. Specifically, JSCC provides graceful degradation and improvement against the cliff effect that refers to a mismatch between the SNR and the adaptation rate, unlike digital transmission.

4.5 Beamforming

The need for large antenna arrays to reap the potential benefits of massive MIMO systems and to compensate for the high propagation losses in mmWave transmission makes beamforming a key technology for future wireless systems. Beamforming is the method to produce constructive interference at some angles and destructive interference at others. Conventionally, beamforming is implemented digitally in the baseband (digital precoder) by controlling the amplitude and phase of each antenna element connected to a separate RF chain.[4] A more practical architectural solution is to share RF chains among antenna elements and perform beamforming in the digital and analog domains since having a dedicated RF chain per antenna element leads to excessive power consumption and device cost.[5]

The selection of synthesized beams after training/alignment, for example, according to reference signal received powers (RSRPs) and their monitoring, tracking, and recovery, is known as beam management. 5G NR adopts a beam management framework relying on beam sweeping, measurement, and reporting at the expense of overhead and latency. Beam sweeping is classically performed over quantized codebooks to search for the best beam. Data-driven solutions can avoid trying never-used beams in non-uniform topologies and prevent wasting resources and increasing latency. To inspect the problem without using a fixed codebook, specifying the underlying optimization problem within a mathematical framework is helpful.

An example transmitter that omits some RF implementation details is illustrated in Figure 4.9 for N_s streams, N_t^{RF} RF chains, and N_t transmit antenna elements such that $N_t^{RF} < N_t$. The RF chain involves a digital-to-analog converter (DAC) for the in-phase and quadrature signals, anti-aliasing filter, local oscillator (LO), and mixer. Analog beamforming can typically only control the signal phases via phase shifters to synthesize different beams. In contrast, digital precoding controls both amplitude and phase. The optimization problem that maximizes

4 In 3GPP systems, this case corresponds to having a separate TXRU for each antenna element, and antenna port virtualization can fulfill the task of a digital precoder.

5 Sub-array partition model and full-connection model have been studied within 3GPP to map TXRUs to antenna elements via TXRU virtualization.

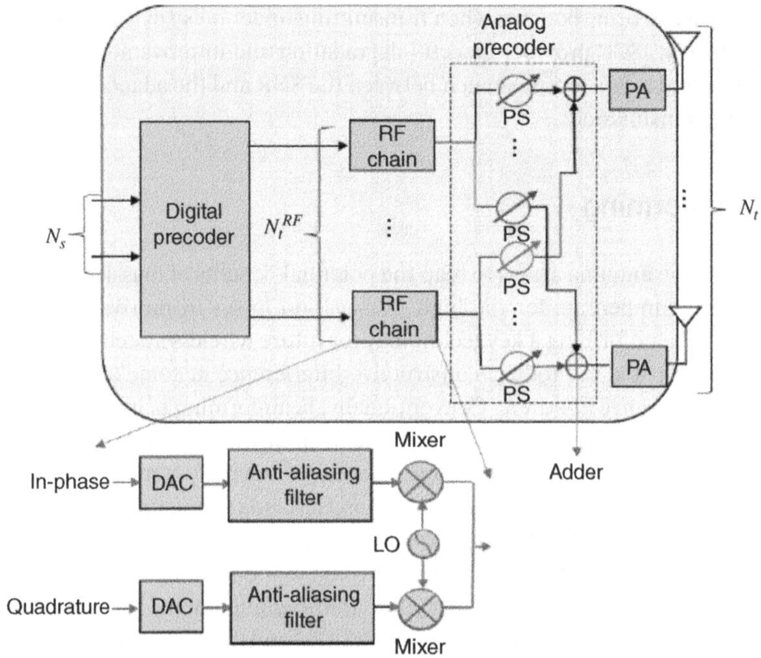

Figure 4.9 A simple illustration of a hybrid beamforming architecture with phase shifter (PS) in the analog domain and power amplifier (PA). *Source:* Eren Balevi.

spectral efficiency over the analog beamformer $\mathbf{F}_{\mathrm{RF}}$ and the digital precoder $\mathbf{F}_{\mathrm{BB}}$ can be expressed as

$$\begin{aligned}(\mathbf{F}_{\mathrm{RF}}^{\mathrm{opt}}, \mathbf{F}_{\mathrm{BB}}^{\mathrm{opt}}) = \underset{\mathbf{F}_{\mathrm{RF}},\mathbf{F}_{\mathrm{BB}}}{\arg\max} \quad & \log_2\left(\left|\mathbf{I}_{N_r} + \frac{\rho}{N_s}\mathbf{H}\mathbf{F}_{\mathrm{RF}}\mathbf{F}_{\mathrm{BB}}\mathbf{F}_{\mathrm{BB}}^H\mathbf{F}_{\mathrm{RF}}^H\mathbf{H}^H\right|\right) \\ \text{subject to} \quad & |(\mathbf{F}_{\mathrm{RF}})_{i,j}|^2 = 1/N_t, \forall i, j \\ & ||\mathbf{F}_{\mathrm{RF}}\mathbf{F}_{\mathrm{BB}}||_F^2 = N_s,\end{aligned} \tag{4.8}$$

where the first and second constraints correspond to the constant modulus due to analog phase shifters and total transmission power, respectively. Additionally, $|(\mathbf{F}_{\mathrm{RF}})_{i,j}|$ is the magnitude of the (i, j)th element of $\mathbf{F}_{\mathrm{RF}}$, $(.)^H$ is the Hermitian operation, and $||.||_F$ shows the Frobenius norm.

Many iterative algorithms, including deep learning-aided methods, have been proposed to solve the problem formulated in Eq. (4.8). Similar to the other physical layer applications, applying generative AI models can enhance spectral efficiency while balancing the inherent additional computational complexity of inference of the neural networks. Unfolding is a promising solution to decrease the computational complexity of iterative algorithms by exploiting domain knowledge [28], [29]. Unfolding can be used to optimize hybrid beamforming over a compressed channel extracted with a generative model. This approach handles the computational complexity and the feedback overhead while minimizing the necessity of CSI by transforming the search space to the low-dimensional channel manifold [7].

The generative AI-based framework with unfolding that separates hybrid beamforming into unrolled upper and lower-level problems is given in Figure 4.10. This framework is composed of L layers, each with an upper and lower-level problem with an update rule, and the low-dimensional input vector $\mathbf{x}$, initialized with the compressed channel acquired by the latent of a generative model. The upper-level problem that outputs the digital precoder and the lower-level problem that provides the analog beamformer are implemented with a CNN-based architecture, and they are trained offline with supervised learning. It is worth noting that supervised learning can be performed with ideal labels if available. Otherwise, one of the algorithms proposed for hybrid beamforming could be utilized to generate the labels. Parameterizing the signal processing grounded algorithms and encoding to CNNs for the beamforming design brings robustness for changing configurations and environments.

The input $\mathbf{x}$ must be processed within the layers to learn the digital and analog precoder per channel coherence time. In the first layer, the upper-level problem takes the compressed channel and the randomly initialized instance of the analog precoder $\mathbf{F}_{RF}^{(0)}$ and outputs the digital precoder $\mathbf{F}_{BB}^{(1)}$ to the lower-level problem. Once the lower level takes its input, it produces the analog precoder $\mathbf{F}_{RF}^{(1)}$ for the first layer. The update rule is defined in terms of the gradient, that is, $-\mu_1 \nabla_{\mathbf{x}} \mathcal{L}(\mathbf{F}_{\mathrm{RF}}^{(1)}, \mathbf{F}_{\mathrm{BB}}^{(1)})$ for the first layer, where μ_1 is the step size and $\mathcal{L}$ is the loss function. Any loss function that maximizes the objective function of the optimization problem in Eq. (4.8) can be used. This process is repeated L times. After L steps, the acquired $\mathbf{x}^*$ is used to determine the digital and analog beamforming.

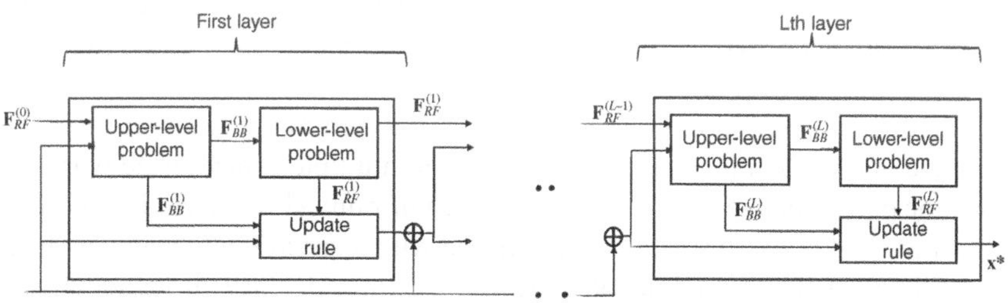

Figure 4.10 The upper- and lower-level problems output the digital and analog beamforming, respectively. Then, the model parameters **x** are updated according to a pre-determined update rule for *L* layers. *Source:* Eren Balevi.

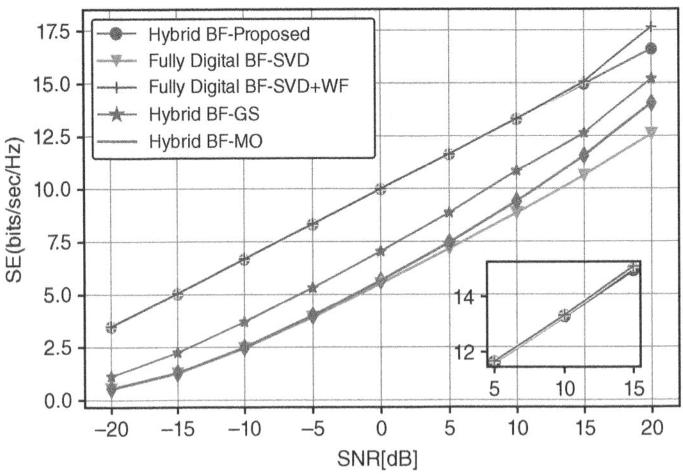

Figure 4.11 The spectral efficiency of the unfolded hybrid beamforming approach compared to the baselines. *Source:* Eren Balevi.

The average spectral efficiency of the unfolded hybrid beamforming algorithm for a single user communication is given by Figure 4.11. All the data used for training the models and obtaining the test results come from a geometric channel model. The comparison baselines are the fully digital precoding based on the singular value decomposition (SVD) with equal power allocation as well as water-filling power allocation and the hybrid precoding algorithms based on Gram-Schmidt (GS) orthogonalization [30] and Manifold Optimization (MO) [31]. Notice that the upper bound of this problem is the fully digital SVD precoding with water-filling power allocation. As can be observed, the unfolded hybrid beamforming achieves the upper bound up to an SNR of 15 dB without the impractical assumption of a perfectly known channel at the transmitter. This means the design gracefully copes with the detrimental impacts of phase shifters and limited channel knowledge. Furthermore, the unfolded beamformer has higher spectral efficiency for all SNRs than the other hybrid beamformer designs and is fully digital beamforming with equal power allocation.

Generative AI can be leveraged for beamforming in many different cases. To illustrate, the GAN framework is generalized to multiple generator networks to cover the rank deficient channels for hybrid beamforming [32]. Accordingly, three generators and a discriminator are optimized with a spectral efficiency module. The spectral efficiency module calculates the spectral efficiency throughout training and provides the input to the discriminator. The other input of the discriminator is the targeted optimum spectral efficiency. With this framework, spectral efficiency is maximized through beamforming. Another application is to learn the dynamics of the strongest beam pair due to mobility and environment via a deep recurrent variational autoencoder (DR-VAE) [33]. To summarize, generative models look promising for capturing beam dynamics and injecting the acquired knowledge into beamforming design. For multiple beam management-related problems, generative AI can lead to many different designs.

4.6 Summary

Generative AI can be helpful in many directions while addressing the scalability issues of future communication systems, emerging mainly because of the broader bandwidths and the increasingly large number of antenna elements to enhance capacity, coverage, and energy efficiency under the hood of immersive communication, massive communication, and hyper-reliable and low-latency communication (HRLLC). Enhancing the power of traditional deep learning models, representative use cases of generative AI at the PHY layer of wireless communication, such as channel modeling, channel estimation, compression, and beamforming, are examined with different types of generative architectures in this chapter. Despite exciting ideas and promising initial evaluations, more research is needed to completely figure out how generative AI works for future communication systems. Diversifying the generative AI applications and selecting one or more of the mentioned use cases for a more detailed analysis and prototyping can constitute the next steps in this fascinating research topic.

References

1 Yang Yang, Yang Li, Wuxiong Zhang, Fei Qin, Pengcheng Zhu, and Cheng-Xiang Wang. Generative-adversarial-network-based wireless channel modeling: Challenges and opportunities. *IEEE Communications Magazine*, 57: 22–27, 2019.

2 Eren Balevi, Akash Doshi, Ajil Jalal, Alexandros Dimakis, and Jeffrey G. Andrews. High dimensional channel estimation using deep generative networks. *IEEE Journal on Selected Areas in Communications*, 39(1): 18–30, 2020.

3 Eren Balevi and Jeffrey G. Andrews. Wideband channel estimation with a generative adversarial network. *IEEE Transactions on Wireless Communications*, 20(5):3049–3060, 2021.

4 Jakob Hoydis, Sebastian Cammerer, Fayçal Ait Aoudia, Avinash Vem, Nikolaus Binder, Guillermo Marcus, and Alexander Keller. Sionna: An open-source library for next-generation physical layer research. *arXiv preprint arXiv:2203.11854*, 2023.

5 Tianyu Hu, Yang Huang, Qiuming Zhu, and Qihui Wu. Channel estimation enhancement with generative adversarial networks. *IEEE Transactions on Cognitive Communications and Networking*, 7(1): 145–156, 2020.

6 Xingqin Lin. An overview of the 3gpp study on artificial intelligence for 5g new radio. *arXiv preprint arXiv:2308.05315*, 2023.

7 Eren Balevi and Jeffrey G. Andrews. Unfolded hybrid beamforming with GAN compressed ultra-low feedback overhead. *IEEE Transactions on Wireless Commun.*, 20(12): 8381–8392, 2021.

8 Jialong Xu, Tze-Yang Tung, Bo Ai, Wei Chen, Yuxuan Sun, and Deniz Gunduz. Deep joint source-channel coding for semantic communications. *IEEE Communications Magazine*, 61(11): 42–48, 2023.

9 H. Jiang, M. Cui, D. W. K. Ng, and L. Dai. Accurate channel prediction based on transformer: Making mobility negligible. *IEEE Journal on Selected Areas in Communications*, 40(9): 2717–2732, 2022.

10 Li Sun, Yuwei Wang, A. Lee Swindlehurst, and Xiao Tang. Generative-adversarial-network enabled signal detection for communication systems with unknown channel models. *IEEE Journal on Selected Areas in Communications*, 39(1): 47–60, 2021.

11 Ke He, Le He, Lisheng Fan, Yansha Deng, George K. Karagiannidis, and Arumugam Nallanathan. Learning-based signal detection for MIMO systems with unknown noise statistics. *IEEE Transactions on Communications*, 69(5): 3025–3038, 2021.

12 Bin Tang, Ya Tu, Zhaoyue Zhang, and Yun Lin. Digital signal modulation classification with data augmentation using generative adversarial nets in cognitive radio networks. *IEEE Access*, 6(1): 15713–15722, 2018.

13 Avi Caciularu and David Burshtein. Unsupervised linear and nonlinear channel equalization and decoding using variational autoencoders. *IEEE Transactions on Cognitive Communications and Networking*, 6(3): 1003–1018, 2020.

14 Lina Bariah, Qiyang Zhao, Hang Zou, Yu Tian, Faouzi Bader, and Merouane Debbah. Large generative AI models for telecom: The next big thing? *IEEE Communications Magazine*, pages 1–7, 2024.

15 Nguyen Van Huynh, Jiacheng Wang, Hongyang Du, Dinh Thai Hoang, Dusit Niyato, Diep N. Nguyen, Dong In Kim, and Khaled B. Letaief. Generative AI for physical layer communications: A survey. *IEEE Transactions on Cognitive Communications and Networking*, 10(3): 706–728, 2024.

16 Yaqi Hu, Mingsheng Yin, William Xia, Sundeep Rangan, and Marco Mezzavilla. Multi-frequency channel modeling for millimeter wave and thz wireless communication via generative adversarial network. *56th Asilomar Conference on Signals, Systems, and Computers*, 2022.

17 William Xia, Sundeep Rangan, Marco Mezzavilla, Angel Lozano, Giovanni Geraci, Vasilii Semkin, and Giuseppe Loianno. Generative neural network channel modeling for millimeter-wave UAV communication. *IEEE Transactions on Wireless Communications*, 21(11): 9417–9431, 2022.

18 Ushnish Sengupta, Chinkuo Jao, Alberto Bernacchia, Sattar Vakili, and Da-shan Shiu. Generative diffusion models for radio wireless channel modelling and sampling. *IEEE Global Communications Conference*, 2023.

19 Taekyun Lee, Juseong Park, Hyeji Kim, and Jeffrey G. Andrews. Generating high dimensional user-specific wireless channels using diffusion models. *arXiv preprint arXiv:2409.03924*, 2024.

20 Michael Baur, Nurettin Turan, Simon Wallner, and Wolfgang Utschick. Evaluation metrics and methods for generative models in the wireless PHY layer. *arXiv preprint arXiv:2408.00634*, 2024.

21 Yang Song and Stefano Ermon. Generative modeling by estimating gradients of the data distribution. In *Proceedings of the 33rd Annual Conference on Neural Information Processing Systems*, 2019.

22 Eren Balevi and Jeffrey G. Andrews. One-bit OFDM receivers via deep learning. *IEEE Transactions on Communications*, 67(6): 4326–4336, 2019.

23 Jeremy P. Vila and Philip Schniter. Expectation-maximization Gaussian-mixture approximate message passing. *IEEE Transactions on Signal Processing*, 61(19): 4658–4672, 2013.

24 Martin Arjovsky, Soumith Chintala, and Léon Bottou. Wasserstein generative adversarial networks. *In Proceedings of the 34th International Conference on Machine Learning*, 70: 214–223, 2017.

25 Chao-Kai Wen, Wan-Ting Shih, and Shi Jin. Deep learning for massive MIMO CSI feedback. *IEEE Wireless Communications Letters*, 7(5): 748–751, 2018.

26 Bassant Tolba, Maha Elsabrouty, Mubarak G. Abdu-Aguye, Haris Gacanin, and Hossam Mohamed Kasem. Massive MIMO CSI feedback based on generative adversarial network. *IEEE Communications Letters*, 24(12): 2805–2808, 2018.

27 Mostafa Hussien, Kim Khoa Nguyen, and Mohamed Cheriet. Prvnet: A novel partially regularized variational autoencoders for massive MIMO CSI feedback. *IEEE Wireless Communications and Networking Conference (WCNC)*, pages 2286–2291, 2022.

28 Karol Gregor and Yann LeCun. Learning fast approximations of sparse coding. In *Proceedings of the International Conference on Machine Learning*, pages 399–406, 2010.

29 J. R. Hershey, J. L. Roux, and F. Weninger. Deep unfolding: Model-based inspiration of novel deep architectures. *arXiv preprint arXiv:1409.2574*, 2014.

30 Ahmed Alkhateeb and Robert W. Heath. Frequency selective hybrid precoding for limited feedback millimeter wave systems. *IEEE Transactions on Communications*, 64(5):1801–1818, 2016.

31 Xianghao Yu, Juei-Chin Shen, Jun Zhang, and Khaled B. Letaief. Alternating minimization algorithms for hybrid precoding in millimeter wave MIMO systems. *IEEE Journal on Selected Topics in Signal Processing*, 10(3): 485–500, 2016.

32 Lihua Pang, Yudong Li, Yang Zhang, Minghao Shang, Yijian Chen, and Anyi Wang. MGGAN-based hybrid beamforming design for massive MIMO systems against rank-deficient channels. *IEEE Communications Letters*, 26(11): 2804–2808, 2022.

33 Muddassar Hussain and Nicolò Michelusi. Learning and adaptation for millimeter-wave beam tracking and training: A dual timescale variational framework. *IEEE Journal on Selected Areas in Communications*, 40(1): 37–53, 2022.

5

Generative AI at the MAC Layer

Kemal Davaslioglu[1,], Ender Ayanoglu[2], and Yalin E. Sagduyu[1]*

[1] *Nexcepta Inc., Gaithersburg, MD, USA*
[2] *Electrical Engineering and Computer Science, Samueli School of Engineering, University of California, Irvine, CA, USA*

*Corresponding author: kdavasli@ieee.org

5.1 Introduction

As the demand for higher data rates and the number of connected devices grow, Medium Access Control (MAC) protocols play a pivotal role in modern communication systems for efficient and conflict-free access to shared wireless spectrum channels. MAC protocols are essential in multiple scenarios. For example, MAC protocols enable Dynamic Spectrum Access (DSA) where multiple users can dynamically access the available spectrum resources in real-time instead of relying on static spectrum allocations. In the face of spectrum scarcity, DSA enhances spectrum efficiency by allowing multiple users to dynamically share the electromagnetic (EM) spectrum while minimizing interference through the management of different hierarchical access levels.

The Citizen Broadband Radio Service (CBRS) in the United States is a great example for DSA. CBRS governs the shared use of the 3.5 GHz band (3550–3700 MHz) [1]. It is designed to maximize spectrum efficiency by facilitating dynamic sharing between different types of users with hierarchical access levels [2].

Another important use case that highlights the importance of MAC protocols is in disconnected, intermittent, and limited (DIL) networks [3]. MAC protocols enable users to maintain connectivity, handle critical transmissions, and employ the limited spectrum efficiently. In congested and contested environments, such as urban areas or large events with high user density, MAC protocols ensure fair and efficient spectrum sharing [4]. They reduce interference and congestion to support reliable communication in these challenging scenarios.

The MAC layer represents a critical component in wireless communication systems. It is responsible for coordinating access to shared communication channels and managing network resources. In next-generation wireless networks, four fundamental challenges persistently limit performance and efficiency: (i) **spectrum scarcity and sharing**, (ii) the **need for adaptive communication strategies**, (iii) **MAC-layer security and potential vulnerabilities**, and (iv) **network performance optimization**.

Generative Artificial Intelligence (AI) offers novel capabilities for the MAC layer by leveraging advanced generative models such as Variational Autoencoders (VAEs), Generative Adversarial Networks (GANs), and diffusion models. These models can enhance spectrum awareness for spectrum sharing by generating precise waveforms tailored to diverse channel conditions, rare scenarios, and anomaly detection. By creating synthetic traffic scenarios, Generative AI enables modeling of complex network behaviors and optimization of scheduling strategies to more efficiently allocate network resources. Additionally, MAC-layer security can benefit from the generative models in simulating common threats such as spoofing, jamming, or replay attacks, thereby facilitating the development and testing of robust countermeasures.

This chapter explores the potential of generative models as a transformative approach to address these challenges. We systematically examine how Generative AI techniques can be integrated within the MAC layer to overcome current limitations and unlock new capabilities in wireless communications. Specifically, we outline the fundamental principles of generative models and analyze their applicability to addressing these four MAC-layer challenges in next-generation wireless communication systems.

The first MAC-layer challenge is **spectrum scarcity and sharing**, which emerges from the exponential growth of wireless devices and applications competing for limited frequency resources [5]. Traditional static spectrum allocation methods have become inadequate to manage the dynamic and increasingly congested wireless communication landscape. Network operators struggle to efficiently allocate and utilize the available spectrum, which leads to inefficient resource utilization, increased interference, and reduced overall network performance [6].

Generative AI models synthesize realistic RF and radar signals and enable domain adaptation through adaptive waveform transformations across different EM environments. By creating high-fidelity synthetic channel models or received signals, Generative AI can support critical MAC-layer applications such as spectrum sensing, modulation recognition, waveform protocol classification, emitter identification, and anomaly detection. The ability to generate and analyze complex signals, channel realizations, or spectrum scenarios allows for more flexible, efficient, and intelligent spectrum management.

The second MAC-layer challenge is the **need for adaptive communication strategies**, which stems from the increasing complexity and heterogeneity of wireless network environments. Traditional communication models struggle to generalize to different network conditions, technologies, and operational scenarios. Recent work has highlighted the importance of incorporating learning-based adaptability at the MAC layer. For instance, Miuccio et al. [7] proposed a generalized wireless MAC protocol learned via abstraction for the MAC protocol to adapt across varying environments, including changes in node density, traffic load, and channel conditions. Xin et al. [8] demonstrated that a deep-learning-based MAC protocol can jointly manage channel access, rate adaptation, and channel switching to dynamically respond to congested and unlicensed spectrum environments. Similarly, DeepWiFi (Davaslioglu et al.) [9] introduced a cognitive MAC protocol that is capable of adapting to interference and jamming through deep neural networks. These studies underscore the critical need for intelligent and adaptive MAC-layer solutions to maintain robust and consistent communication performance in the face of evolving wireless conditions.

Generative models excel in facilitating adaptability through advanced domain adaptation capabilities. By learning underlying data distributions, these models support semi-supervised learning and unsupervised clustering, enabling knowledge transfer across different wireless environments and conditions. This approach allows for the development of more generalized models that can perform effectively under varied communication scenarios. Such models can capture complex data representations and generate synthetic training data to develop more robust and flexible wireless communication technologies.

The third MAC-layer challenge is **MAC-layer security vulnerabilities**, which represent a critical and evolving threat to wireless communication systems. MAC layer is particularly susceptible to sophisticated attacks such as jamming, spoofing, replay attacks, and advanced signal manipulation techniques. Traditional rule-based security mechanisms struggle to respond to these evolving threats in real time, which leave systems exposed to degradation, unauthorized access, or denial of service. Recent work by Shi et al. [10] illustrates how adversarial deep learning (DL) can be used to generate adaptive jamming strategies that target cognitive radio systems. In response, Davaslioglu et al. [11] proposed a continual deep reinforcement learning approach to enable MAC-layer protocols to adapt and retain anti-jamming policies without catastrophic forgetting, even in the presence of generative attacks. These findings underscore the dual role of Generative AI in wireless systems, as both a threat vector and a critical enabler of adaptive and resilient MAC-layer defenses capable of responding to AI-generated attacks in dynamic environments.

Generative AI models can provide a proactive and adaptive approach to MAC-layer security by simulating various attack scenarios and generating robust defensive strategies. These models can learn from historical attack patterns, create synthetic threat signatures, and develop dynamic countermeasures. By generating realistic adversarial scenarios, generative models enable the training of more sophisticated security algorithms that can anticipate, detect, and neutralize emerging threat landscapes. This approach transforms security from a reactive mechanism to a predictive one to significantly enhance the resilience of wireless communication systems.

The fourth MAC-layer challenge is **network performance optimization**, which confronts the persistent need to maintain persistent wireless communication under rapidly changing environmental conditions. The MAC layer must continuously balance competing objectives such as throughput, latency, and reliability, particularly in heterogeneous and high-mobility scenarios. However, traditional optimization techniques often rely on static rules and predefined thresholds that fail to adapt to the dynamic and non-stationary nature of modern wireless environments. Recent advances in deep reinforcement learning (DRL) offer promising alternatives. For example, Xu et al. [12] demonstrated how DRL agents can learn optimal channel access strategies in real time by interacting with the environment. Such approaches allow MAC protocols to move beyond reactive control toward proactive, learning-based optimization, and improve performance across a wide range of network conditions.

Generative models can be used to optimize network performance by synthesizing complex network utilization scenarios and generating adaptive scheduling strategies. These models can create probabilistic models of network behavior, predict traffic patterns, and dynamically generate optimal transmission parameters. By leveraging generative models, wireless systems can efficiently allocate resources in challenging environments such as Reflective Intelligent Surfaces (RIS) and Terahertz (THz) Airborne Networks (TAN). The ability to generate synthetic traffic scenarios and predict performance bottlenecks using generative models enables more responsive and intelligent network management.

In the following sections, we provide an in-depth analysis of various generative modeling techniques, explore potential use cases for Generative AI in the MAC layer, and highlight their applications in wireless communications. Figure 5.1 illustrates three categories of Generative AI use cases in the MAC layer: spectrum awareness, scheduling optimization, and MAC-layer security. We begin

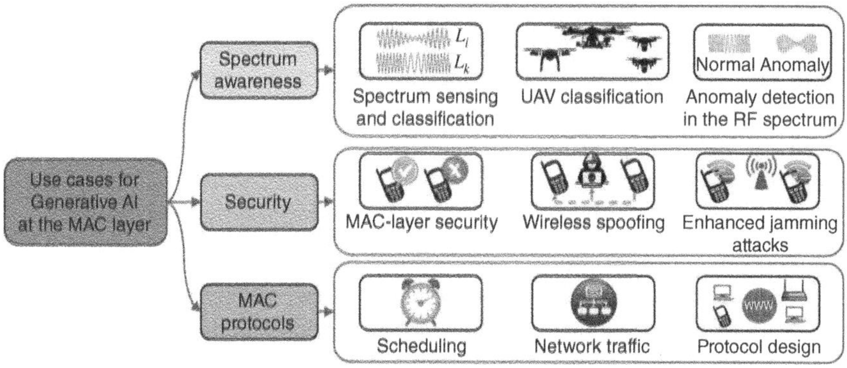

Figure 5.1 Use cases for Generative AI in MAC layer.

with an overview of generative models in Section 5.2, where we introduce VAEs, GANs, and diffusion models, discussing their architectures and training methodologies. In Section 5.3, we focus on spectrum awareness and present different applications of generative models in spectrum sensing, unmanned aerial vehicle (UAV) classification, and anomaly detection. Section 5.4 explores RF spectrum security applications using generative models and covers RF fingerprinting, wireless spoofing, and jamming attacks. Section 5.5 examines how these models enhance scheduling strategies, particularly in emerging network topologies such as RIS and TAN. In Section 5.6, we discuss open problems and future research directions, highlighting areas where further exploration could yield significant advancements. Finally, we conclude the chapter in Section 5.7 where we summarize the key insights and highlight the potential impact of generative models on the future of wireless network design.

5.2 Generative Models

Generative models discover and learn the regularities or patterns in input data automatically, without any supervision. Therefore, they are classified as an unsupervised learning task in machine learning. These models can generate synthetic samples that cannot be distinguished from the real samples they are trained on. These models have many applications. Some examples of these applications are creating images, audio, and video; or drug discovery, medical imaging, or genomic analysis; weather forecasts or natural disaster prediction. There are a number of frameworks to create generative models. In general, these frameworks belong to three different categories. The first category is explicit density models, which explicitly derive the probabilistic distribution of data. Gaussian mixture models and kernel density estimation belong to this category. The second category is latent (hidden) variable models, which use latent variables to model complex distributions. Variational autoencoders and Boltzmann machines belong to this category. The third category is implicit density models, which learn to generate data without explicitly defining a probability distribution. GANs and diffusion models belong to this category.

Generative models can be employed in different ways, as detailed below.

1. **Data Augmentation:** Generative models can be used to synthesize additional realistic waveforms that can help the spectrum classification models learn better feature representations across various conditions. This capability can substantially reduce the amount of labeled data needed for training by making better use of the available data.
   In wireless communications, data augmentation with generative models enables the creation of diverse waveform instances by simulating channel effects, modulation schemes, and interference conditions. This helps spectrum classification models generalize better to unseen environments and reduces reliance on large labeled datasets, which are often costly or impractical to collect over the air.
2. **Semi-supervised Learning:** Semi-supervised learning bridges supervised and unsupervised learning by leveraging both labeled and unlabeled data to train AI models for downstream tasks, such as classification and regression. This approach is particularly valuable in scenarios where obtaining sufficient labeled data is prohibitively difficult or expensive, while large amounts of unlabeled data are readily available. In such cases, neither fully supervised nor unsupervised methods are adequate. As an example, in spectrum awareness applications, semi-supervised learning can be used to pretrain a neural network on unlabeled data and subsequently fine-tune it using a limited amount of labeled data.
   The fact that labeled data is costly and time-consuming to obtain means that in wireless, manual labeling of signal types or network events is needed. On the other hand, unlabeled data can be available, for example, a wireless network can collect a large amount of signals without labeling. There are several applications of semi-supervised learning in wireless communication have emerged, for example, wireless interference identification, wireless signal recognition, and integrated sensing and communications.
3. **Domain Adaptation:** A model trained in one environment, known as the *source domain*, can be adapted to perform effectively in a different environment, known as the *target domain*, by aligning feature distributions between the two domains to mitigate environmental discrepancies without extensive retraining. By leveraging generative models for domain adaptation, models

can achieve greater robustness and generalizability, where models can efficiently transfer and adapt to previously unseen environments or those with limited labeled data.

In wireless applications, domain adaptation is employed because of the fact that a wireless environment is dynamic and can change due to several factors as well as because of the difficulty and cost of collecting wireless data. As examples of the applications it can be used in wireless communications, one can count employing ML models for different network deployment scenarios, using models for accurate user localization, and adapting models for different channel conditions and signal types.

4. **Unsupervised Clustering:** The learned features of the generative model can be used to perform clustering on the unlabeled data. This can help identify natural groupings within the data, which can later be used to help humans label the data more efficiently. Additionally, these clusters can help automate the labeling process by assigning provisional labels based on the identified groupings, streamlining the overall RF waveform annotation workflow.

   In wireless communications, unsupervised clustering is used for tasks such as network traffic analysis, resource allocation, and anomaly detection. In these tasks the data is unlabeled and patterns need to be discovered automatically. Some of the applications where unsupervised clustering is used in wireless communications include network traffic analysis, RF fingerprinting, which can be used to identify wireless devices with respect to their RF and hardware characteristics, user clustering, and spectrum analysis.
5. **Anomaly Detection:** Generative models can help identify outliers or anomalies in the data by learning the distribution of the normal data. This can assist in data labeling by highlighting unusual or rare data points that may require special attention or different labels.

   In wireless applications, anomaly detection is used to detect unauthorized access and adversarial attacks for performance degradation such as reduced throughput, increased delay, and dropped packets. The degradation performance can be caused by interference, hardware failure, or network congestion.

In the following, we provide a brief overview of commonly used generative models.

5.2.1 Variational Autoencoders

Variational Autoencoders (VAEs) [13] learn complex distributions in an unsupervised manner. As shown in Figure 5.2, a VAE consists of an encoder that maps input data (e.g., RF waveforms) to a latent space distribution and a decoder that reconstructs the data. Unlike standard autoencoders, VAEs model the latent

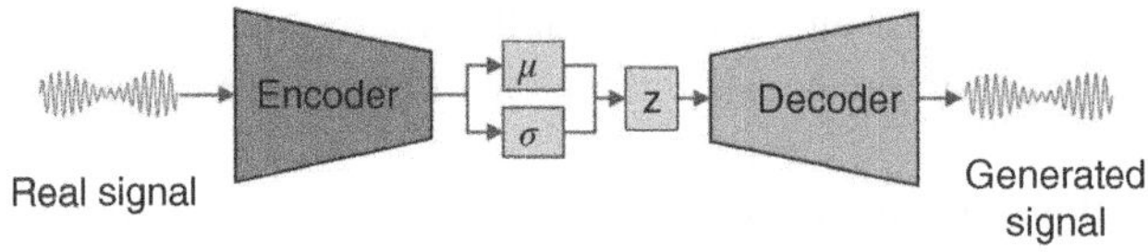

Figure 5.2 VAE generative model.

space as a multivariate statistically independent normal distribution, parameterized by two vectors, a learned mean μ and variance σ^2. This probabilistic approach enables uncertainty modeling and the generation of diverse samples. VAEs are trained using reconstruction loss (ensuring input fidelity) and Kullback-Leibler (KL) divergence (encouraging a structured latent space for smooth interpolation). This facilitates high-fidelity RF signal generation.

Vector Quantized VAEs (VQ-VAEs) [14, 15] extend VAEs by learning a discrete latent space, improving structure and interpretability for tasks such as RF waveform synthesis and classification. They mitigate posterior collapse, where the latents are ignored due to an overpowered decoder. Alongside reconstruction loss, VQ-VAEs introduce vector quantization loss to align encoder outputs with a discrete codebook and commitment loss to ensure effective codebook usage. These losses improve training stability and output quality.

5.2.2 Generative Adversarial Networks

Generative Adversarial Networks (GANs), introduced in 2014 [16, 17], are part of the class of deep-learning-based generative models. Generative modeling belongs to the category of unsupervised ML. The tasks in this category automatically discover the regularities or patterns in input data. This is done so that the model can generate new samples that cannot be discriminated from the original dataset.

GANs decompose the problem into two sub-models, as depicted in Figure 5.3. The first of these sub-models is a generator network trained to generate new examples, while the second is a discriminator network that tries to classify examples as either real (from the domain) or fake (generated). These two models are jointly trained. This process continues until the generator network learns to synthesize plausible samples. It is important that plausible examples generated by the GAN are not distinguished from real data.

GANs and their variations have been used in a variety of applications. Some examples are as follows: creating synthetic human faces [18, 19], transforming images from one medium (e.g., photographs of real scenery) to another (paintings by famous painters) [20], making image-to-image translation including conversion of horse pictures to pictures of zebras and vice versa, pictures of summer

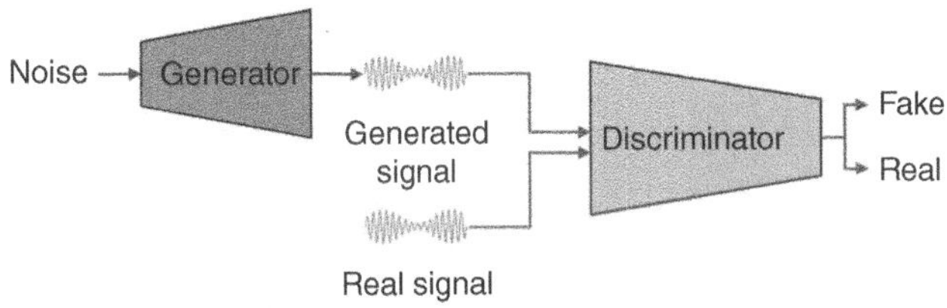

Figure 5.3 Generative Adversarial Network (GAN) with RF data generation application.

scenery to pictures of the same scenery in winter [20], etc. Among other applications, one can count face aging [21], creating super-resolution images from those of low resolution [22], generating music [23], audio synthesis [24], and video synthesis [25]. GANs have also been applied to problems in spectrum sensing and security in wireless networks, for example, Refs. [26–33].

Additional information that is correlated with the input data, such as class labels, can be used to improve GAN performance, either in the form of more stable or faster training, or generated images that have better quality. These GANs are called conditional GANs (CGANs). They are trained in such a way that both the generator and the discriminator models are conditioned on the class label, so that when the trained generator is used as a standalone model to generate samples in the domain, samples of a given type can be generated [34]. As an example, in the synthesis of faces one could focus on generating a female face, or one could convert a summer scene into a winter scene, etc. Popular face aging applications are also based on this principle.

In the literature, there are different GAN formulations beyond those mentioned above. Examples are Least-Squares GAN (LS-GAN) [35], Wasserstein GAN (WGAN) [36], and WGAN with Gradient Penalty (WGAN-GP) [37].

5.2.3 Diffusion Models

Diffusion models, introduced in Ref. [38], have emerged as a powerful approach to unsupervised learning, employing deep learning techniques inspired by thermodynamic diffusion. They offer high-quality data generation, flexibility in modeling complex distributions, and simpler implementation compared to GANs.

Unlike GANs, which rely on adversarial training, diffusion models employ a denoising network that iteratively refines noisy samples. The process consists of forward and reverse diffusion [39], illustrated in Figure 5.4. In forward diffusion, Gaussian noise is incrementally added, degrading the original structure. In reverse diffusion, the model learns to denoise step by step, reconstructing the original sample and enabling realistic data generation across domains such as image synthesis, audio generation, and RF waveform modeling.

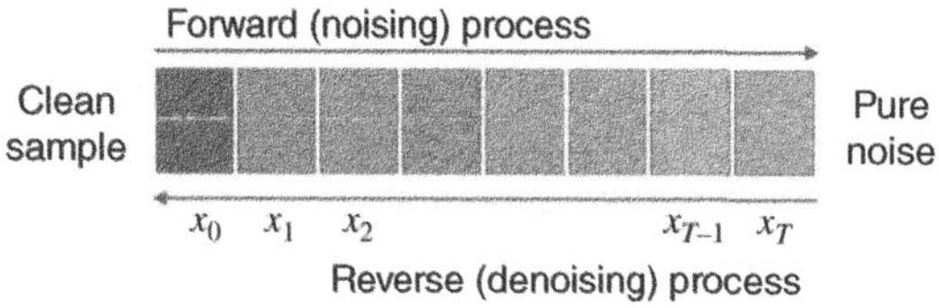

Figure 5.4 Diffusion process for an RF signal. The color version of this figure is available on Wiley.com.

Recent advances have improved adaptive noise schedules and sampling techniques to improve training efficiency and output quality. Latent Diffusion Models (LDMs) [40] optimize this process by operating in a compressed latent space rather than a high-dimensional data space and significantly reduce computational overhead.

These generative models are finding innovative applications in wireless communications to tackle unique challenges. For example, VAEs excel at learning a low-dimensional, latent representation of complex and rapidly changing wireless channels. This ability is ideal for creating compact and realistic channel state information feedback, which is crucial for efficient communication and link adaptation. GANs, on the other hand, have proven powerful for applications in wireless security. The adversarial training process allows GANs to generate highly realistic synthetic signals to augment limited training data, which is invaluable for training robust discriminators to detect sophisticated jamming or adversarial attacks. Finally, diffusion models have made a significant impact on signal processing. They are excellent at modeling complex noise profiles and can reverse this process to recover high-fidelity signal information from noisy measurements. This capability is being leveraged to improve the accuracy of channel estimation and to design novel waveforms for applications like physical layer security.

The wireless domain is characterized by intricate and highly dynamic data, including rapidly changing channel conditions, complex signal patterns, and the need for robust security. In Generative AI models, the underlying principles of learning and generating complex and high-dimensional data are applicable to the field of wireless communications. By leveraging generative models, researchers can address these unique challenges, moving beyond traditional statistical models to create more adaptive, intelligent, and efficient wireless communication systems.

5.3 Spectrum Awareness Applications

Spectrum scarcity and sharing require intelligent spectrum awareness capabilities to efficiently manage and optimize access to the EM spectrum. Spectrum awareness encompasses the ability of communication systems to monitor and understand the EM spectrum environment. It involves sensing the spectrum, detecting the signal of interest (SOI), isolating the signal, classifying the signal, and managing possible interference. In some cases, it also involves geolocating the emitters when multiple antennas or multiple cooperating nodes are used. Spectrum awareness enables systems to optimize spectrum access in multiple node settings, ensures regulatory compliance, and improves security through anomaly detection and cooperative spectrum awareness efforts. Additionally, it allows for efficient allocation and management of communication resources to improve overall system performance.

5.3.1 Data Augmentation and Synthetic Data Generation

As ML becomes integral to spectrum awareness, a critical challenge arises from the scarcity and domain-specific nature of training data, which motivates the use of data augmentation and synthetic data generation to increase the model robustness and generalization. Davaslioglu et al. [26] presented one of the first studies that applied generative models to improve spectrum awareness, where they developed the Spectrum Augmentation/Adaptation with GAN (SAGA) framework. SAGA framework addressed two important challenges: First, ML-based spectrum sensing and waveform classification models require significant amounts of labeled data to capture complex channel and emitter characteristics and train an ML model. Second, the training data collected from one spectrum environment cannot generally be used for another one. For example, Line-of-Sight (LOS) channel characteristics are significantly different from Non-Line-of-Sight (NLOS) channel conditions. Due to the discrepancy in channel distributions, classifiers trained in one environment may suffer performance degradation in another channel environment. Therefore, signals collected in one environment may need to be adapted to other channel conditions before being used in the training of AI models.

To address the first challenge of data augmentation, generative models can be used to synthesize additional data. This framework is illustrated in Figure 5.5, where over-the-air captured data or simulated data can be used to train a generative network and obtain an augmented dataset, which is then used to train an ML classifier. Along these lines, the SAGA framework in Ref. [26] used a conditional GAN to generate synthetic training data and improve signal classification accuracy. They evaluated three ML classifiers: one trained with limited real training data, another using real and GAN-augmented data, and a third using adapted training data to detect signals. Their results showed that training data augmentation using a conditional GAN significantly enhanced classifier accuracy.

For the second challenge of domain adaptation, SAGA employed a bidirectional GAN [41], where the generator learned to map latent samples to the RF data distribution and also learned an inverse mapping from the data to the latent representation. This enabled the SAGA framework to adapt the training data to changing spectrum dynamics. This is illustrated in Figure 5.6. It was demonstrated in

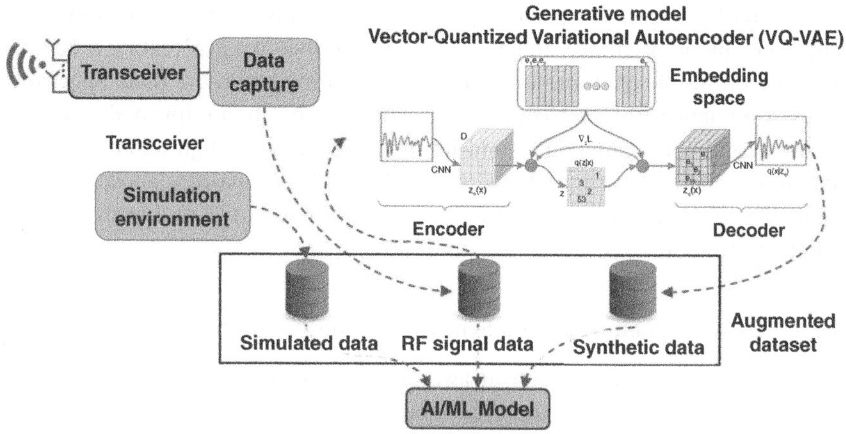

Figure 5.5 Data augmentation for RF spectrum awareness.

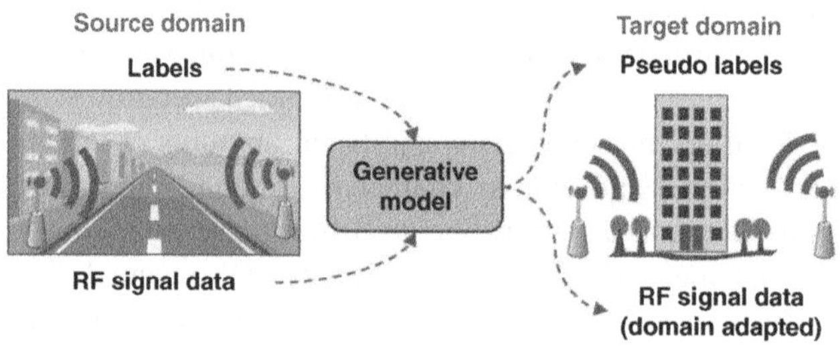

Figure 5.6 Domain adaptation for RF spectrum awareness.

Ref. [26] that as the spectrum conditions change, classification performance was maintained. Overall, the gains in SAGA were attributed to the fact that GAN-augmented data facilitated the creation of rare or unusual signal conditions, which are critical for testing the robustness of classifiers in diverse environments. This approach enhanced the adaptability of models and improved their generalization across varying spectrum scenarios. Similar to Ref. [26], Patel et al. [42] also proposed using a conditional GAN to provide an effective solution to the limited data problem. They trained a CNN-based modulation classifier using the augmented dataset. They demonstrated that as they added more synthetic data samples, the validation loss of the modulation classifier substantially decreased.

Liu et al. [43] used a Deep Convolutional GAN (DCGAN) to augment spectrum sensing signals. They calculated the covariance matrices for the real and imaginary parts of the complex signal separately. To structure the data in a form resembling the three-channel representation of a red, green, and blue (RGB) image, they extended these covariance matrices by appending a zero matrix of the same dimensions. This structured data was then used as input to train a DCGAN, leveraging the convolutional layers optimized for image-like data. Similar to Ref. [26], Liu et al. employed the trained GAN to generate synthetic data to improve the performance of the spectrum classifier.

Kompella et al. [44] employed a Vector-Quantized Variational Autoencoder (VQ-VAE) model designed to enhance wireless signal classification by augmenting training data with synthetic Radio Frequency (RF) signals. The VQ-VAE architecture, shown in Figure 5.7, comprises encoder and decoder networks that compress input signals into a latent space and subsequently reconstruct them back to the original signal space. A key feature of this model is the quantizer, which discretizes latent embeddings and learns a codebook within the latent space. The model was trained on 18 different High Frequency (HF) communication waveforms to generate high-fidelity synthetic RF signals that capture the complex variations inherent in these communications. Compared to the baseline model trained only on real data, the model trained with augmented data achieved a 4.06% absolute increase in classification accuracy across all SNRs and a 15.86% improvement at −10 dB SNR, the latter being very significant. These findings suggest that the VQ-VAE model can be used to effectively increase the diversity and fidelity of training datasets, and improve the generalization and robustness of wireless signal classifiers, especially under low SNR conditions.

A recently emerging communication system that has garnered significant attention is autoencoder-based communication. This system does not rely on an explicit analytic model of channel impairments and can be used when the channel response is unknown or cannot be easily modeled with a closed-form analytic expression. Generative modeling has been applied to enhance the performance of autoencoder-based communications [45–47]. For example, Ye et al. [47] used the received pilot signals as conditioning information to train a conditional GAN

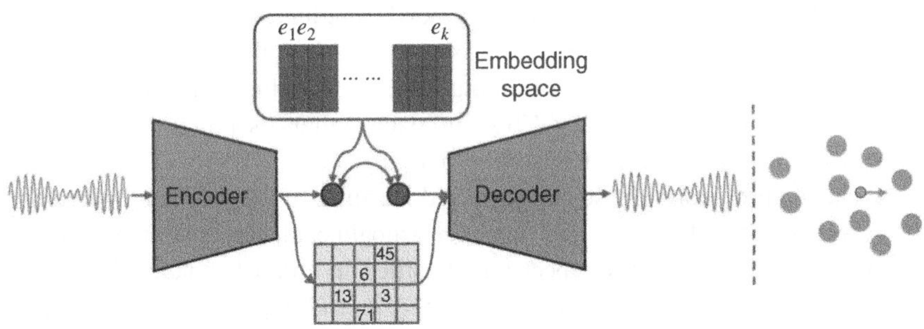

Figure 5.7 VQ-VAE model to augment RF waveforms.

and created realistic time-varying channels that are then used to train an autoencoder. Similarly, O'Shea et al. [46] proposed a method called CommGAN to fully adapt the transmitter and receiver to a physical model-free channel response using a hardware-in-the-loop generative adversarial channel autoencoder network over-the-air. CommGAN directly learned on an unknown and uncharacterized physical channel and approximated the channel, which was shown to be sufficient to backpropagate and adapt both encoder and decoder networks. In a similar approach, Davaslioglu et al. [45, 48] applied WGAN-GP to learn the received signal distribution and synthesized received signal samples that were used in the training of the autoencoder communication network for (intentional or unintentional) interference suppression. The authors investigated the performance under different conditions: data-limited regions and data-abundant regimes. They showed that a classifier that is trained with GAN-augmented data in the data-limited region achieved much lower bit error rate (BER) compared to the one that was trained in both data-limited and data-abundant regimes. A subtle but important distinction between these similar studies [45–47] lies in the nature of the inputs to the generative models. While all three references applied GANs to augment the data of autoencoder-based communication networks, the type of input data differs. Specifically, Ye et al. [47] used different channel realizations as inputs, whereas O'Shea et al. [46] and Davaslioglu et al. [45] employed the signals received at the receiver.

Spectrum coexistence is critical for next-generation (NextG) systems to share bandwidth with incumbent users while meeting rising demand. For instance, in the 3.5 GHz CBRS band, NextG systems must opportunistically access the spectrum when incumbents (e.g., radar) are inactive, requiring a high-fidelity deep neural network-based classifier to ensure low misdetection (protecting incumbents) and low false alarms (maximizing throughput). In dynamic wireless environments, the classifier's use is constrained by the coherence time, split between sensing (to collect data and train the classifier) and transmission. This creates a tradeoff: longer sensing improves accuracy but reduces transmission time. To address this, a GAN-based approach was employed in Ref. [49] to generate synthetic sensing data, augmenting training datasets to reduce sensing time while maintaining classifier accuracy. This approach enhances protection for high-priority users and boosts NextG user throughput.

A related study applying semi-supervised learning to RF spectrum awareness was presented in Ref. [50]. In this work, Davaslioglu et al. demonstrated the effectiveness of pretraining an RF spectrum classifier on unlabeled data using a Self-supervised Learning (SSL) framework. The SSL pretraining architecture, depicted in Figure 5.8, incorporated two encoders: a query encoder and a momentum encoder. The query encoder includes the classifier backbone, a projection head,

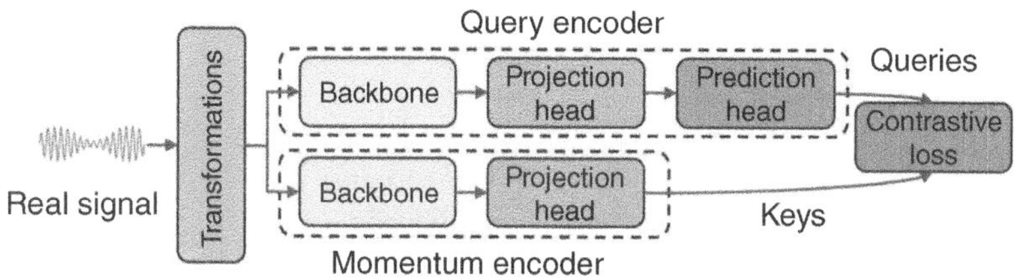

Figure 5.8 Self-supervised learning to pretrain a neural network for RF classification.

and a prediction head. The backbone typically comprises a convolutional neural network (CNN), while the projection and prediction heads consist of fully connected feed forward network layers. The momentum encoder, as shown in Figure 5.8, mirrors the backbone and projection head of the query encoder, but it is updated differently to maintain stability. Specifically, the momentum encoder is updated via a moving average of the query encoder network. During SSL pretraining, each real RF signal is subjected to two tailored augmentations for RF data and two augmented versions of the same waveform are generated. These augmented pairs form positive pairs, while other signals within the training batch form negative pairs. A contrastive loss function is used to train the model, which makes sure that the embeddings of positive pairs are closer together while pushing negative pairs apart in the feature space. The query encoder is updated directly using backpropagation, whereas the momentum encoder is updated indirectly to maintain consistency across training. The results in Ref. [50] underscored the power of SSL in improving sample efficiency. The RF spectrum classifier pretrained on unlabeled data and fine-tuned with just 0.5% of labeled data achieved performance comparable to a randomly initialized model fine-tuned with 5% of labeled data. This represents nearly an order-of-magnitude improvement in label efficiency and highlights the significant potential of SSL for RF spectrum awareness applications.

5.3.2 Signal Classification Applications – UAV Classification

Unmanned Aerial Vehicle (UAV) classification is crucial for enhancing security by distinguishing between benign and potentially malicious drones, and preventing unauthorized activities. It supports efficient air traffic management by tracking and controlling drone movements, reducing collision risks, and ensuring safe airspace use. In addition, accurate UAV classification enhances surveillance and monitoring capabilities, aiding in regulatory compliance and general situational awareness in urban environments. In this context, Figure 5.9 illustrates Generative AI applications that can be used to augment UAV data captures or adapt them from one environment to another to increase classification robustness.

Zhao et al. [51] considered a passive sensing scenario in which a sensor receives UAV transmissions and categorizes them into four UAV types. They proposed

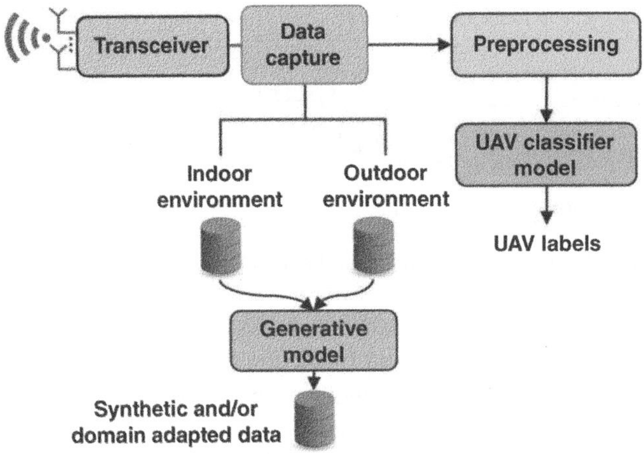

Figure 5.9 Generative AI applications in the UAV classification problem.

using Auxiliary Classifier Wasserstein GANs (AC-WGANs) to address the limitations of traditional classification methods such as Support Vector Machines (SVM) and k-Nearest Neighbors (KNN), which requires large datasets for effective feature extraction. While similar to cGAN, AC-WGAN integrates an auxiliary classifier to enhance data quality, class discrimination, training stability, and interpretability. Their system achieved a 95% recognition accuracy in indoor environments and shows potential for outdoor applications, highlighting the effectiveness of generative models in improving UAV classification which is a key aspect of spectrum awareness and security in modern wireless communication systems.

5.3.3 Anomaly Detection in RF Spectrum

Generative modeling can be applied to detect anomalies or patterns in the data that do not conform to "normal behavior." In this context, the normal behavior can be modeled using generative models, which can then be used to calculate an anomaly score. This score can serve as a basis for identifying deviations from expected patterns. For RF spectrum awareness, anomaly detection plays an important role because it helps identify unauthorized or malicious transmissions, interference, and irregularities that could disrupt communication networks, compromise security, or lead to inefficiencies in spectrum utilization. Ayanoglu et al. [52] addressed this problem and used GANs to detect anomalies in the RF spectrum. They used the Fast AnoGAN (F-AnoGAN) approach [53] that extends the AnoGAN [54] by using a more stable GAN architecture.

F-AnoGAN consists of three neural network models: an encoder, generator, and discriminator, as illustrated in Figure 5.10. The generative model is trained

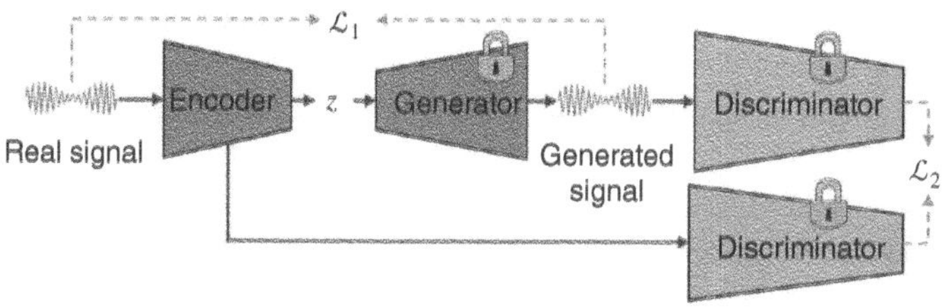

Figure 5.10 Inference steps of f-AnoGAN model for anomaly scoring.

exclusively on known SOIs, while the encoder is trained to map new data efficiently to the latent space of the GAN. The anomaly score is computed as the sum of two mean squared errors (MSEs). The first term measures the reconstruction error between the input data and its generator-based reconstruction, expressed as the mean of $||\mathbf{x} - G(E(\mathbf{z}))||_2^2$ over all the samples. The second term quantifies the discrepancy in discriminator feature representations between the original and reconstructed signals, given by the mean of $||D_f(\mathbf{x}) - D_f(G(E(\mathbf{x})))||_2^2$ over all the samples. The final anomaly score is a weighted combination of these terms, where the second term is scaled by a factor κ.

The F-AnoGAN training process typically consists of three steps:

1. Train the generator and discriminator to learn SOI distributions.
2. Train the encoder to learn an efficient mapping of new data into the latent space of the GAN.
3. Evaluate anomaly detection performance by computing the anomaly score for each sample.

Ayanoglu et al. [52] trained an F-AnoGAN model using the RFMLS 2016a dataset [55], which contains eleven different modulation types. The model was trained on a single known modulation, designated as the inlier, and evaluated across all eleven modulations, where other modulation types were expected to be classified as outliers. Across different fidelity metrics, the f-AnoGAN model achieved an area under the receiver operating characteristic (ROC) curve (AUROC) score of 0.706 on average over all eleven modulations. Notably, the AUROC scores varied significantly depending on the similarity between modulation types, ranging from 0.088 to 0.970.

5.4 RF Spectrum Security Applications

5.4.1 Emitter Identification

Specific emitter identification (SEI) is a technique used to identify the unique hardware characteristics of wireless transmitters, primarily for intelligence, surveillance, and security applications. Subtle attributes such as I/Q imbalance, phase imbalance, and carrier frequency offset create a distinctive RF signature for each transmitting device. These RF fingerprints play a critical role in enhancing wireless communication security to detect and identify unauthorized or rogue devices.

The significance of RF SEI lies in its ability to provide an additional layer of security, particularly in scenarios where traditional identification methods may fail. As illustrated in Figure 5.11, generative modeling can improve RF fingerprinting by generating realistic synthetic data, allowing models to better distinguish between legitimate and illegitimate transmitters.

Yang and Yan [56] proposed a GAN-based approach for RF fingerprint recognition, where I/Q data was first preprocessed using wavelet transform to highlight the unique characteristics of RF fingerprints. The GAN model then generated pseudo-data distributions, which were used to train the discriminator to recognize subtle differences between trusted and rogue transmitters. Experimental results demonstrated that this GAN-based method achieved a 98.1% accuracy in distinguishing between trusted and unauthorized transmitters, outperforming traditional CNNs and deep neural networks (DNNs).

Another paper that studies the use of generative modeling in RF fingerprinting application is by Tan et al. [57], where the authors proposed a semi-supervised algorithm for SEI using bispectrum feature extraction integrated with a CGAN (BFE-CGAN). In this approach, the bispectrum estimation of the signal serves as the feature representation of the emitters, while label information is embedded into the input latent layer to guide the GAN's training. Unlike traditional SEI models that only handle labeled data, this framework assigns uniform categories to unlabeled samples for semi-supervised training, which allows the shared weights between the unsupervised and supervised components to help the discriminator capture a more comprehensive data distribution, and thereby improves the efficiency of using unlabeled samples.

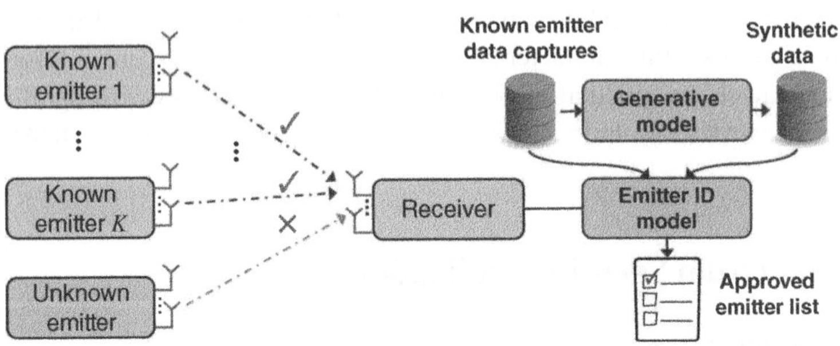

Figure 5.11 Emitter identification problem in wireless applications.

5.4.2 Wireless Spoofing

While generative models can improve RF fingerprinting classification, they also introduce security risks, as adversaries may exploit them to bypass PHY-layer signal authentication. Statistical techniques, such as Kernel Density Estimation (KDE), allow adversaries to generate rogue device signals by approximating the probability density function (PDF) of received signals [58, 59].

As an extension of Generative AI, Shi et al. [60] investigated deep learning-based spoofing attacks, where synthetic wireless signals were generated to be indistinguishable from legitimate transmissions. In this approach, illustrated in Figure 5.12, the adversary operated as a GAN-based system, where a transmitter and receiver acted as the generator and discriminator, respectively, playing a minimax game over the air. The adversarial transmitter leveraged a DNN to craft spoofing signals capable of deceiving the defense mechanisms at the adversarial receiver, which was also modeled as a DNN.

The attack captured waveform, channel, and radio hardware effects, ensuring that the generated signals closely mimicked the targeted legitimate transmissions. Compared to spoofing using random or replayed signals, GAN-based spoofing significantly increased the probability of misclassification, with its effectiveness influenced by network topology and mobility patterns. The success probability of the spoofing attacks improved when the adversary employed multiple antennas, whereas it diminished when the defending receiver employed multiple antennas, highlighting the role of spatial diversity in mitigating such adversarial threats.

GANs can also be used to generate synthetic Fifth Generation (5G) cellular wireless signals and then transmit them to access the 5G-enabled services [49]. Suppose that a classifier at the 5G gNodeB detects 5G user equipment (UEs) by analyzing I/Q signal data, which includes power and phase, to differentiate legitimate user signals from noise and provide authenticated services. Adversaries can exploit this system by sensing the spectrum to collect I/Q signal samples and determine if signals are classified as legitimate, based on feedback or observed communication with the gNodeB. Using these samples, adversaries employ GANs to

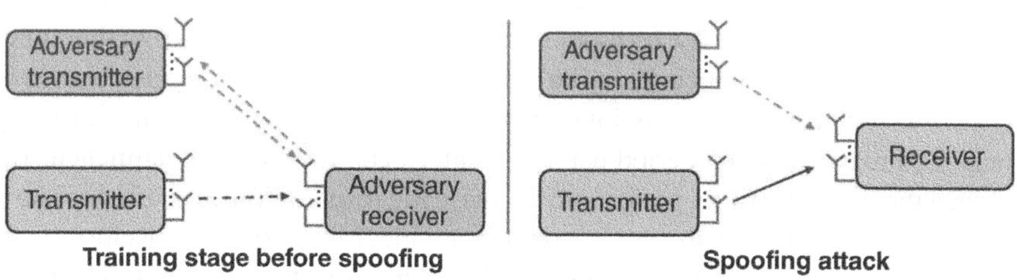

Figure 5.12 Adversary transmitter, equipped with a generative model, learns to spoof legitimate transmitter signals.

generate spoofed 5G signals. The GAN's generator and discriminator are trained collaboratively, with the adversary transmitter sending labeled signals to the adversary receiver, which provides feedback for refining the spoofed signals. This enables adversaries to bypass authentication and access 5G-enabled services.

A GAN-based attack strategy, IoTGAN, was devised in Ref. [61] to manipulate IoT device traffic and evade machine learning-based identification. The key challenges are: (i) obtaining the discriminative model in a black-box setting and (ii) introducing traffic perturbations to evade identification without disrupting IoT device functionality. To address these, IoTGAN employed a neural network-based substitute model to approximate the target model and served as the discriminative component. A manipulative model was then generated adversarial perturbations to evade this substitute model.

In this subsection, so far, we have discussed how Generative AI can be employed to create wireless spoofing attacks. While this is true, in fact Generative AI can also be used to defend against such attacks. Generative AI, particularly GANs, can be trained on datasets of clean wireless signals and known spoofing attack patterns. Generative AI can then point to the anomalies that may have the signature of a spoofing attack. It is important to understand that the kind of spoofing attack scenarios can be very varied, from more conventional ones to even new and previously unseen ones. One standard solution is to use Generative AI to establish a basis of normal behavior. Then, Generative AI can determine potential spoofing attacks in the anomaly detection mode. It is known that GANs, in particular, have been used very successfully for anomaly detection, for both regular security applications such as concealed weapon detection in traveler luggage as well as wireless applications.

5.4.3 Enhanced Jamming Attacks

Erpek et al. [62] applied generative modeling to enhance jamming attacks, focusing on a reactive jammer that senses the spectrum and predicts whether a successful transmission will occur in a given time slot (indicated by an acknowledgment signal, ACK) or not (no feedback). If the jammer predicted a successful transmission, it jammed the transmission with a certain power P_j during that time slot; otherwise, it remained idle. The jammer must learn to distinguish between time slots with active transmissions and those that are idle. This learning process may require a prolonged period before the attack can be effectively launched. To reduce the initial learning period and decrease the response time required to start jamming, the jammer can train a generative model to learn the waveform and channel distributions from the sensed real data in a shorter timeframe. The generative model can then generate synthetic data samples to augment training data and reduce the probabilities of false alarms and missed detections.

As in the previous subsection, Generative AI can protect against jamming attacks in wireless communication networks by enhancing the capabilities of anti-jamming systems. In particular, AEs and VAEs can learn the distribution of normal network traffic and identify anomalies indicative of jamming attacks. Generative AI can analyze historical and real-time data. By doing so, it can generate effective strategies and plans. This can reduce response time and minimize downtime. Generative AI can enable communication systems to adapt their transmission strategies. These can be in the form of adjusting transmit power, changing frequency, or adopting different modulation schemes to avoid or mitigate jamming effects. As in their typical applications, GANs can be used to generate synthetic jamming data to train models that are more robust and adaptive to new or unknown jamming patterns.

5.5 Scheduling Applications

Generative AI can greatly improve MAC-layer scheduling, especially in wireless communication systems where optimal utilization of the shared medium is essential. It can be used to create adaptive schedules based on the network conditions and traffic demands to optimize the process of how devices access the shared medium. In this subsection, we explore key ways in which Generative AI can improve MAC-layer scheduling, with illustrations provided in Figure 5.13.

5.5.1 Traffic Prediction and Pattern Generation

Generative AI plays a critical role in MAC-layer scheduling through traffic prediction and pattern generation. By learning from historical network data, generative models can synthesize realistic traffic patterns and predict future traffic conditions, as demonstrated in Refs. [63–65]. These conditions may include idle periods, congestion levels, and demand surges.

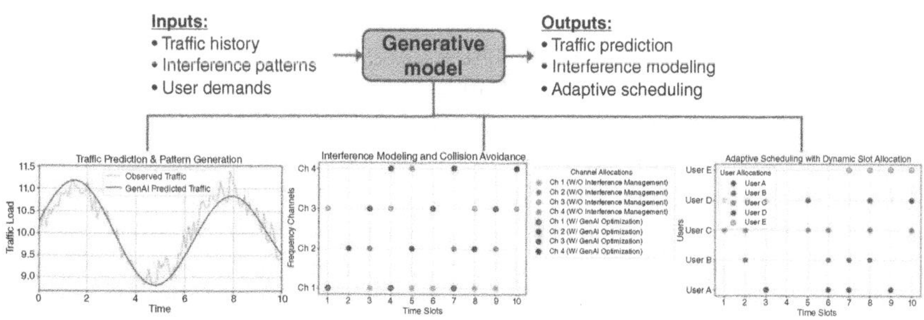

Figure 5.13 Generative AI applications in MAC-layer scheduling.

By identifying such patterns, AI-driven scheduling algorithms can optimize access times, minimize delays, and prevent unnecessary transmissions. Additionally, Generative AI can simulate synthetic traffic scenarios, enabling the evaluation of scheduling algorithms without the need for costly and time-intensive real-world experiments. This capability is particularly valuable for modeling rare events that are difficult to observe in practice, ensuring more robust and adaptive scheduling strategies.

5.5.2 Adaptive Scheduling Algorithms

Beyond static or semi-static approaches, Generative AI can create adaptive scheduling algorithms that dynamically respond to fluctuating network conditions, traffic variations, and interference patterns. This adaptability can be achieved through mechanisms such as dynamic channel allocation, flexible transmission window placement, and adaptive time slot assignment. These strategies, discussed in Ref. [66], allow MAC-layer protocols to adjust in real time, ensuring optimal performance even in unpredictable wireless environments.

5.5.3 Interference Patterns

Generative AI can model wireless interference patterns among devices and cells. By learning these patterns, it can generate optimized schedules to minimize overlap and therefore reduce the occurrence of collisions. In particular, Generative AI can be used to design cooperative scheduling plans where neighboring units or cells work together to coordinate access times and frequency allocations. As a result, interference among users or cells can be reduced.

5.5.4 Fairness and QoS

Two additional areas where Generative AI can be applied in MAC scheduling are fairness and quality of service (QoS), which are often closely related. In terms of fairness, Generative AI can help strike a balance between equitable resource allocation and overall network throughput by generating schedules that are both fair and efficient. Additionally, Generative AI can learn user profiles and service requirements to create schedules that meet individual QoS demands while maintaining fairness across the network.

Other areas where Generative AI can be employed for better scheduling in MAC are efficient spectrum and channel management, scalability for large networks, improving efficiency in high-density networks, and energy efficiency.

5.5.5 Millimeter-wave Networks

A novel application of generative models is to help improve the MAC protocol design for TAN. MAC protocol for THz communication has some fundamental differences from conventional low-frequency (<15 GHz) MAC protocols. In general, the purpose of MAC protocol design is to guarantee collision-free, high-rate communications among the dynamic one-hop neighborhood. For example, Wi-Fi products can simply use omnidirectional antennas to radiate signals within a range of 360 degrees and do not need to concentrate the signals on a small angle as what directional antennas do. In the TAN, both the transmitter and receiver are equipped with directional antennas. This shifts the design focus from channel access collision avoidance in the conventional MAC schemes to the precise neighbor coordination in terms of antenna orientation and transmission (Tx)/ reception (Rx) schedule. However, the THz channel is extremely sensitive to antenna misalignment. Delaying transmission rate adjustments until a THz link outage is detected can result in significant packet loss. Therefore, accurate prediction of node position/posture, which determines the antenna orientation changes, is important to proactive mitigation of link degradation.

He et al. [67] proposed to address these problems by training a generative model that used historical data to predict node state in TAN. A predictive, spatiotemporal node state tracking scheme based on deep learning (DL) model called Deep Neighbor Adaptation (DNA) was proposed. This model was designed to accurately track and predict the next-time node position/posture. The GAN model was used to handle the sparse nature of the network feature matrix and overcome limited TAN training dataset problems. Note that, in this application, the generative model was trained using a spatiotemporal graph representing the TAN topology. By this means, the generative model learned the distribution of this graph and synthesized the next-phase TAN topology.

5.6 Open Problems and Future Research Directions

As generative modeling continues to evolve and its integration into wireless networks expands, several open problems and future research directions emerge, particularly in areas such as RIS-assisted networks, spectrum sharing in the presence of interference, Integrated Sensing and Communications (ISAC), and enhancing wireless MAC-layer security, presenting both challenges and opportunities for further advancements in this promising technology.

5.6.1 Reconfigurable Intelligent Surface (RIS)-assisted Networks

Unlike traditional channel models discussed earlier in this chapter, the RIS-assisted channel consists of a cascade of three subchannels: the transmitter-to-receiver (Tx–Rx), transmitter-to-RIS (Tx-RIS), and RIS-to-receiver (RIS-Rx). We will discuss the case of an RIS whose elements are passive, as in most of the RISs considered in the literature. The unique characteristics of this channel depend on the specific behavior of these three subchannels. However, due to the difficulties of practical measurements at the RIS elements, it is challenging to separately obtain the responses of Tx-RIS and RIS-Rx subchannels using traditional channel measurements. Furthermore, the lack of publicly available RIS-assisted channel datasets hinders the development and application of AI-enabled RIS channel modeling. Currently, the scarcity of publicly available RIS-assisted channel datasets poses a major barrier to AI-driven RIS channel modeling and estimation. However, recent efforts have started to address this gap. Rossanese et al. [68] provide experimental sub-6 GHz measurement datasets from an RIS testbed, while Tewes et al. [69] offer large-scale physical 5 GHz channel measurement data.

In Ref. [70], it was recognized that accurate channel estimation is a fundamental technical prerequisite to achieve the huge performance gains from RIS. Towards that end, by leveraging the low-rank structure of RIS channels, three practical residual neural networks, named convolutional blind denoising network, convolutional denoising GAN, and multiple residual dense networks, are proposed to obtain accurate channel state information. Simulation results reveal the evolution direction of these three methods and illustrate their superior performance compared with existing benchmark schemes.

In Ref. [71], the extensive data upload and download demands of AI-generated content were identified as major challenges to spectrum efficiency, with data security emerging as a critical concern. To address these issues, Duan et al. [71] proposes an active RIS-assisted physical layer security (PLS) scheme to improve both spectrum efficiency and secrecy performance. Unlike passive RIS schemes, it integrates an active reflected power amplifier circuit within the RIS elements to mitigate multiplicative fading and amplify signals to improve channel gain. To optimize secrecy capacity, an efficient alternative optimization algorithm is used for RIS element selection and power allocation. The simulation results demonstrate that the active RIS scheme effectively mitigates the fading effects of passive RIS and achieves superior secrecy capacity compared to random allocation and existing RIS-assisted methods.

While a legitimate RIS can enhance physical layer secret key generation (PL-SKG) by increasing channel randomness, a malicious RIS can compromise legitimate channels and undermine most existing PL-SKGs. The study in Ref. [72] proposed an adversarial learning framework between legitimate parties to counter man-in-the-middle malicious RIS (MITM-RIS) eavesdropping. First, the theoretical mutual information gap between legitimate pairs and MITM-RIS is derived. Then, the two legitimate parties leveraged GANs to train a common feature surface that minimizes mutual information overlap with MITM-RIS. Next, the interpretability of black-box neural networks for signal processing is increased by employing a symbolic explainable AI (xAI) representation. These symbolic terms of dominant neurons were used for feature-engineering validation and the design of PLS common feature spaces. Simulation results in Ref. [72] demonstrated that their GAN-based and symbolic-based PL-SKGs achieved high-key agreement rates among legitimate users and remained resistant to MITM-RIS eavesdropping, even when Eve possessed knowledge of legitimate feature generation (NNs or formulas). This work establishes a foundation for securing wireless communications with untrusted reflective devices in future 6G networks.

In Ref. [73], a novel joint sensing, communication, and AI framework was proposed to optimize extended reality (XR) experiences over THz wireless systems. The proposed framework consisted of three main components. First, a tensor decomposition framework was proposed to extract unique sensing parameters for XR users and their environment by exploiting the THz channel sparsity. Essentially, the quasi-opticality nature of THz band was used and the sensing parameters were extracted from the uplink communication signal to use the same waveform, spectrum, and hardware for both communication and sensing functionalities. Then, the Cramer-Rao lower bound was derived to assess the accuracy of the estimated sensing parameters. Second, a non-autoregressive multi-resolution Generative AI framework integrated with an adversarial transformer was proposed to predict missing and future sensing information. The proposed framework offered robust and comprehensive historical sensing information and anticipatory forecasts of future environmental changes, which were generalizable to fluctuations in both known and unforeseen user behaviors and environmental conditions. Third, a multi-agent deep recurrent hysteretic Q-neural network was developed to control the handover policy of RIS subarrays in order to leverage the informative nature of sensing information to minimize handover cost, maximize the individual quality of personal experiences (QoPEs), and improve the robustness and resilience of THz links. Simulation results showed a high generalizability of the proposed unsupervised Generative AI framework to fluctuations in user behavior and velocity, leading to a 61% improvement in instantaneous reliability compared to schemes with known channel state information.

5.6.2 Spectrum Sharing in the Presence of Interference

Most publicly available datasets include SOIs collected in pristine conditions, and datasets where multiple waveforms are present in a single sample are scarce. Signals can coexist with each other in time, frequency, coding, or space domains. Depending on the interference domain, the SOI needs to be filtered accordingly.

For example, if we consider signals are orthogonal in time domain, signals need to be isolated by simply selecting the I/Q samples to the transmission time of the SOI, which is sometimes called as *signal isolation*. Signals occupying different carrier frequencies without any overlapping bandwidths can be digitally frequency shifted to the center in the complex baseband and passed through a lowpass filter (e.g., Butterworth filtering) with a cutoff frequency equal to the bandwidth of the signal [74]. For spatial domain coexistence, signals can be transmitted from different spatial directions and can be captured individually with an antenna array. A common example is the uniform linear array (ULA). For this case, signals can be isolated using model order estimation (see Algorithm 1 in Ref. [74] for details).

Despite these traditional filtering and separation techniques, real-world interference scenarios are often highly dynamic and non-stationary, which make it challenging to rely solely on rule-based methods or over-the-air collected data. Generative AI can be used to learn complex interference patterns and generate synthetic datasets to improve the robustness of spectrum sensing and interference mitigation algorithms. Specifically, Generative AI can enhance spectrum sharing in the presence of interference in the following ways:

- **Synthetic Data Generation with Interference for Robust Training:** Publicly available datasets often lack interference conditions. The lack of such effects makes it difficult to train robust spectrum sensing models. Generative modeling can generate synthetic datasets that include realistic interference either in the time, frequency, coding, or spatial domains, and help improve the generalization of AI-based classifiers and signal processing algorithms.
- **Denoising and Signal Enhancement:** Conventional filtering methods struggle when interference is non-stationary or highly correlated with the SOI. Generative AI can learn the statistical characteristics of both clean and interference signals such that the intended clean signal can be reconstructed in challenging conditions where traditional denoising methods may fail.
- **Interference-aware Signal Separation:** Traditional source separation methods such as model order estimation require prior domain knowledge and may not work well in highly congested environments. Generative AI models can learn to separate overlapping signals adaptively, to optimize their parameters dynamically based on observed spectrum conditions.
- **Adaptive Waveform Design for Coexistence:** Instead of just mitigating interference, Generative AI can be used to proactively optimize the waveform for coexistence. By predicting interference patterns, the transmission parameters (e.g., frequency allocation, time slots, beamforming angles) can be adjusted using Generative AI to minimize collisions and maximize spectrum efficiency.

- **Spectrum Anomaly Detection and Policy Compliance:** Unauthorized transmissions, policy violations, and unexpected co-channel interference are difficult to detect with rule-based methods. Generative AI algorithms can model normal spectrum usage patterns and detect anomalous behaviors to assist in regulatory compliance and real-time spectrum monitoring.

5.6.3 Integrated Sensing and Communications (ISAC)

The purpose of ISAC is to combine wireless communication and sensing functionalities into a unified network infrastructure for simultaneous communication and environmental sensing. With ISAC, moving targets within a coverage area can be detected and tracked while concurrently meeting the communication needs of other users. In 5G networks, specific physical resource blocks (PRBs) are allocated for sensing purposes, such as those used for Positioning Reference Signal (PRS) waveforms. PRS employs unique sequences with high autocorrelation and low cross-correlation characteristics to accurately detect and estimate the range and velocity of targets. A recent application of ISAC in 5G networks was studied in Ref. [75], where Demir et al. (2024) demonstrated the use of PRS to detect moving targets within the coverage range of a gNodeB while employing the remaining subcarriers in each resource block for communication.

Despite these benefits, there is currently a lack of publicly available datasets with real-world ISAC data, and no existing work has incorporated generative modeling to enhance data fidelity and scalability. Nevertheless, one can speculate that generative modeling offers promising capabilities to improve ISAC systems in the following ways:

- **Synthetic Data Generation for ISAC Model Training:** Real-world ISAC datasets are limited due to the difficulty of collecting joint sensing-communication data under various environmental conditions. Traditional data-driven generative models may fail to capture the intricate physical interactions of EM waves with the environment, which can create unrealistic signal representations. By incorporating physics-informed Generative AI, synthetic ISAC datasets can be generated such that realistic multipath propagation, Doppler shifts, and noise conditions are captured. This physics-aware approach could enhance model generalization and robustness of ISAC algorithms that perform well in real-world deployments.
- **Enhancing Sensing Fidelity with Generative Modeling:** PRS-based sensing relies on highly correlated signal reflections, which can be difficult to model under dynamic conditions. Generative models can learn complex channel and reflection characteristics, and generate synthetic sensing data to improve ISAC target detection and tracking.

- **Denoising and Interference Mitigation for Sensing Signals:** ISAC signals are often corrupted by multi-user interference (MUI) and noise, which can degrade sensing accuracy. Generative models can be used to reconstruct sensing signals and improve detection accuracy in low-SNR environments.
- **Improved Target Classification and Anomaly Detection:** Traditional ISAC systems primarily focus on detecting moving targets, but classifying different types of objects or detecting anomalous behavior remains a challenging task. Physics-based Generative AI algorithms can be used to generate realistic radar signatures constrained by wave propagation principles to categorize moving targets (e.g., humans, vehicles, and drones) while improving anomaly detection in dynamic environments.

5.6.4 Link Scheduling in Large Networks

Efficient link scheduling is critical in large-scale networks, such as TAN, where communication links can be represented as graphs. Representing the network as a graph allows for the analysis of link dependencies, interference, and scheduling constraints in a structured manner. However, data scarcity remains a significant challenge, as real-world datasets are often limited in size and diversity. Generative AI tailored for graph-based representations offers several promising solutions to address these challenges:

- **Synthetic Graph Data Generation:** Real-world network datasets are often scarce and may not sufficiently cover the diverse operational scenarios. Graph-based Generative AI models can generate synthetic communication graphs that preserve statistical properties of real networks, which can be used to develop diverse datasets.
- **Augmenting Training Data for Link Scheduling Models:** Machine learning models for link scheduling require large datasets to generalize effectively. Generative AI can create diverse network topologies, interference conditions, and traffic patterns to improve the robustness of scheduling algorithms.
- **Data-efficient Reinforcement Learning for Scheduling:** Reinforcement learning (RL) algorithms for link scheduling may suffer from long convergence times. Generative modeling can be used to generate realistic network state transitions to accelerate convergence for RL-based scheduling policies.

5.6.5 Enhancing Wireless MAC-layer Security

Enhancing security at the MAC layer in wireless networks is critical to safeguard communication systems against a variety of threats such as signal spoofing, jamming, and unauthorized access. These vulnerabilities, if exploited, can disrupt communication, degrade network performance, and compromise data integrity and confidentiality. While existing approaches have made progress in addressing these issues, several open challenges remain, particularly as networks become more complex, and adversaries become increasingly sophisticated. Generative AI presents a promising approach to address these challenges in the following ways:

- **Detecting Dynamic and Intelligent Attacks:** Many existing techniques rely on static signatures or predefined rules, which fail against adaptive adversaries capable of evolving their attack strategies, such as covert signal spoofing or adaptive jamming. Generative models can synthesize adversarial examples to train detection models that recognize novel and evolving attack patterns, improving network resilience.
- **Real-time Countermeasures for Adaptive Threats:** In critical applications such as autonomous vehicles and military communications, rapid counter-responses to attacks are essential. Generative models can simulate diverse attack scenarios, enabling reinforcement learning-based agents to develop adaptive real-time countermeasures against jamming and spoofing threats.
- **Addressing Security Vulnerabilities in Emerging Technologies:** New wireless technologies such as THz communication, RIS, and dense IoT deployments introduce unique MAC-layer vulnerabilities that traditional security methods cannot easily address. Generative models can help explore these vulnerabilities by generating synthetic attack models, aiding in the design of robust and adaptive MAC-layer security protocols.
- **Privacy Preservation in Adversarial Environments:** MAC-layer privacy is increasingly crucial to protect user identities and data from adversaries using signal analysis or traffic fingerprinting. Generative models such as VAEs can obfuscate transmission patterns while maintaining communication efficiency, reducing the risk of passive eavesdropping.
- **Optimizing Resource-efficient Security Countermeasures:** Traditional countermeasures such as frequency hopping and retransmissions consume valuable bandwidth, power, and processing resources. Generative models can optimize resource allocation by predicting attack likelihoods and tailoring security responses to minimize overhead, ensuring efficient threat mitigation.
- **Generalizing Security Solutions Across Heterogeneous Networks:** Wireless networks often consist of diverse devices, protocols, and environments, making it difficult for security models to generalize across different network conditions. Domain adaptation techniques in Generative AI can enable security solutions to dynamically adjust and maintain effectiveness across heterogeneous network deployments.

5.7 Concluding Remarks

The integration of generative models within the MAC layer of wireless communications represents a transformative approach to enhance network performance, security, and adaptability. By leveraging advanced techniques such as VAEs, GANs, and diffusion models, this chapter has discussed studies from the literature that employed generative models to significantly improve spectrum awareness, scheduling, and overall network efficiency. The exploration of applications in data augmentation, semi-supervised learning, domain adaptation, unsupervised clustering, and anomaly detection underscores the versatility and potential of these models to address both existing challenges and emerging demands in wireless networks. Moreover, the analysis of generative models in advancing spectrum awareness, PHY security, and the performance of innovative topologies such as RIS and TANs highlights their critical role in the future of wireless communication.

However, as promising as these advancements are, they also open up new avenues for research, particularly in refining these models to meet the unique challenges of wireless environments. Continued exploration in this area is essential to fully realize the potential of generative modeling in shaping the next generation of wireless networks. By embracing these technologies, the wireless communications community can pave the way for more intelligent, secure, and resilient networks that can adapt to the dynamic demands of the future.

References

1 Federal Communications Commission. FCC releases rules for innovative spectrum sharing in 3.5 GHz band. https://www.fcc.gov/document/fcc-releases-rules-innovative-spectrum-sharing-35-ghz-band, 2015. Accessed on July 30, 2025.

2 U.S. Government Publishing Office. 47 CFR Part 96—Citizens Broad-band Radio Service. https://www.ecfr.gov/current/title-47/chapter-I/subchapter-D/part-96, 2015. Accessed on July 30, 2025.

3 Claudia Cormio and Kaushik R. Chowdhury. A survey on mac protocols for cognitive radio networks. *Ad Hoc Networks*, 7(7):1315–1329, 2009. ISSN 1570-8705. https://doi.org/10.1016/j.adhoc.2009.01.002.

4 Ali Mamadou Mamadou, Joël Toussaint, and Gérard Chalhoub. Survey on wireless networks coexistence: Resource sharing in the 5G era. *Mobile Networks and Applications*, 25(5):1749–1764, October 2020. https://doi.org/10.1007/s11036-020-01564-w.

5 Akhil Gupta, Raabia Kausar, Sudeep Tanwar, Abdulatif Alabdulatif, Vrince Vimal, and Srinivas Aluvala. Efficient spectrum sharing in 5g and beyond network: a survey. *Telecommunication Systems*, 88(1): 1–37, March 2025. https://doi.org/10.1007/s11235-025-01261-7

6 U.S. Government Accountability Office. Spectrum management: Key practices could help address challenges to improving receiver performance. Report to Congressional Committees GAO-24-106325, U.S. Government Accountability Office, 2024. https://www.gao.gov/assets/gao-24-106325.pdf. Accessed July 25, 2025.

7 Luciano Miuccio, Salvatore Riolo, Sumudu Samarakoon, Daniela Panno, and Mehdi Bennis. Learning generalized wireless mac communication protocols via abstraction. In *GLOBECOM 2022 - 2022 IEEE Global Communications Conference*, pages 2322–2327, 2022. https://doi.org/ 10.1109/ GLOBECOM48099.2022.10000805.

8 Jiantao Xin, Wei Xu, Bin Cao, Taotao Wang, and Shengli Zhang. A deep-learning-based mac for integrating channel access, rate adaptation, and channel switch. *Digital Communications and Networks*, 2024. https://doi.org/10.1016/ j.dcan.2024.10.010.

9 Kemal Davaslioglu, Sohraab Soltani, Tugba Erpek, and Yalin E. Sagduyu. DeepWiFi: Cognitive WiFi with deep learning. *IEEE Transactions on Mobile Computing*, 20(2): 429–444, 2021.

10 Yi Shi, Yalin E. Sagduyu, Tugba Erpek, Kemal Davaslioglu, Zhuo Lu, and Jason H. Li. Adversarial deep learning for cognitive radio security: Jamming attack and defense strategies. In *2018 IEEE International Conference on Communications Workshops (ICC Workshops)*, pages 1–6, 2018. https://doi.org/10.1109/ICCW. 2018.8403655.

11 Kemal Davaslioglu, Srihari Kompella, Tugba Erpek, and Yalin E. Sagduyu. Continual deep reinforcement learning to prevent catastrophic forgetting in jamming mitigation. In *Proceedings of the IEEE Military Communications Conference (MILCOM)*, Washington, D.C., USA, 2024.

12 Yue Xu, Jianyuan Yu, William C. Headley, and R. Michael Buehrer. Deep reinforcement learning for dynamic spectrum access in wireless networks. In *MILCOM 2018 - 2018 IEEE Military Communications Conference (MILCOM)*, pages 207–212, 2018. https://doi.org/10.1109/MILCOM.2018.8599723.

13 Diederik P. Kingma. Auto-encoding variational Bayes. (Preprint) *arXiv:1312.6114*,2013.

14 Ali Razavi, Aaron Van den Oord, and Oriol Vinyals. Generating diverse high-fidelity images with VQ-VAE2-2. *Advances in Neural Information Processing Systems*, 32, 2019.

15 Aaron Van Den Oord and Oriol Vinyals. Neural discrete representation learning. *Advances in Neural Information Processing Systems*, 30, 2017.

16 Ian J. Goodfellow, Jean Pouget-Abadie, Mehdi Mirza, Bing Xu, David Warde-Farley, Sherjil Ozair, Aaron Courville, and Yoshua Bengio. Generative adversarial nets. In *Advances in Neural Information Processing Systems*, pages 2672–2680, 2014.

17 Thalles Silva. An intuitive introduction to generative adversarial networks (GANs). https://www.freecodecamp.org/news/an-intuitive-introduction-to-generative-adversarial-networks-gans-7a2264a81394/, 2018.

18 Michael Kan. NVIDIA's AI creates artificial human faces from celebrity photos. https://www.pcmag.com/news/357068/nvidias-ai-creates-artificial-human-faces-from-celebrity-ph, 2017.

19 Tero Karras, Timo Aila, Samuli Laine, and Jaakko Lehtinen. Progressive growing of GANs for improved quality, stability, and variation. In *International Conference on Learning Representations (ICLR)*, pages 1–26, 2018.

20 Jun-Yan Zhu, Taesung Park, Phillip Isola, and Alexei A. Efros. Unpaired image-to-image translation using cycle-consistent adversarial networks. In *International Conference on Vision and Visualization (ICCV)*, pages 1–18, 2017.

21 Grigory Antipov, Moez Baccouche, and Jean-Luc Dugelay. Face aging with conditional generative adversarial networks. (Preprint) *arXiv:1702.01983v2*, 2017.

22 Christian Ledig, Lucas Theis, Ferenc Huszár, Jose Caballero, Andrew Cunningham, Alejandro Acosta, Andrew Aitken, Alykhan Tejani, Johannes Totz, Zehan Wang, and Wenzhe Shi. Photo-realistic single image super-resolution using a generative adversarial network. (Preprint) *arXiv:1609.04802v5*, 2017.

23 Li-Chia Yang, Szu-Yu Chou, and Yi-Hsuan Yang. MiDiNet: A convolutional generative adversarial network for symbolic-domain music generation. (Preprint) *arXiv:1703.10847v2*,2017.

24 Chris Donahue, Julian McAuley, and Miller Puckette. Adversarial audio synthesis. (Preprint) *arXiv:1802.04208v3*, 2018.

25 Ting-Chun Wang, Ming-Yu Liu, Jun-Yan Zhu, Guilin Liu, Andrew Tao, Jan Kautz, and Bryan Catanzaro. Video-to-video synthesis. (Preprint) *arXiv:1808.06601v2*, 2018.

26 Kemal Davaslioglu and Yalin E. Sagduyu. Generative adversarial learning for spectrum sensing. In *2018 IEEE International Conference on Communications (ICC)*, pages 1–6, 2018.

27 Debashri oy, Tathagata Mukherjee, Mainak Chatterjee, and Eduardo L. Pasiliao Jr. Detection of rogue RF transmitters using generative adversarial nets. In *Proceedings of IEEE Wireless Communications and Networking Conference (WCNC)*, pages 1–7, 2019.

28 Tamoghna Roy, Tim O'Shea, and Nathan West. Generative adversarial radio spectrum networks. In *Proceedings of ACM Workshop on Wireless Security and Machine Learning (WiseML)*, pages 12–15, 2019.

29 Yi Shi, Kemal Davaslioglu, and Yalin E. Sagduyu. Generative adversarial network for wireless signal spoofing. In *Proceedings of the ACM Workshop on Wireless Security and Machine Learning*, pages 55–60, 2019.

30 Yi Shi and Yalin E. Sagduyu. Sensing-throughput tradeoffs with generative adversarial networks for NextG spectrum sharing. In *IEEE Military Communications Conference (MILCOM)*, 2022.

31 Dung N. Tran, Trac D. Tran, and Lam Nguyen. Generative adversarial networks for recovering missing spectral information. In *2018 IEEE Radar Conference (RadarConf18)*, pages 1223–1227, 2018.

32 Muhammad Usama, Muhammad Asim, Siddique Latif, and Junaid Qadir. Generative adversarial networks for launching and thwarting adversarial attacks on network intrusion detection systems. In *International Wireless Communications Mobile Computing Conference (IWCMC)*, pages 78–83, Jun. 2019.

33 Bo Yang and Min Liu. Attack-resilient connectivity game for UAV networks using generative adversarial learning. In *Proc. International Conference on Autonomous Agents and Multiagent Systems (AAMAS)*, pages 1743–1751, May 2019.

34 Mehdi Mirza and Simon Osindero. Conditional generative adversarial nets. (Preprint) *arXiv:1411.1784*,2014.

35 Xudong Mao, Qing Li, Haoran Xie, Raymond YK Lau, Zhen Wang, and Stephen Paul Smolley. Least squares generative adversarial networks. In *Proc. IEEE International Conference on Computer Vision (ICCV)*, pages 2813–2821, 2017.

36 Martin Arjovsky, Soumith Chintala, and Léon Bottou. Wasserstein generative adversarial networks. In *Proc. International Conference on Machine Learning (ICML)*, pages 214–223, 2017.

37 Ishaan Gulrajani, Faruk Ahmed, Martin Arjovsky, Vincent Dumoulin, and Aaron C. Courville. Improved training of Wasserstein GANs. In *Advances in Neural Information Processing Systems*, pages 5767–5777, 2017.

38 Jascha Sohl-Dickstein, Eric Weiss, Niru Maheswaranathan, and Surya Ganguli. Deep unsupervised learning using nonequilibrium thermodynamics. In *International Conference on Machine Learning*, pages 2256–2265. PMLR, 2015.

39 Jonathan Ho, Ajay Jain, and Pieter Abbeel. Denoising diffusion probabilistic models. *Advances in Neural Information Processing Systems*, 33: 6840–6851, 2020.

40 Robin Rombach, Andreas Blattmann, Dominik Lorenz, Patrick Esser, and Björn Ommer. High-resolution image synthesis with latent diffusion models. In *Proceedings of the IEEE/CVF Conference on Computer Vision and Pattern Recognition*, pages 10684–10695, 2022.

41 Jeff Donahue, Philipp Krähenbühl, and Trevor Darrell. Adversarial feature learning. (Preprint) *arXiv:1605.09782*, 2016.

42 Mansi Patel, Xuyu Wang, and Shiwen Mao. Data augmentation with conditional GAN for automatic modulation classification. In *Proceedings of the 2nd ACM Workshop on Wireless Security and Machine Learning*, WiseML ’20, page 31–36, New York, NY, USA, 2020. Association for Computing Machinery. ISBN 9781450380072. https://doi.org/10.1145/3395352.3402622.

43 Zheng Liu, Xiaojun Jing, Ronghui Zhang, and Junsheng Mu. Spectrum sensing based on deep convolutional generative adversarial networks. In *2021 International Wireless Communications and Mobile Computing (IWCMC)*, pages 796–801, 2021.

44 Srihari Kamesh Kompella, Kemal Davaslioglu, Yalin E. Sagduyu, and Sastry Kompella. Augmenting training data with vector-quantized variational autoencoder for classifying RF signals. In *2024 IEEE Military Communications Conference (MILCOM)*, 2024.

45 Kemal Davaslioglu, Tugba Erpek, and Yalin Sagduyu. End-to-end autoencoder communications with optimized interference suppression. *Physical-Layer Security for 6G*, pages 153–184, 2024.

46 Timothy J. O'Shea, Tamoghna Roy, Nathan West, and Benjamin C. Hilburn. Physical layer communications system design over-the-air using adversarial networks. (Preprint) *arXiv:1803.03145*, 2018.

47 Hao Ye, Geoffrey Ye Li, Biing-Hwang Fred Juang, and Kathiravetpillai Sivanesan. Channel agnostic end-to-end learning based communication systems with conditional GAN. In *2018 IEEE Globecom Workshops (GC Wkshps)*, pages 1–5, 2018.

48 Kemal Davaslioglu, Tugba Erpek, and Yalin E. Sagduyu. Autoencoder communications with optimized interference suppression for NextG RAN. In *2022 IEEE Future Networks World Forum (FNWF)*, pages 164–169. IEEE, 2022.

49 Yalin E. Sagduyu, Tugba Erpek, and Yi Shi. Adversarial machine learning for 5G communications security. In Charles A Kamhoua, Christopher D Kiekintveld, Fei Fang, Quanyan Zhu editors, *Game Theory and Machine Learning for Cyber Security*, pages 270–288, Wiley Online library 2021.

50 Kemal Davaslioglu, Serdar Boztaş, Mehmet Can Ertem, Yalin E. Sagduyu, and Ender Ayanoglu. Self-supervised RF signal representation learning for NextG signal classification with deep learning. *IEEE Wireless Communications Letters*, 12(1): 65–69, 2022.

51 Caidan Zhao, Caiyun Chen, Zhibiao Cai, Mingxian Shi, Xiaojiang Du, and Mohsen Guizani. Classification of small UAVs based on auxiliary classifier Wasserstein GANs. In *2018 IEEE Global Communications Conference (GLOBECOM)*, pages 206–212, 2018.

52 Ender Ayanoglu, Kemal Davaslioglu, and Yalin E. Sagduyu. Machine learning in NextG networks via generative adversarial networks. *IEEE Transactions on Cognitive Communications and Networking*, 8(2): 480–501, 2022.

53 Thomas Schlegl, Philipp Seeböck, Sebastian M. Waldstein, Georg Langs, and Ursula Schmidt-Erfurth. f-AnoGAN: Fast unsupervised anomaly detection with generative adversarial networks. *Medical Image Analysis*, 54: 30–44, 2019.

54 Thomas Schlegl, Philipp Seeböck, Sebastian M. Waldstein, Ursula Schmidt-Erfurth, and Georg Langs. Unsupervised anomaly detection with generative adversarial networks to guide marker discovery. In *International Conference on Information Processing in Medical Imaging*, pages 146–157. Springer, 2017.

55 Timothy J. O'Shea, Johnathan Corgan, and T. Charles Clancy. Convolutional radio modulation recognition networks. In *17th International Conference on Engineering Applications of Neural Networks*, pages 213–226. Springer, 2016.

56 Yixuan Yang, and Tianfeng Yan. Radio frequency fingerprint recognition method based on generative adversarial net. In *2021 13th International Conference on Communication Software and Networks (ICCSN)*, pages 361–364, 2021.

57 Kaiwen Tan, Wenjun Yan, Limin Zhang, Qing Ling, and Congan Xu. Semi-supervised specific emitter identification based on bispectrum feature extraction CGAN in multiple communication scenarios. *IEEE Transactions on Aerospace and Electronic Systems*, 59(1): 292–310, 2022.

58 Yalin E. Sagduyu and Tugba Erpek. Adversarial attacks on LoRa device identification and rogue signal detection with deep learning. In *IEEE Military Communications Conference (MILCOM)*, pages 385–390, 2023.

59 Yalin E. Sagduyu and Tugba Erpek. Adversarial attack and defense for LoRa device identification and authentication via deep learning. *IEEE Internet of Things Journal*, 2025.

60 Yi Shi, Kemal Davaslioglu, and Yalin E. Sagduyu. Generative adversarial network in the air: Deep adversarial learning for wireless signal spoofing. *IEEE Transactions on Cognitive Communications and Networking*, 7(1): 294–303, 2021.

61 Tao Hou, Tao Wang, Zhuo Lu, Yao Liu, and Yalin Sagduyu. IoTGAN: GAN powered camouflage against machine learning based IOT device identification. In *IEEE International Symposium on Dynamic Spectrum Access Networks (DySPAN)*, pages 280–287, 2021.

62 Tugba Erpek, Yalin E. Sagduyu, and Yi Shi. Deep learning for launching and mitigating wireless jamming attacks. *IEEE Transactions on Cognitive Communications and Networking*, 5(1): 2–14, 2019.

63 Alireza Koochali, Peter Schichtel, Andreas Dengel, and Sheraz Ahmed. Probabilistic forecasting of sensory data with generative adversarial networks – ForGAN. *IEEE Access*, 7: 63868–63880, 2019.

64 Laisen Nie, Xiaojie Wang, Qinglin Zhao, Zhigang Shang, Li Feng, and Guojun Li. Digital twin for transportation big data: A reinforcement learning-based network traffic prediction approach. *IEEE Transactions on Intelligent Transportation Systems*, 25(1): 896–906, 2024.

65 Hongling Zheng, Xiang Li, Yongfeng Li, Ziqin Yan, and Tinghong Li. GCN-GAN: Integrating graph convolutional network and generative adversarial network for traffic flow prediction. *IEEE Access*, 10: 94051–94062, 2022.

66 Juhwan Song, Yujin Nam, Hyungtae Kwon, Ilsong Sim, Seung Joo Maeng, and Seowoo Jang. Adaptive generalized proportional fair scheduling with deep reinforcement learning. In *2022 IEEE Globecom Workshops (GC Wkshps)*, pages 759–764, 2022.

67 Linsheng He, Fei Hu, Zhe Chu, Jiamiao Zhao, Yalin Sagduyu, Ngwe Thawdar, and Sunil Kumar. Intelligent Terahertz medium access control (MAC) for highly dynamic airborne networks. *IEEE Transactions on Aerospace and Electronic Systems*, 59(3): 2494–2512, 2023.

68 Marco Rossanese, Placido Mursia, Andres Garcia-Saavedra, Vincenzo Sciancalepore, Arash Asadi, and Xavier Costa-Perez. Open experimental measurements of sub-6ghz reconfigurable intelligent surfaces. *IEEE Internet Computing*, 28(2): 19–28, 2024. https://doi.org/10.1109/MIC.2024.3376772.

69 Simon Tewes, Markus Heinrichs, Kevin Weinberger, Rainer Kronberger, and Aydin Sezgin. A comprehensive dataset of RIS-based channel measurements in the 5ghz band. In *2023 IEEE 97th Vehicular Technology Conference (VTC2023-Spring)*, pages 1–5, 2023. https://doi.org/10.1109/VTC2023-Spring57618.2023.10200973.

70 Yu Jin, Jiayi Zhang, Chongwen Huang, Liang Yang, Huahua Xiao, Bo Ai, and Zhiqin Wang. Multiple residual dense networks for reconfigurable intelligent surfaces cascaded channel estimation. *IEEE Transactions on Vehicular Technology*, 71(2): 2134–2139, 2022.

71 Junhao Duan, Ying Zhang, Jinyuan Gu, Lei Zhang, and Wei Duan. Physical layer security for AI-generated content: Power and elements allocation for active RIS. *IEEE Transactions on Consumer Electronics*, 70(1): 839–848, 2024.

72 Zhuangkun Wei, Wenxiu Hu, Junqing Zhang, Weisi Guo, and Julie McCann. Explainable adversarial learning framework on physical layer secret keys combating malicious reconfigurable intelligent surface. (Preprint) *arXiv:2402.06663*, 2024.

73 Christina Chaccour, Walid Saad, Mérouane Debbah and H. Vincent Poor. Joint sensing, communication, and AI: A trifecta for resilient THz user experiences. *IEEE Transactions on Wireless Communications,* 23 (9): 11444–11460, 2024.

74 Samuel R. Shebert, Benjamin H. Kirk, and R. Michael Buehrer. Open set wireless signal classification: Augmenting deep learning with expert feature classifiers. (Preprint) *arXiv:2302.03749*, 2023.

75 Utku Demir, Kemal Davaslioglu, Yalin E. Sagduyu, Tugba Erpek, Gustave Anderson, and Sastry Kompella. I-SCOUT: Integrated sensing and communications to uncover moving targets in NextG networks. In *2024 IEEE Military Communications (MILCOM)*, pages 1–6, 2024.

6

Generative AI at Network Layer

Athanasios Karapantelakis[1,*], *Pegah Alizadeh*[2], *Abdulrahman Alabbasi*[1], *Kaushik Dey*[3], *and Alexandros Nikou*[1]

[1] *Ericsson Research, Ericsson AB, Kista, Stockholm, Sweden*
[2] *Ericsson Research, Ericsson France, Massy, France*
[3] *Ericsson Research, Ericsson India, Perungudi, Chennai, Tamil Nadu, India*

*Corresponding author: athanasios.karapantelakis@ericsson.com

6.1 Introduction

In packet-switched computer networks, the network layer manages all aspects of data packet transfer between two endpoints. In mobile networks, this packet transfer includes the radio access network (RAN), backhaul and core network. RAN consists of geographically distributed radio base stations, which, in 5G terminology, are known as 5G Node-Bs (gNBs). RAN provides wireless network access to mobile devices, which are known as user equipment (UE), using the new radio (NR) interface. The backhaul network connects RAN with the core network. The latter is a collection of network functions (NFs), managing connection between UE and between UE and other services that can reside outside of the mobile network (e.g., third-party cloud services and applications). As shown in Figure 6.1, internet protocol (IP) is the network protocol that is commonly used throughout the mobile network, with the exception of control plane signaling in NR. Due to the extensive adoption of IP, this chapter will explore generative AI (GenAI) approaches to IP in a broad context, extending beyond mobile networks.

Every component of the network layer faces distinct challenges. In RAN, challenges are signal interference at the wireless (or so-called "air") interface, dynamic user mobility, and optimal radio resource allocation. Today, these challenges are

Generative AI for Communications Systems: Fundamentals, Applications, and Prospects,
First Edition. Edited by Diep N. Nguyen et al.

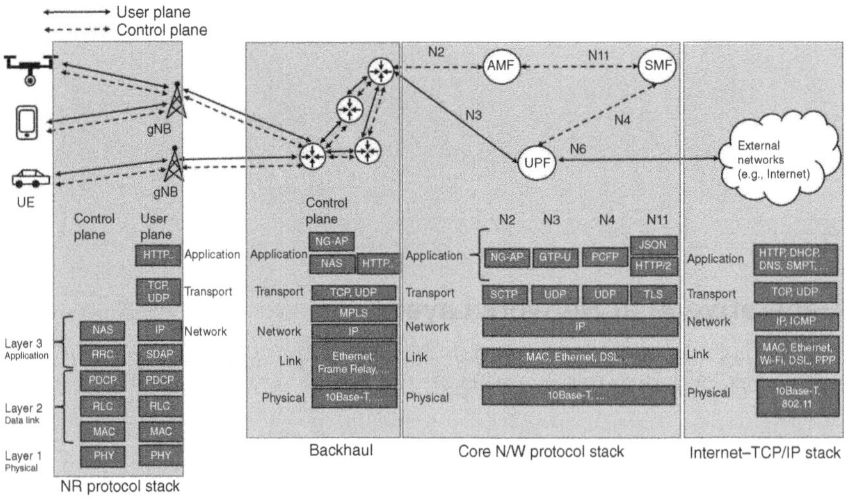

Figure 6.1 Data flow and protocol stacks of 5G as specified in 3GPP (Third Generation Partnership Project) TS series 23.5xx for the Core Network and 3GPP, for example [1], TS series 38.3xx for RAN (for example [2]).

addressed through adaptive modulation schemes, scheduling and partitioning algorithms, and handover procedures for seamless transitions between cells.

In the backhaul network, maintaining high throughput and low latency amidst fluctuating traffic loads is important for many communication services. Existing solutions make use of high-capacity physical layer transport, such as optical fiber links or microwave transmissions. They also use dynamic routing protocols in order to manage data distribution and reduce network congestion.

In the core network, integrating and coordinating network functions—such as the access management function (AMF), service management function (SMF), and user plane function (UPF)—requires scalable, resilient network infrastructure technologies. This is achieved with cloud computing technologies such as network function virtualization (NFV) and software-defined networks (SDNs).

Additionally, the pervasive use of IP brings its own set of challenges, including packet loss, routing inefficiencies, and congestion control. These issues are traditionally mitigated by implementing robust routing protocols, quality of service (QoS) policies, and traffic steering techniques that dynamically adapt to changing network conditions.

These mechanisms play a crucial role in managing the different types of traffic found in 3GPP mobile networks. Specifically, traffic in these networks is divided in two categories: control plane traffic, which manages network operations and signaling tasks (such as session management, user authentication, and mobility control), and user plane traffic, which carries the data packets for end-user applications. Control plane traffic is routed to NFs such as the AMF and SMF, while user plane traffic is directed to NFs such as the UPF. These two types of traffic are often routed over different networks and have different requirements from the network infrastructure. For instance, control plane traffic generally requires lower latency for efficient connection and mobility management, whereas user plane traffic demands consistent throughput and, depending on the application, may tolerate higher latency. Moreover, control plane traffic requires less bandwidth and occurs sporadically, while user plane traffic is continuous and requires more bandwidth—especially for data-intensive applications. These differing requirements also influence which existing technologies are best suited to manage each type of traffic.

In mobile networks, routing and addressing complexities grow with network size and device diversity and are traditionally managed through static routing tables and predefined paths that require manual updates. Data packet delivery faces the challenge of ensuring timeliness and minimal loss, particularly under heavy traffic conditions, which are addressed using load balancing and traffic prioritization to manage congestion and maintain service quality. The efficiency of the control plane, crucial for rapid management of signaling and operational tasks, becomes strained during peak times; this is mitigated by distributing control plane functions across multiple NFs to reduce bottlenecks [1]. Meanwhile, maintaining high throughput for user plane traffic, especially vital for data-intensive applications in densely populated areas, is achieved through strategies such as caching data near users and dynamically optimizing data paths based on real-time conditions. Additionally, network security is fortified against sophisticated cyber threats through multilayered security protocols and continuous traffic monitoring.

These challenges above highlight the potential areas where GenAI can play a transformative role. GenAI could dynamically generate and adapt routing protocols to improve efficiency, predict and simulate traffic patterns for better load management, enhance anomaly detection in network security, and optimize resource allocation in real-time, making the network more adaptive and efficient. This hint at GenAI's capabilities sets the stage for a deeper exploration of its specific applications in enhancing network layer functions.

The remainder of this chapter is organized into four subsequent sections. Section 6.2 presents more details in the network layer across the different protocol stacks used in mobile network. Section 6.3 describes areas in the network layer where GenAI algorithms can contribute. Subsequently, Section 6.4 outlines limitations of GenAI algorithms in context of the network layer and suggest future opportunities in GenAI research to address these limitations. We end with Section 6.5, which summarizes key points of this chapter.

6.2 Network Layer in Mobile Networks

This section provides an overview of the network layer protocol and the nodes responsible for decisions impacting network layer functionality in mobile networks. The description is framed within the context of 5G, although similar nodes and protocols are present in previous generations and are anticipated to remain comparable in 6G.

6.2.1 Radio Access Network

6.2.1.1 Network Layer Protocols of the Air Interface

In this section, we describe the responsibilities, functionalities, and challenges of the higher RAN layers that are relevant to the network layer, that is, radio link control (RLC), packet data convergence protocol (PDCP), and service data adaptation protocol (SDAP), partly illustrated in Figure 6.1. The connection between RAN and the core network is as follows. Control plane traffic, UE (and RAN protocols), connect to the core network through NAS. For data transmission from the user plane, gNB connects to the core network through UPF.

The RLC protocol, as described in Ref. [3], is responsible for several key tasks, including reliable data transfer between the UE and the network, error correction, and flow control. It ensures that data is accurately transmitted, even in the presence of errors that cannot be handled by the medium access control (MAC) layer. Additionally, the RLC protocol manages the flow of data to prevent buffer overflow, ensuring smooth communication between the UE and the network. The RLC protocol also has specific functions that enhance data transfer efficiency. One key function is segmentation and reassembly, where large packets are broken down into smaller segments for transmission and reassembled at the receiver. The automatic repeat request (ARQ) mechanism is used to manage retransmissions for error correction. Furthermore, RLC handles buffer management, ensuring that data is properly buffered to maintain smooth data flow during transmission.

The PDCP protocol, as described in Ref. [4], is primarily responsible for ensuring data integrity and security. It achieves this through encryption and integrity protection, safeguarding the confidentiality and accuracy of the transmitted data. Additionally, it utilizes header compression techniques to reduce the size of headers, thus improving transmission efficiency.

In terms of functionality, PDCP performs ciphering and deciphering to encrypt and decrypt user data, ensuring that the data remains confidential during transmission. It also provides integrity protection by verifying the integrity of the data to prevent tampering. Furthermore, PDCP uses robust header compression (ROHC) for header compression, which reduces the overhead of IP headers, improving overall transmission efficiency.

The functionalities of the SDAP layer, as described in Ref. [4], include:

- **QoS Enforcement:** The SDAP layer enforces the QoS rules provided by the 5G core network, ensuring that each traffic flow is treated according to its QoS requirements. This could involve prioritizing certain types of traffic, reserving resources for critical applications, or applying traffic-shaping techniques.
- **Traffic Prioritization:** SDAP enables traffic prioritization by mapping QoS flows into different radio bearers. Higher priority traffic is placed on bearers with guaranteed resources and low latency, while lower priority traffic uses best-effort bearers.
- **Bearer Aggregation:** In scenarios where multiple bearers are available (e.g., carrier aggregation or dual connectivity), SDAP can split traffic across multiple bearers, ensuring optimal use of available radio resources.
- **Flow Control:** The SDAP layer helps regulate the flow of packets to match the available radio resources. If there is network congestion or limited bandwidth, SDAP can prioritize certain QoS flows to prevent packet loss or degradation in critical services.

The RRC layer is a critical protocol layer within the control plane of the 5G NR, operating between the UE and the gNB [5]. It is responsible for managing the overall control and configuration of both UE and network resources. Key responsibilities of the RRC layer includes connection management, where RRC establish, maintain, and release the RRC connection between the UE and the gNB. RRC manages the transitions between various states, such as RRC_IDLE, RRC_CONNECTED, and RRC_INACTIVE. It handles connection establishment procedures when the UE needs to send or receive data, ensuring that resources are allocated for the connection. In addition, RRC is responsible for mobility management, including the handover process, to ensure seamless connectivity as a UE moves between different cells or gNB. Coordinates the signaling required to initiate, execute, and complete handover operations. Cell reselection, in RRC_IDLE state, is another RRC responsibility. RRC also is responsible for broadcasting system information, such as frequency bands, available services, access restrictions, configuration details, measurement and reporting, and paging. Additionally, RRC is responsible for setting up, modifying, and releasing radio bearers. These bearers define how data will be transferred between the UE and the gNB, both for user data (user plane) and control signaling. Different bearers are established for different QoS requirements, such as voice, video, or general data.

6.2.1.2 Interactions Between RAN Layers

In the context of network layer functionality, the RAN protocols in the NR protocol stack collaborate to manage data transmission, ensuring efficient routing, error correction, and quality of service. This section details how the physical, MAC,

RLC, PDCP, SDAP, and service layers interact to support the network layer's role in data delivery.

- **PHY to MAC:** The PHY layer provides the MAC layer with information about channel conditions and available resources. The MAC layer then decides how to allocate these resources.
- **MAC to RLC:** The MAC layer sends data to the RLC layer, which ensures that the data is error free and properly sequenced.
- **RLC to PDCP:** The RLC layer delivers error-corrected and ordered data to the PDCP layer, which handles encryption and header compression.
- **PDCP to Service Layer:** The PDCP layer provides secure and compressed data to the service layer, which then manages the delivery of specific services to the user.
- **SDAP and PDCP:** SDAP sits above the PDCP layer and works with it to ensure that the packets are encrypted, compressed (if necessary), and transmitted with the appropriate QoS.
- **SDAP and RLC:** The RLC layer handles segmentation and reassembly of the data received from SDAP, ensuring that the packets are split into smaller units that can be sent over the radio interface.
- **SDAP and MAC:** The MAC layer schedules the transmission of data based on the resources available in the PHY. SDAP informs the MAC layer of the QoS requirements so that MAC can prioritize certain flows and ensure timely transmission.

6.2.1.3 RAN Challenges

This section outlines several challenges in RAN that are closely related to network layer functionality. One of the primary challenges is dynamic spectrum management, which is crucial for efficiently allocating spectrum resources as users move and traffic patterns change. This directly impacts network layer routing and resource optimization. Beamforming optimization is also essential, as it directs signal energy toward UE for optimal performance, particularly in high-mobility scenarios, where strong connections are necessary.

Interference mitigation is another key challenge, as managing interference across users, cells, and spectrum bands is vital for ensuring clear communication and minimizing packet loss at the network layer. Additionally, massive multiple input and multiple output (MIMO) optimization plays a significant role in improving coverage and capacity by efficiently managing large antenna arrays, supporting the network layer's need for high throughput. The MAC layer also plays an important role, as it schedules resources that affect real-time data flow decisions and ensure that QoS requirements are met for diverse traffic types.

Traffic congestion must be mitigated during peak loads by optimizing routing and ensuring that critical traffic is prioritized to maintain service reliability.

Reducing latency is critical for ultra-low-latency services such as ultra-reliable low-latency communication (URLLC), which support time-sensitive applications. Ensuring efficient data transmission with minimal delays is a priority in this regard. Moreover, optimizing error recovery and retransmission strategies is necessary to balance error correction and latency, ensuring reliable real-time data delivery.

Maximizing throughput while minimizing delays and errors is essential for supporting high-performance services like video streaming. Efficient segmentation and reassembly of large data packets are crucial for handling varying network conditions without causing disruptions. At the same time, latency-sensitive data encryption must be implemented to ensure secure, fast data transmission without introducing excessive delays. Header compression efficiency is also important, as it reduces IP header overhead while maintaining effective communication, which optimizes network resource usage.

Mobility and handover management are critical for ensuring seamless data sessions during handovers, maintaining low latency, and preventing packet loss for continuous service. Managing QoS flows for different traffic types is equally important, as it ensures that varying service requirements are met across the network layer. Flow differentiation, alongside traffic prioritization, plays a significant role in maintaining optimal service delivery, particularly when traffic types are competing for resources.

Self-organizing networks (SON) in RAN enable the network to autonomously adapt to changing conditions, optimizing resources, and improving the network layer's responsiveness and efficiency. Energy efficiency is another challenge, as dynamic adjustment of resources based on usage patterns can reduce power consumption while enhancing network layer performance. Lastly, anomaly detection and fault prediction allow the network to proactively identify and address issues such as traffic changes or hardware failures, ensuring the network layer can adjust and maintain service quality.

6.2.2 Core Network

6.2.2.1 Introduction

The core network is an important part of 5G, and it controls data and call sessions, performs subscription and mobility management, and routes data to and from other networks. It also handles security, QoS, and billing. Figure 6.1 illustrates the architecture of the 5G network, including the core network. The core network contains NFs, which are discrete software components that provide a specific network-related functionality. These NFs are part of 5G's service-based architecture (SBA). From the perspective of the network layer, some key NFs of the core network include AMF, SMF, and UPF.

6.2.2.2 Core Network Entities

AMF is a central control-plane NF in the core network. Specifically, AMF manages signaling protocols like nonaccess stratum (NAS) to facilitate communication between the RAN and the core network, supporting key network layer functions such as registration, connection establishment, and mobility management, which are essential for routing and maintaining session continuity. On a user-plane level, AMF interacts with SMF and UPF to ensure that data paths are optimized based on mobility and network policy. Specifically, AMF communicates with the SMF to handle session establishment, modification, and release. SMF then interacts with UPF to establish user-plane tunnels and manage data routing.

SMF is also a key NF in 5G core network, as it is responsible for session management, policy enforcement, mobility management, and interfacing with other network functions, such as AMF, UPF, and PCF, via N11, N4, N7, respectively. Functionalities of SMF relevant to network layer functionality include:

- Initiating and managing protocol data unit (PDU) sessions between UE and the data network and modifying session attributes.
- Assigning IP addresses to UE either dynamically or based on predefined policies, while handling the corresponding dynamic host configuration protocol (DHCP) functions.
- Ensuring that QoS key performance indicators (KPIs) are met for different traffic types, while steering traffic to the appropriate UPF based on QoS requirements and network conditions.
- Applying charging rules and policies (e.g., data caps, throttling, and prioritization) as per the user's subscription.
- Managing network slices and its impact on PDU sessions and ensuring isolation and optimized performance for each slice.

UPF NF is responsible for managing user-plane data traffic between UE and external networks. The data traffic belongs to three different classes, namely, enhanced mobile broadband (eMBB), URLLC, and massive machine-type communication (mMTC). UPF interfaces with RAN via N3, with SMF via N4, and with external nodes via N6. UPF focuses on:

- Handling user data traffic with minimal latency and high throughput, considering both uplink and downlink transmission.
- Ensuring service-level agreement (SLA) is met for high-priority traffic.
- Enabling traffic optimization techniques such as traffic shaping, steering, and caching (and edge computing), in addition to routing optimizations to improve relevant KPIs.

Aside from the aforementioned entities and functionalities of the core network (i.e., SMF, AMF, UPF), there are other NFs that handle network-related tasks.

Such tasks include policy enforcement related to QoS management, charging control, authentication, user identification, network slicing, routing, load balancing (across multiple instances of network functions), and traffic management.

6.2.2.3 Core Network Challenges

While RAN primarily deals with lower-layer protocols, the core network focuses on higher layers and IP-based functionalities. These include routing, session management, and interfacing with external networks, which are critical for delivering the promised QoS, mobility support, and security in 5G.

To perform these functions efficiently, the core network faces several challenges that could be addressed by GenAI algorithms:

- Ensuring scalability and adaptability in routing decisions, as optimal paths may vary across different regions of the network due to topology, load distribution, and service demands.
- Supporting decision-making when real network data is unavailable or insufficient.
- Efficiently detecting anomalous behaviors at the network layer, such as deviations in TCP or UDP traffic patterns.

6.3 Generative AI in the Network Layer

6.3.1 Introduction

In this section, we begin by describing the general advantages of GenAI models over their nongenerative counterparts. We also present network layer-related use-cases that could benefit from the application of GenAI algorithms. These areas are categorized by their timeline for the adoption of GenAI solutions—specifically, "short-term" and "long-term." Short-term areas can be addressed using current GenAI technologies, while long-term areas will require advancements in both the performance and accuracy of GenAI models [6]. Furthermore, widespread adoption will depend on collaboration with network equipment vendors and the standardization community to grant GenAI models the authority to design communication protocols and optimize network packet routing.

6.3.2 Advantages of GenAI Models

Nongenerative AI models are already extensively applied in mobile networks. While nongenerative models focus on analyzing available data for predictions, classifications, or decision making, GenAI models generate novel, previously unseen content. We distinguish between three key areas that GenAI models have over their predecessors: their adaptability to diverse tasks without the need for extensive retraining, their ability to generate new data, and their ability to efficiently detect anomalies in input data. We describe these three aspects in more detail below.

Ease of Adaptation and Reduced Lead-time: Generative AI (GenAI models are typically easier to adapt to new domains and use cases than non-GenAI models. This is due to their ability to handle a wide range of tasks with minimal or no additional training. On the other hand, nongenerative models often require extensive task-specific feature engineering, data preparation, and retraining to generalize across domains. In contrast, GenAI models, such as large language model (LLM), leverage their pretraining on large datasets to generalize across diverse domains with minimal additional fine-tuning. Techniques like zero-shot or few-shot prompting [7], chain-of-thought (CoT) [8], and retrieval-augmented generation (RAG) [9] can be used—individually or in combination, to reduce the time required to tailor a model for a specific use case.

Moreover, this ease of adaptation extends to network configuration automation, where GenAI can automate the setup and adjustment of complex network configuration with minimal manual intervention [10]. One example of such configuration could be generating new routing rules for routing tables. Generative models can create alternative versions of these routing rules previously unseen, by simulating network traffic scenarios. By automating such configuration processes, GenAI reduces the operational burden on network engineers, accelerates deployment timelines, and minimizes human error. Moreover, the ability to rapidly generate and validate configurations ensures that networks can be adapted for diverse environments and requirements without extensive lead times. This is especially important as mobile network deployments are expected to increase, as private or enterprise mobile networks become more and more popular.

Additionally, network-automated optimization benefits significantly from GenAI's adaptability, particularly in scenarios where traditional AI algorithms may falter. GenAI excels in handling dynamic and nonlinear network behaviors, supporting tasks such as rotation, generalization, and scalability of input data. The ability to infer optimal configurations and adjustments—even in edge cases—ensures robust performance and resilience. This adaptability is critical in addressing challenges like network congestion, resource allocation, and fault recovery, enabling networks to maintain high performance and reliability under varying conditions.

Together, these capabilities reduce the lead time for network operations and optimization when compared to the use of other types of AI models.

Data Generation Capabilities: GenAI models, such as generative adversarial networks (GANs), variational autoencoders (VAEs), and diffusion models can learn the probability value distribution of input features [11]. This enables them to generate not only realistic data but also diverse variations of training data that may be rare in real-world scenarios, costly to produce manually, or impossible to obtain due to risk factors and network constraints [12].

Consider a scenario of dynamic routing optimization, where a model must determine how to route traffic across multiple network hops. The decision may be influenced by factors such as congestion on different network paths, but also the traffic type, which can itself be influenced by network policies. These policies, which define various QoS parameters like relative priority among traffic flows, guaranteed throughput, packet drop rate ceiling, and maximum latency, are known as Policy and Charging Control (PCC) rules in a 5G network and are configured in the Policy Control Function (PCF). A non-GenAI model must be trained on data that captures all possible permutations of PCC rules based on current or predicted congestion in different parts of the network. Not only does this dataset need to represent all potential data variations that reflect real-world conditions, but it also carries a risk of introducing bias for a specific condition over others. Producing such a dataset manually for supervised learning is labor intensive as data needs to mitigate bias and exhibit variance that is possible to encounter in the real world. Similarly, using an approach such as reinforcement learning (RL) or online learning risks waiting for all different conditions to happen in the real network so that the training algorithm can observe them and train a model.

On the other hand, a GenAI model can capture this knowledge as it can learn the underlying probability value distribution and generate permutations of data on its own, augmenting real network observations. They can also help reduce noise in real-world collected data by learning an underlying distribution, as seen in models like diffusion models.

In another scenario, when a network link fails and a secondary link has yet to be identified, a generative model can produce synthetic data to temporarily substitute for actual transmission data. This approach can be particularly beneficial for applications like extended reality (XR) [13].

Network Layer Security: Encoder-decoder type of models such as VAEs have been used in the past for anomaly detection in the image domain [14], particle physics [15], and health [16]. Such models capture data distribution and, given an input, can highlight deviations from normal patterns in a resource-efficient, low-latency manner. Similarly, diffusion models, which model data distribution through iterative processes, have shown promise in anomaly detection [17] by generating synthetic data to simulate rare events and identifying deviations more robustly. In the context of the network layer, the GenAI solution may focus on identifying unusual patterns or behaviors in transport layer protocols, such as Transmission Control Protocol (TCP) or User Datagram Protocol (UDP).

6.3.3 Short-term Applications (GenAI for Network Layer)

6.3.3.1 Authentication and Access Control

The authentication and access control (AAC) process determines whether new traffic flows can be admitted into the networks based on the current network condition, QoS and quality of experience (QoE) requirements. One advantage of using the machine learning (ML) approaches is predicting traffic spikes, whereas the traditional AAC mechanism cannot adapt to dynamic changes of resource requirements during unpredicted traffic spikes. On the other hand, GenAI models can also learn the network condition distribution and better predict the traffic spikes, thanks to their strengths in learning the contextual information using architectures such as transformers and attention mechanisms [18].

However, non-GenAI models may not be able to consider external context, which significantly impacts QoE[1] such as for example the profile of the UE (e.g., hardware capabilities, installed software), the profile of the applications being used (e.g., video streaming, web browsing, etc.), and the profile of the user (in terms of preferences, demographics, etc.). All this information has a significant impact on QoE.

6.3.3.2 Traffic Management for Multiple RATs

GenAI can improve decision-making processes for complex network coordination tasks, such as choosing which information to route over which RAT in multiconnectivity deployments. Unlike traditional models, which typically rely on preprovisioned rules, GenAI-based solutions adapt in real-time to changing and unseen network conditions by generating new strategies that balance load across different networks.

For example, in NTN scenarios where satellite links are involved, generative models can simulate various environmental factors, predict channel quality, and recommend optimal configurations for traffic steering between terrestrial and satellite networks. This flexibility is especially important for supporting the diverse use cases anticipated for 5G and beyond, where the need for low latency, high throughput, and reliable connections across a mix of network types is paramount.

In another example, GenAI models may be used to support access traffic steering, switching, and splitting (ATSSS), a functionality that allows devices to use multiple access networks (e.g., cellular and WiFi) simultaneously or interchangeably for enhanced performance, reliability, and user experience [19]. In order to decide how to manage traffic flows, ATSSS relies on policies defined as PCC rules at PCF. SMF in the core network is responsible for interpreting these rules and

1 QoE (quality of experience) is subjective to the user of the UE and to the mobile device capabilities, and, as opposed to QoS is dependent upon the preferences of the individual user.

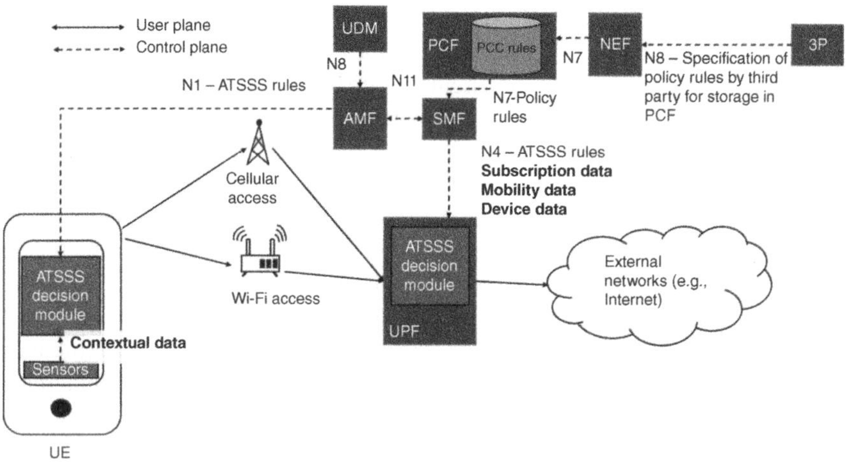

Figure 6.2 ATSSS implemented using GenAI decision modules that integrate context beyond policies (PCC rules).

instructing UPF and UE (via AMF), to split, steer, and switch traffic between different access networks based on these rules. In Figure 6.2, we show how it is possible to integrate context, in addition to PCC rules.

As the figure shows, two ATSSS decision modules, one in UE and one in UPF integrate additional context such as mobility information and device information (e.g., information from device sensors such as location, weather, velocity, etc.). Depending on the capabilities of a device, different contextual information may be provided to the ATSSS decision module, which would every time have to adapt its decisions to the capabilities of the device. As presented in Section 6.3.2, one of the advantages of transformer-based GenAI models is their flexibility. They would, therefore, make them ideal for such applications where dynamic adaptation to different device capabilities is required. In the illustrated example, PCC rules, which are translated to ATSSS rules at SMF, are either configured by the network operator, or can also be set by a 3P (third-party), via the Network Exposure Function (NEF).

6.3.3.3 Network Slicing Configuration

GenAI models have the potential to significantly enhance network slicing in future wireless networks, especially for 5G and 6G. Network slicing involves creating multiple virtual networks on top of a shared physical infrastructure, allowing customized services with performance guarantees. GenAI can optimize this process by predicting traffic patterns, dynamically adjusting network resources, and even designing network slices tailored to specific application needs, such as low-latency for gaming or high-throughput for video streaming [20]. By utilizing data from various network components, GenAI can anticipate shifts in service demands and automatically allocate resources to maintain optimal performance without manual intervention.

Moreover, GenAI can enhance the flexibility and responsiveness of network slicing by simulating different network conditions and predicting how slices might need to be adjusted over time. This allows for proactive management, including real-time optimization, slice creation, and lifecycle management. In 6G, where network configurations become more complex, generative models can also assist in handling diverse use cases, from immersive communications to machine-type communications, ensuring each slice meets the specific requirements of emerging technologies.

6.3.3.4 Mobility Management

GenAI models can significantly improve mobility management in 5G networks, particularly at the network layer, by enhancing predictive capabilities and optimizing network resources. These models can predict user equipment UE trajectories, anticipate mobility patterns, and dynamically adjust resource allocation for optimal handovers and network congestion management. By simulating various mobility scenarios, GenAI can fine-tune parameters such as handover thresholds, beamforming strategies, and resource block allocation in real-time to accommodate shifting user locations and varying network loads. This level of foresight and adaptability surpasses traditional systems, which often rely on predefined thresholds and static predictions.

Moreover, GenAI models can optimize the placement and management of cells in dense urban environments or high-demand areas, ensuring efficient load balancing. The AI can simulate potential traffic spikes and adapt the network's topology and handover mechanisms before congestion occurs, reducing latency and improving overall network performance. This predictive capability is especially crucial for managing high-density traffic during peak times or large events. With these advances, GenAI can enable a more efficient and proactive mobility management strategy, improving the user experience and network efficiency in 5G environments.

6.3.3.5 Adaptive Routing

GenAI can significantly improve adaptive routing in RAN by optimizing resource allocation and traffic flow. This could be applicable in different areas, for example, in integrated access and backhaul (IAB) and distributed-MIMO (D-MIMO).

Traditional methods of routing rely on pre-defined algorithms and static optimization, which often fail to account for the highly variable and unpredictable conditions in mobile networks. With ML and mostly with GenAI, the routing algorithms can be trained to adapt in real-time and network dynamics, considering a variety of network parameters, including traffic load, device mobility, and channel conditions. For instance, by simulating a wide range of potential scenarios, GenAI can predict and adapt to congestion or signal interference more effectively than traditional routing methods. This is particularly useful for D-MIMO, where GenAI can optimize beamforming strategies based on real-time conditions, improving spectral efficiency and system capacity.

Furthermore, GenAI can help optimize routing decisions between devices, enhancing the overall user experience by ensuring efficient communication paths based on chosen device locations, traffic demands, and network states. Generative models can also simulate and recommend optimal configurations for different network layers and devices, improving coordination between base stations, access points, and end-user devices.

One example of application of GenAI models in adaptive routing in RAN is to be used as a decision criterion for choosing routing paths in IAB. Figure 6.3 shows an example of an IAB topology.

In IAB, wireless nodes that have no backhaul connectivity to the core network (IAB-nodes) can use part of the radio access spectrum, to connect to a node that has backhaul access, known as IAB-donor. The idea is that IAB is an inexpensive way to expand the mobile network, as a backhaul connection to the core network is expensive relative to the cost of the access nodes. The IAB-donor node is responsible for configuring routing for data packets between UE and itself with one or more IAB-nodes as intermediate hops. In order to make this decision, IAB-donor node receives network load related information from all IAB-Nodes. As was the

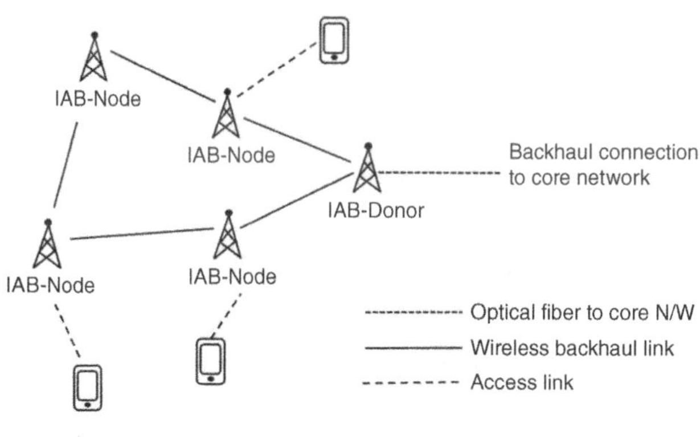

Figure 6.3 An example of an IAB topology in a 5G network.

case with previous examples, the IAB-donor node can benefit from GenAI, models which would be able to integrate more contextual sources in the routing decisions, but also adapt to a changing topology, as the number of IAB-nodes can change over time.

6.3.3.6 Synthetic Data Packet Forwarding

Generative AI models, such as GANs, can significantly enhance synthetic data packet forwarding in networking by compensating for gaps in real-time data. When real network data is unavailable or delayed at a router, GANs can generate synthetic data that closely mimics actual traffic conditions. This synthetic data helps maintain continuous network operation, allowing the router to process and forward packets as if real data were available. For example, GANs can simulate network traffic spikes, security breaches, or bandwidth constraints, enabling more accurate decision-making when real-time data is missing. This approach is particularly valuable in complex scenarios where traditional methods struggle to predict or simulate network behaviors, ensuring robust packet forwarding and reducing latency in critical applications.

6.3.3.7 Anomaly Detection and Network Security

In the context of network security and anomaly detection, GenAI models, particularly VAEs, offer a promising approach. The VAEs are powerful for identifying outliers or anomalies in network data, which could signal security threats such as intrusions or malfunctioning devices. The VAE architecture allows for the efficient learning of complex data patterns and can flag anomalies by comparing reconstructed inputs to original data. If the reconstruction error exceeds a predefined threshold, the data point is considered anomalous.

In practice, VAEs are trained to encode network traffic or system behavior into a latent space, where normal patterns are learned. Any deviation from these learned patterns, such as an unusual network access request or unexpected packet flow, can be identified as an anomaly. Custom thresholds can be set based on the reconstruction loss to fine-tune the model's sensitivity to anomalies, thus balancing false positives and negatives. This ability to detect subtle deviations is particularly useful for real-time security monitoring and proactive network protection.

6.3.4 Long-term Applications (Network Layer for GenAI)

GenAI models can play a transformative role in the long-term development of mobile networks, especially in the design and optimization of network protocols. By leveraging advanced agents and multiagent systems, these models can help tailor network protocols to meet diverse use-case requirements such as high-speed data transfer, low power consumption, and enhanced security. These AI-driven agents can collaborate and share knowledge, adapting protocols dynamically based on real-time network conditions and specific user needs. For instance, protocols could be designed to prioritize certain traffic flows or adjust based on the current user environment. Furthermore, generative agents can take a more proactive role by anticipating future network demands, learning the network dynamics, and optimizing protocol performance for both current and future use cases. This approach is exemplified by multiagent architectures such as LangGraph or TaskWeaver, where different agents specialize in handling specific tasks, allowing for the efficient allocation of resources and ensuring network performance is always optimized.

Such AI-driven frameworks could facilitate smarter, more autonomous network operation. For example, agents could dynamically adjust protocols to handle new technologies like 5G/6G or hybrid networks involving both terrestrial and NTNs. This level of flexibility, adaptation, and self-optimization, enabled by GenAI, is crucial for maintaining high-quality service in increasingly complex, heterogeneous mobile networks.

6.4 Challenges and Opportunities for GenAI in the Network Layer

This section outlines key research challenges and opportunities for applying GenAI in the network layer, with a forward-looking perspective on potential advancements beyond 2030.

6.4.1 Challenges

Here, we outline several challenges in using GenAI models at the network layer:

- **Hallucinations:** GenAI models can produce outputs that are factually incorrect or nonsensical, despite appearing plausible [21]. These errors arise from reliance on training patterns without fact verification, posing risks to reliability. Addressing this requires fallback mechanisms to ensure data accuracy and maintain network integrity.
- **Scalability:** Deploying GenAI, particularly LLMs, in the network layer faces several constraints:
  - **Resource Demands:** High storage, computational, and transport requirements limit widespread adoption.
  - **Data Privacy Compliance:** Regulations like the General Data Protection Regulation (GDPR) restrict centralized training on sensitive network data.
  - **Lifecycle Management:** Efficient machine learning operations (MLOps) frameworks are needed to manage updates, monitoring, and maintenance at scale.

- **Inference Time:** Often inference latency required for large models [22] exceeds the maximum latency allowed by the system. In the RAN network, the latency is often less than 1 ms in places like MAC, PDCP, and RRC, while the latency required in the core network like in UPF for packet priority changes may be close to one second. Large models often tend to have an inference exceeding 100 ms, and hence there is a challenge incorporating GenAI models for problems requiring low-latency inference.
- **Robustness:** It refers to the sensitivity of LLMs to small changes in input, which can lead to disproportionately large and unexpected changes in their outputs [23]. The robustness issue arises because LLMs rely heavily on statistical patterns in training data rather than a true understanding of the context of the input or the semantics of the input. The main issue in the lack of robustness is due to training data bias: the vast training data often contains inconsistencies, which the model learns and replicates.

6.4.2 Research Opportunities

6.4.2.1 Model Verification and Validation

Ensuring GenAI reliability and safety requires robust verification and validation frameworks. Verification confirms model performance within predefined parameters (e.g., via controlled tests), while validation assesses alignment with real-world expectations, leveraging human-in-the-loop assessments, benchmarking, and pilot testing. Incorporating fallback mechanisms addresses potential errors and biases, mitigating risks such as hallucinations and ethical concerns.

6.4.2.2 Distributed Training

Distributed training across multiple nodes accelerates convergence, improves scalability, and preserves data privacy by avoiding centralized processing. This approach also supports compliance with data protection laws, making it essential for secure GenAI development in networked environments.

6.4.2.3 Model Optimization

Techniques like distillation, pruning, and quantization can reduce GenAI resource demands while preserving performance [24]. Deploying optimized models in distributed systems may involve tailored strategies based on specific use cases and hardware constraints, enabling efficient and scalable GenAI applications in the network layer.

6.4.2.4 Inference Time Efficiency

Inference latency may be reduced by identifying attention heads that provide contextual sparsity [25] or efficient management of key-value (KV) cache [26] by dynamically adapting to resource usage variances between phases of requests.

However, further research is required before the GenAI models can fit to the stringent requirements of some of the network domains.

6.5 Summary

This chapter explores how GenAI can improve the network layer of mobile networks. The network layer spans across the entirety of the mobile network, including the RAN, backhaul, and core network, and is responsible for managing the transmission of data packets across the network. From a mobile network perspective, data packets are part of either control plane traffic, which is responsible for network operations, or user plane traffic, which carries user data.

We describe several potential applications of GenAI in the network layer, including adaptive routing, mobility management, and network security. These applications leverage GenAI's ability to learn complex patterns, generate new data, and detect anomalies, offering significant advantages over traditional methods. For example, GenAI can be used to predict traffic congestion and optimize resource allocation in real-time. It can also be used to detect and mitigate security threats.

We also note that there are limitations to be addressed. One of the key challenges is the risk of GenAI models producing factually incorrect or nonsensical outputs, known as "hallucinations." To mitigate this, we suggest implementing fallback mechanisms and focusing on model verification and validation. The authors also highlight the need for efficient distributed training strategies to manage the computational demands of GenAI and to protect data privacy.

Overall, the chapter provides a comprehensive overview of the potential of GenAI to transform the network layer in mobile networks. While there are still some challenges to be addressed, we are optimistic that GenAI will play a key role in the development of more efficient, resilient, and intelligent mobile networks.

References

1 3GPP. 3rd Generation Partnership Project; Technical Specification Group Services and System Aspects; System architecture for the 5G System (5GS); Stage 2 (Release 19). Technical Specification (TS) 23.501, 3rd Generation Partnership Project (3GPP), 2025. URL https://www.3gpp.org/ftp/Specs/archive/23_series/23.501/23501-j40.zip. Version 19.4.0.

2 3GPP. 3rd Generation Partnership Project; Technical Specification Group Radio Access Network; NR; NR and NG-RAN Overall Description; Stage 2 (Release 18). Technical Specification (TS) 38.300, 3rd Generation Partnership Project (3GPP), 2025. URL https://www.3gpp.org/ftp/Specs/archive/23_series/23.501/23501-j40.zip. Version 18.6.0.

3 3GPP. 3rd Generation Partnership Project; Technical Specification Group Radio Access Network; NR; Radio Link Control (RLC) protocol specification; Stage 2 (Release 18). Technical Specification (TS) 38.322, 3rd Generation Partnership Project (3GPP), 2025. URL https://www.3gpp.org/ftp/Specs/archive/38_series/38.322/. Version 18.6.0.

4 3GPP. 3rd Generation Partnership Project; Technical Specification Group Radio Access Network; NR; Packet Data Convergence Protocol (PDCP) specification; Stage 2 (Release 18). Technical Specification (TS) 38.323, 3rd Generation Partnership Project (3GPP), 2025.URL https://www.3gpp.org/ftp/Specs/archive/38_series/38.323/. Version 18.6.0.

5 3GPP. 3rd Generation Partnership Project; Technical Specification Group Radio Access Network; NR; Radio Resource Control (RRC) protocol specification; Stage 3 (Release 18). Technical Specification (TS) 38.331, 3rd Generation Partnership Project (3GPP), 2025.URL https://www.3gpp.org/ftp/Specs/archive/38_series/38.331/. Version 18.6.0.

6 Athanasios Karapantelakis, Pegah Alizadeh, Abdulrahman Alabbasi, Kaushik Dey, and Alexandros Nikou. Generative AI in mobile networks: A survey. *Annals of Telecommunications*, 79(1): 15–33, 2024.

7 Tom Brown, Benjamin Mann, Nick Ryder, Melanie Subbiah, Jared D Kaplan, Prafulla Dhariwal, Arvind Neelakantan, Pranav Shyam, Girish Sastry, Amanda Askell, Sandhini Agarwal, Ariel Herbert-Voss, Gretchen Krueger, Tom Henighan, Rewon Child, Aditya Ramesh, Daniel Ziegler, Jeffrey Wu, Clemens Winter, Chris Hesse, Mark Chen, Eric Sigler, Mateusz Litwin, Scott Gray, Benjamin Chess, Jack Clark, Christopher Berner, Sam McCandlish, Alec Radford, Ilya Sutskever, and Dario Amodei. Language models are few-shot learners. In H. Larochelle, M. Ranzato, R. Hadsell, M.F. Balcan, and H. Lin, editors, *Advances in Neural Information Processing Systems*, volume 33, pages 1877–1901. Curran Associates, Inc., 2020. URL https://proceedings.neurips.cc/paper_files/paper/2020/file/1457c0d6bfcb4967418bfb8ac142f64a-Paper.pdf.

8 Jason Wei, Xuezhi Wang, Dale Schuurmans, Maarten Bosma, Brian Ichter, Fei Xia, Ed H. Chi, Quoc V. Le, and Denny Zhou. Chain-of-thought prompting elicits reasoning in large language models. In *Proceedings of the 36th International Conference on Neural Information Processing Systems*, NIPS '22, Red Hook, NY, USA, 2024. Curran Associates Inc. ISBN 9781713871088.

9 Patrick Lewis, Ethan Perez, Aleksandra Piktus, Fabio Petroni, Vladimir Karpukhin, Naman Goyal, Heinrich Küttler, Mike Lewis, Wen-tau Yih, Tim Rocktäschel, Sebastian Riedel, and Douwe Kiela. Retrieval-augmented generation for knowledge-intensive NLP. tasks. In *Proceedings of the 34th International Conference on Neural Information Processing Systems*, NIPS '20, Red Hook, NY, USA, 2020. Curran Associates Inc. ISBN 9781713829546.

10 Gordon Owusu Boateng, Hani Sami, Ahmed Alagha, Hanae Elmekki, Ahmad Hammoud, Rabeb Mizouni, Azzam Mourad, Hadi Otrok, Jamal Bentahar, Sami Muhaidat, Chamseddine Talhi, Zbigniew Dziong, and Mohsen Guizani. A survey on large language models for communication, network, and service management: Application insights, challenges, and future directions, 2024. URL https://doi.org/10.48550/arXiv.2412.19823. arXiv preprint arXiv:2412.19823.

11 Alvaro Figueira and Bruno Vaz. Survey on synthetic data generation, evaluation methods and GAN. *Mathematics*, 10(15): 2733, 2022. doi: 10.3390/math10152733. URL https://www.mdpi.com/2227-7390/10/15/2733.

12 Shoucai Jing, Yuyu Zhao, Xiangmo Zhao, Fei Hui, and Asad J. Khattak. An efficient high-risk lane-changing scenario edge cases generation method for autonomous vehicle safety testing. *IEEE Transactions on Intelligent Vehicles*, 10(1): 459–471, 2025. doi: 10.1109/TIV.2024.3414828.

13 Christina Chaccour, Athanasios Karapantelakis, Timothy Murphy, and Mischa Dohler. Telecom's artificial general intelligence (AGI) vision: Beyond the GenAI frontier. *IEEE Network*, 38(5): 21–28, 2024. doi: 10.1109/MNET.2024.3425594.

14 Huy Hoang Nguyen, Cuong Nhat Nguyen, Xuan Tung Dao, Quoc Trung Duong, Dzung Pham Thi Kim, and Minh-Tan Pham. Variational autoencoder for anomaly detection: A comparative study, 2024. URL https://arxiv.org/abs/2408.13561.

15 Adrian Alan Pol, Victor Berger, Cecile Germain, Gianluca Cerminara, and Maurizio Pierini. Anomaly detection with conditional variational autoencoders. In *2019 18th IEEE International Conference on Machine Learning and Applications (ICMLA)*, pages 1651–1657, 2019. doi: 10.1109/ICMLA.2019.00270.

16 Siwei Guan, Zhiwei He, Shenhui Ma, and Mingyu Gao. Multivariate time series anomaly detection with variational autoencoder and spatial–temporal graph network. *Computers & Security*, 142: 103877, 2024. ISSN 0167-4048. doi: https://doi.org/10.1016/j.cose.2024.103877. URL https://www.sciencedirect.com/science/article/pii/S0167404824001780.

17 Jing Liu, Zhenchao Ma, Zepu Wang, Yang Liu, Zehua Wang, Peng Sun, Liang Song, Bo Hu, Azzedine Boukerche, and Victor Leung. A survey on diffusion models for anomaly detection. *arXiv preprint arXiv:2501.11430*, 2025.

18 Jianxiang Zhou, Erdong Liu, Wei Chen, Siru Zhong, and Yuxuan Liang. Navigating spatio-temporal heterogeneity: A graph transformer approach for traffic forecasting. *arXiv preprint arXiv:2408.10822*, Aug 2024.https://arxiv.org/abs/2408.10822.

19 3GPP. Study on Access Traffic Steering, Switching and Splitting (ATSSS) support in the 5G System Architecture. Technical Report TR 23.793, 3rd Generation Partnership Project (3GPP), 2018.

20 Mourice O Ojijo and Olabisi E Falowo. A survey on slice admission control strategies and optimization schemes in 5G network. *IEEE Access*, 8: 14977–14990, 2020.

21 Patrice Béchard and Orlando Marquez Ayala. Reducing hallucination in structured outputs via retrieval-augmented generation. *arXiv preprint arXiv:2404.08189*, 2024.

22 Zixuan Zhou, Xuefei Ning, Ke Hong, Tianyu Fu, Jiaming Xu, Shiyao Li, Yuming Lou, Luning Wang, Zhihang Yuan, Xiuhong Li, Shengen Yan, Guohao Dai, Xiao-Ping Zhang, Yuhan Dong, Yu Wang. A survey on efficient inference for large language models. *arXiv preprint arXiv:2404.14294*, 2024.

23 Boxin Wang, Weixin Chen, Hengzhi Pei, Chulin Xie, Mintong Kang, Chenhui Zhang, Chejian Xu, Zidi Xiong, Ritik Dutta, Rylan Schaeffer, Sang T. Truong, Simran Arora, Mantas Mazeika, Dan Hendrycks, Zinan Lin, Yu Cheng, Sanmi Koyejo, Dawn Song, and Bo Li. Decodingtrust: A comprehensive assessment of trustworthiness in GPT models, 2023. URL https://doi.org/10.48550/arXiv.2306.11698. arXiv preprint arXiv:2306.11698.

24 Babak Rokh, Ali Azarpeyvand, and Alireza Khanteymoori. A comprehensive survey on model quantization for deep neural networks in image classification. *ACM Transactions on Intelligent Systems and Technology*, 14(6): 1–50, 2023.

25 Zichang Liu, Jue Wang, Tri Dao, Tianyi Zhou, Binhang Yuan, Zhao Song, Anshumali Shrivastava, Ce Zhang, Yuandong Tian, Christopher Re, Beidi Chen. Deja vu: Contextual sparsity for efficient LLMs at inference time. In *International Conference on Machine Learning*, pages 22137–22176. PMLR, 2023.

26 Bingyang Wu, Shengyu Liu, Yinmin Zhong, Peng Sun, Xuanzhe Liu, and Xin Jin. Loongserve: Efficiently serving long-context large language models with elastic sequence parallelism, 2024. URL https://arxiv.org/abs/2404.09526.

7

Generative AI at Application Layer: Mobile AI-generated Content

Paria Mohammadzadeh Hesar, Amirhossein Mohammadi, and Hina Tabassum*

Lassonde School of Engineering, York University, Toronto, ON, Canada

*Corresponding author: pariamdz@yorku.ca

7.1 Introduction to AIGC

7.1.1 General Overview

AI-generated content (AIGC) has reshaped content development by using generative models to produce personalized, high-quality text, images, audio, video, augmented training data, and interactive 3D media [1]. Within the broader Web3.0/Metaverse ecosystem, it now stands alongside professionally generated content (PGC) and user-generated content (UGC) as a scalable and cost-effective modality for creating diverse media [2]. However, the deployment assumptions that underpin cloud-centric AIGC—abundant centralized compute and stable, high-bandwidth backhaul—often fail in mobile and edge settings [3]. Interactive mobile experiences such as live translation, AR overlays, and camera assistants impose stringent latency and jitter budgets; client devices are heterogeneous and thermally constrained; uplinks for multimodal streams can be intermittent or costly; and privacy, compliance, and localization requirements frequently prohibit raw context (camera, microphone, location) from leaving the device or jurisdiction. These factors motivate mobile/edge-native AIGC that places inference and adaptation closer to users, leveraging multiaccess edge computing to reduce round-trip delay and bandwidth while enabling policy-compliant personalization.

At the model level, capabilities have increased significantly with large language models such as GPT-3 and GPT-4, which excel in few-shot and zero-shot performance across domains [4]. However, simply expanding data and parameters does not resolve the constraints of mobile and edge settings. GPT-3 was trained using hundreds of billions of tokens, and GPT-4 improves accuracy even more [5]. However, deploying generation closer to users necessitates a different approach: compact model variants, careful compression and task- or user-specific adaptation, more frugal generation strategies, and privacy-preserving learning that keeps sensitive context on the device. Combined with execution plans that divide work across the device, nearby edge servers, and the cloud, these measures help systems meet practical targets for responsiveness, battery and compute budgets, and data costs while maintaining quality—goals that are difficult to achieve with a purely cloud-based approach [6, 7].

7.1.2 AIGC in the Application Layer

AIGC is transforming content creation across text, visual media, audio, and 3D environments—enhancing productivity and engagement across diverse sectors. Importantly, these advancements are driving innovative application-layer solutions that deliver real-time, context-aware, and highly personalized user experiences.

- **Text Generation:** Recent advancements in Natural Language Processing (NLP) enable producing AI-generated text that is nearly indistinguishable from human writing, thus driving widespread adoption—especially in mobile networks [8]. Models such as OpenAI's ChatGPT [9] and Google's Bard [10] leverage large language models and advanced NLP techniques to simulate human communication, thereby enhancing tasks across engineering, programming, marketing, and specialized search applications. Their integration into platforms such as New Bing and Microsoft's Office 365 Copilot exemplifies how these technologies deliver personalized, summarized responses and automate document generation, data analysis, and task management, respectively. Furthermore reinforcing these innovations are lightweight models such as ALBERT and MobileBERT [11, 12], which are optimized for low-resource and mobile environments. By reducing computational and memory demands, these models enable real-time, privacy-preserving applications on smartphones, IoT devices, and edge networks. This ensures low-latency performance for conversational assistants, chatbots, and context-aware systems, expanding access to high-quality, personalized information. Together, these developments underscore the critical role of AIGC at the application layer, where the integration of advanced text generation capabilities is reshaping user engagement and productivity across diverse platforms.

- **Audio Generation:** AI-generated audio is increasingly vital in mobile networks, enhancing user experience through improved efficiency, security, and personalization at the application layer [13]. For instance, AIGC-driven speech synthesis boosts call quality, while speech recognition and compression lower data transmission overheads and automate tasks like transcription. Voice biometrics further enhance security by using unique voiceprints. A notable example is Audio Albert [14], a streamlined adaptation of BERT for audio tasks. It achieves competitive performance in speech recognition, speaker identification, and music genre classification, while reducing inference time by about 20%. Its lightweight design makes it ideal for resource-constrained edge devices, delivering faster responses, lower latency, and reduced power consumption.
- **Image Generation:** AI-generated visual media is transforming mobile applications by enhancing image quality, efficiency, and creative workflows at the application layer. These advancements enable critical tasks such as image enhancement in low-light conditions, efficient compression for reduced data usage, and robust image recognition for object and facial detection. Moreover, text-to-image generation—exemplified by models like Make-a-Scene and techniques like SPADE [15]—supports dynamic visual storytelling, inpainting, and style transfer, while innovations in video generation, as seen in OpenAI's DALL-E2/DALL-E3 [16], use diffusion-based models and cross-modal techniques to produce high-resolution, coherent visuals from textual inputs. Additionally, Adobe Sensei [17] automates complex editing tasks, further streamlining creative processes for both amateurs and professionals, although the rise of deep fake technology poses ethical and legal challenges that warrant careful consideration.
- **Video Generation:** AI-generated videos enhance mobile applications by offering immersive, dynamic content that goes beyond static images. They support functions like video compression, enhancement, summarization, and synthesis, while allowing for tailored adjustments in style, resolution, and frame rate to suit specific purposes such as advertising, education, or entertainment. For example, Imagen Video—a text-conditioned system—uses a cascade of video diffusion models and super-resolution techniques to produce high-definition videos with exceptional controllability and diverse artistic styles [18]. This approach not only creates engaging visual experiences but also opens new avenues for storytelling by synthesizing new content from existing videos, images, text, or audio.
- **3D Content Creation:** AI-generated 3D content is mainly used in AR/VR, and the metaverse, by automating the creation of intricate, realistic models from textual inputs and eliminating manual design efforts. Techniques in this field are capable of generating detailed 3D shapes and textures from

2D images, enabling applications such as object reconstruction and scene understanding. Complementary approaches such as diffusion models and their variants further enhance 3D shape synthesis, completion, and offering fine-grained control over diverse shapes, and generating high-quality models directly from natural language descriptions. These innovations not only drive dynamic virtual environments, avatars, and immersive experiences in the metaverse, but also benefit industries such as e-commerce through improved product visualization and education via interactive simulations. Moreover, by optimizing performance on resource-limited devices like AR/VR headsets, AIGC for 3D content enables real-time prototyping, immersive storytelling, and accessible asset production, ultimately reducing costs and accelerating development across diverse sectors.

7.1.3 AIGC Product Lifecycle

The generation of AIGC involves several key stages, including content creation, distribution, and exchange. Initially, producers collaborate with edge service providers (ESPs) to assist in content creation, incurring associated expenses. Once the content is created, creators retain full ownership and can choose to publish it on dedicated AIGC platforms or social media.

The AIGC product lifecycle consists of several critical stages, including data collection, pretraining, fine-tuning, inference, and product management, each contributing to AIGC's overall effectiveness and precision (Figure 7.1).

- **Data Collection:** AIGC quality and diversity are determined by the crucial step of data gathering. AI models learn patterns and correlations influenced by the data they are trained with. For instance, online platforms like Amazon Mechanical Turk and Datatang offer specialized services and high-quality datasets to improve model performance [19]. Moreover, edge devices like mobile phones and smart city sensors collect environmental data, such as GPS records and wireless sensing data [20], to develop individualized AI models. These data are then used to train GenAI models, ensuring the accuracy and diversity of AI-generated material [21].
- **Pretraining:** Central servers with robust processing capabilities are used for training after data collection. Training involves understanding the fundamental pattern and structure of the collected data to generate the desired response. In this regard, selecting a GAI model depends on the task, resources, desired outcome, and computational hardware. Therefore, significant computational resources are required for pretraining GenAI models. In one case, Stability AI pretrained the Stable Diffusion model on 256 Nvidia A100 GPUs for 150,000 hours, costing $600,000. After training, the models can be fine-tuned and deployed to network edge servers to process user requests locally [2].

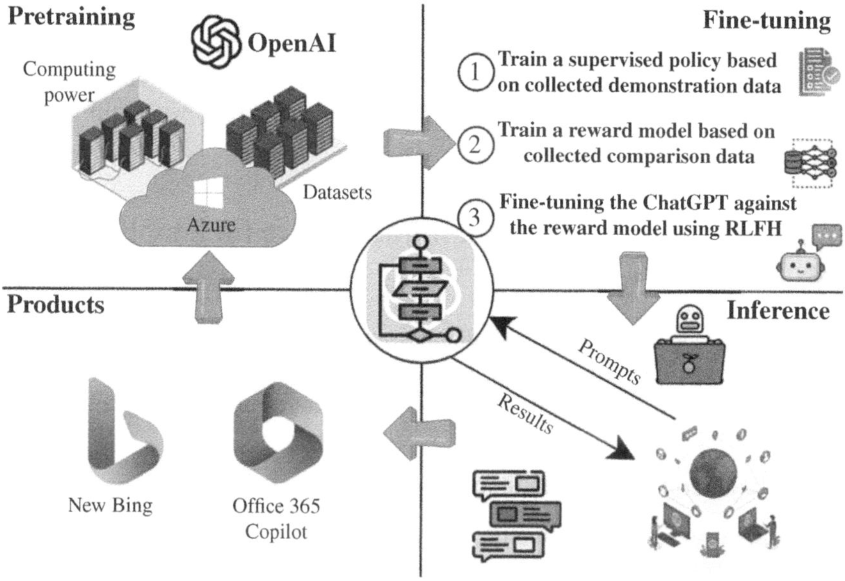

Figure 7.1 The lifecycle of AIGC [2]/IEEE.

This process and the infrastructure required to support such operations will be discussed further in the next section on infrastructure for GenAI.

- **Fine-tuning:** Fine-tuning involves retraining a machine learning model on a particular dataset to preserve its fundamental capabilities while adapting it for specific applications. It entails incorporating supplementary data to adjust a pretrained GenAI model for new tasks or domains. This approach improves the model's efficacy by calibrating its parameters to new data. Edge networks in mobile systems can execute accurate modifications by employing the limited datasets mobile users provide.
- **Inference:** Inference involves generating the intended output by employing the acquired GenAI model. Inference in wireless networks involves using GenAI models to analyze observable data to make conclusions about network states or behaviors. This can include optimization and graphical models to estimate parameters, deduce topologies, or improve performance. Historically, GenAI models were managed via centralized servers, but a growing trend is to implement AIGC services on edge networks for improved computational capabilities and user proximity.
- **Product Management:** Product management in AIGC involves overseeing the entire lifecycle of AI-powered graphic production technologies, understanding customer needs, and designing products to address specific user

tasks. AIGC products are essential due to their unique ownership, copyright, and value. Edge networks enable producers like organizations or individual mobile users to distribute content to AIGC platforms or social media. Additionally, AIGC products are tradable, facilitating fund ownership transfer between parties.

7.2 Collaborative Network Infrastructure for Enabling GenAI Services

We first introduce the critical challenges and key components of the network infrastructure, which is necessary to enable GenAI services for mobile networks. We then discuss the significance of collaborative cloud-edge infrastructure.

7.2.1 Enabling AIGC – Challenges

As the demand for AIGC in mobile networks increases, new challenges also arise. The emergence of new technologies utilizing this service, such as Virtual Reality (VR), Augmented Reality (AR), and Mixed Reality (MR), which have low latency requirements, underscores the need for a robust infrastructure to generate real-time responses. The challenges of deploying current GenAI services in mobile network can be summarized as follows:

7.2.1.1 Resource Scalability

Currently, the employment of generative models is primarily cloud based. Cloud-based solutions leverage the substantial computational power available in cloud services to handle the intensive computations required for training and inference of generative models. Although these solutions are scalable to some extent, they cannot satisfy the concurrent requests of a massive number of users in terms of both bandwidth and computation. Even with significant financial investment to massively scale the infrastructure, the physical distance of cloud services from users does not meet the low latency requirements of some applications like the Metaverse.

Therefore, researchers are turning their attention towards edge-based solutions. Two alternative options are mobile edge computing (MEC) and edge cloud computing. MEC has emerged as a technology that extends cloud computing capabilities to the edge of the network, closer to the users. This approach is beneficial in terms of latency, energy consumption, and privacy, as it leverages the proximity of edge servers to data sources, facilitating more efficient and scalable training [22, 23]. Despite its advantages in latency and scalability, several challenges arise within this framework. For instance, edge servers are resource-constrained devices, limiting their computational capacity significantly. Additionally, resource management, considering communication overhead and latency, poses a substantial challenge [24, 25].

Another alternative is utilizing both edge and cloud cooperatively to handle users' requests. Combining resources from edge and cloud improves user quality of service by offloading less critical and computationally intensive tasks to the cloud, while handling sensitive tasks requiring low latency at the edge, closer to users [26–28]. The integration of edge and cloud resources is especially useful in mobile AIGC frameworks, where pretraining of generative models is computationally expensive and not feasible on edge devices alone, while inference can be done on edge servers [2, 29]. This integration provides satisfactory QoS for GenAI-related applications such as Metaverse. One solution to integrate AIGC models within mobile networks is the adaptation of AIGC as a service [30], which we will discuss in more detail later. However, there is also a significant challenge in orchestrating each network component to find the lowest latency to respond to a request.

7.2.1.2 End-to-end Latency

It is noteworthy when we talk about latency, it is essential to adopt a broader view beyond just network latency. Network latency is the delay in moving packets over a network, but compute latency—the time from when a server receives a request to when it responds—is equally critical. GenAI infrastructure requires optimizing for lower latency [31] in all forms to ensure exceptional user experience.

Different AIGC applications have varying sensitivity to these latency components. For instance, GenAI applications like image or text generation are more impacted by computational latency, while real-time applications like the metaverse or AVs require minimizing both computational and network latency to operate effectively. Studies show that while 58% of end-users can reach a nearby edge server in less than 10 ms, only 29% can achieve similar latency with cloud servers [32]. This disparity highlights the importance of strategic deployment of edge servers to complement cloud services, ensuring optimal response times for AIGC applications.

The total latency in AIGC services must also account for potential bottlenecks in the network path, such as congestion in backhaul networks and processing delays at intermediate nodes. This becomes particularly challenging when trying to scale services to handle numerous concurrent requests while maintaining consistent performance levels.

7.2.1.3 Security and Privacy

Security and privacy are also concerns for users when using cloud servers for computational tasks. Users prefer to keep their data as close as possible to their devices and ideally process everything locally. However, this is not always feasible since some tasks cannot be handled with on-device computation. Due to the powerful capabilities of cloud servers, some computations must inevitably be performed there, as neither edge servers nor user devices may be able to handle such requests [9].

The edge-cloud architecture can further enhance privacy by implementing a hierarchical security model. Edge servers can act as privacy-preserving intermediaries, filtering sensitive information before data reaches the cloud. This allows for a balance between maintaining data privacy and leveraging the computational power of cloud infrastructure. However, this requires careful orchestration of security policies across different layers of the infrastructure and clear user consent mechanisms for data handling at each level [33, 34].

7.2.2 Infrastructure Components and Capabilities

To understand mobile networks, we need to familiarize ourselves with each component in the network and their ability to contribute to delivering the best quality of service to users. By examining the capabilities and limitations of cloud servers, edge computing resources, and user devices (Figure 7.2), we can develop effective strategies for deploying GenAI services in mobile networks. In this section, we will explore how these components can work together to tackle the challenges of resource scalability, end-to-end latency, and security and privacy, ultimately optimizing the delivery of high-quality GenAI services to users.

7.2.2.1 Cloud Infrastructure

Cloud servers are the primary locations for performing the bulk of computations. These servers offer scalable, on-demand access to vast amounts of computational power, which is crucial for training GenAI models. Training GenAI models, for instance, LLMs and diffusion models, requires considerable computational power, including GPUs and TPUs [35]. Cloud services offer access to these specialized hardware resources, enabling large-scale model training that requires extensive computational resources [36]. These servers can handle the significant memory and processing demands of such models, which involve billions of parameters and require optimized parallel processing capabilities [29].

For example, Meta's Llama 3 model includes versions with 8 and 70 billion parameters, and 405 billion parameters trained on 15.6 trillion tokens [37]. The most efficient implementation achieved compute utilization of over 430 TFLOPS per GPU, utilizing up to 16,000 H100 GPUs simultaneously. These efforts were supported by custom-built GPU clusters and an advanced training stack that ensured high efficiency and reliability [38].

The computational demands vary significantly based on the type of generated content. Text generation with LLMs like Llama 3 requires around 280GB VRAM for a 70 billion parameter model. Image generation models such as Stable Diffusion demand up to 400 GB VRAM for high-resolution outputs, while video generation models require even more resources, often necessitating multiple GPUs working in parallel. This is due to the inherently larger output dimensions of visual content - a single 1080p video frame requires about 6 MB of data, significantly more than traditional AI classification task outputs [2].

To handle these demands, cloud providers implement sophisticated request management systems. A typical cloud setup can process thousands of concurrent requests through request queuing and load distribution across multiple data centers. However, this comes at a significant operational cost - running inference on a 70 billion parameter model can cost between $0.5 and 2 per hour per GPU, depending on the provider and usage pattern. As the number of concurrent users increases, providers must balance computational capacity, cost efficiency, and service quality [39].

7.2.2.2 User Devices

With recent advancements in hardware capabilities, some GenAI applications can now be performed directly on user devices without the need to send requests to cloud or edge servers. The foremost benefit of on-device computation is enhanced privacy and security. Sensitive information remains on the device, minimizing the risk of data breaches and unauthorized access. By keeping data local, users gain greater control over their information, reducing the potential for data tracking and manipulation [40]. Additionally, on-device computation eliminates communication latency, as there is no need to transmit large amounts of data over the network. This can benefit the overall communication network by reducing congestion. However, computational latency on user devices is typically higher than on cloud servers. This trade-off can be managed based on the sensitivity of the task, determining whether it is more appropriate to process data locally or send it to the cloud for faster computation.

Compared to cloud servers with at least 20TB of memory, user devices typically have less than 64GB of memory. This limitation prevents inference with models containing billions of parameters. However, optimized versions of these models can perform simpler tasks on the device. Researchers use techniques like quantization and knowledge distillation to create smaller versions of these models. For example, Qualcomm AI Research has optimized the Stable Diffusion model for mobile devices using quantization and hardware acceleration. Using such methods they have enabled photorealistic image generation directly on smartphones in under 15 seconds, achieving performance comparable to cloud-based systems [36, 41–44].

On-device GenAI enables the creation of highly personalized AI experiences. User devices can leverage local data, such as usage patterns, preferences, and environmental context, to tailor AI responses and functionalities to individual users. This personalization extends to various applications, from personalized fitness coaching based on health data to customized virtual assistants that adapt to a user's communication style. The ability to provide personalized experiences enhances user satisfaction and engagement, making GenAI more relevant and useful in daily life [45].

Cloud computing, while capable of handling large-scale computations, incurs high operational costs and energy consumption. By offloading some of these tasks to user devices, the overall demand on cloud infrastructure is reduced, leading to cost savings for service providers. Additionally, user devices, optimized for energy-efficient performance, can execute GenAI tasks with lower power consumption compared to cloud servers. This not only reduces the environmental impact but also aligns with sustainability goals [46].

7.2.2.3 Edge Computing

Edge servers occupy a crucial middle ground in the network hierarchy, bridging the gap between user devices and cloud servers. While they may not match the computational power of cloud servers, they offer significant advantages in terms of latency, privacy, and personalization, making them a vital component in delivering high-quality GenAI services.

Edge servers are strategically positioned closer to end-users, often at the network edge or within local data centers. This proximity enables them to provide lower communication latency compared to cloud servers. For latency-sensitive applications such as AR, VR, and AVs, the reduced latency of edge servers is essential for real-time responsiveness and user experience. Edge servers can provide communication latency usually less than 10 ms compared to the 60 ms latency of cloud servers, making their usage essential in latency-sensitive applications like Metaverse scene generation [32]. Without the deployment of edge servers, the high latency associated with cloud servers would severely hinder the immersive and interactive experiences crucial for Metaverse applications.

From a computational perspective, edge servers offer a balanced solution. They possess more computational power than user devices, enabling them to handle moderately complex tasks and offload workloads from user devices. This is particularly beneficial for tasks that require more resources than a user device can provide but do not necessitate the extensive capabilities of a cloud server. For example, edge servers can perform intermediate processing of data before sending it to the cloud for further analysis, thus optimizing the overall computational workflow.

Privacy is another critical advantage of edge servers. By processing data closer to the source, edge servers reduce the need to transmit sensitive information over

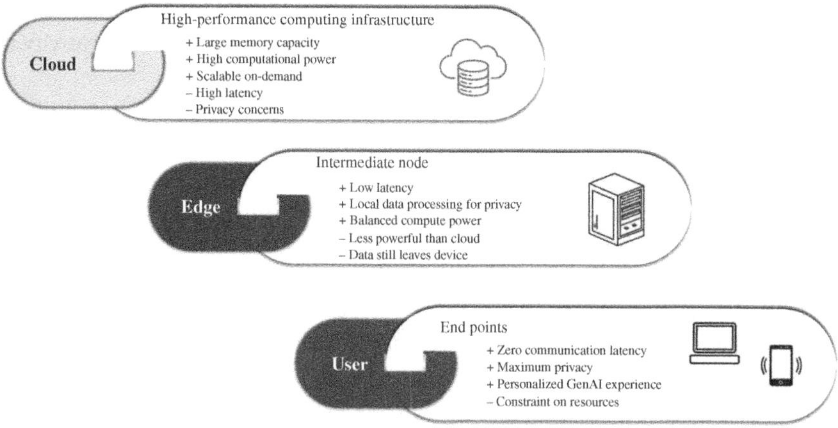

Figure 7.2 Hierarchical infrastructure for GenAI service delivery: Cloud infrastructure for high-performance computing, edge nodes for intermediate processing, and user devices for on-device executions.

long distances, thereby minimizing the risk of data breaches. This localized processing also allows for greater control over data privacy, as users can choose to process certain tasks at the edge rather than sending all data to the cloud. However, it is important to note that while edge servers offer improved privacy over cloud servers, they do not match the level of privacy achievable with on-device computation, as data must still leave the user's device.

Personalization is a key strength of edge servers relative to cloud servers. By leveraging local data and contextual information, edge servers can provide tailored AI experiences that are more relevant to individual users. This capability is particularly valuable for applications such as personalized healthcare, where edge servers can analyze local health data to provide customized recommendations and interventions. Additionally, edge servers can support adaptive learning models that evolve based on user interactions, enhancing the overall effectiveness and user satisfaction of GenAI services.

7.2.3 Collaborative Edge-cloud Infrastructure

As we discussed earlier, for effective deployment of GenAI services, we need to exploit all components of the network to be able to serve users in the best way. Although depending on the specific applications we might need to slightly change our infrastructure, yet there are common properties that all of different applications must follow to deliver reliable services in mobile network. GenAI-as-a-mobile service (GAIaaMS) refers to using edge-cloud-based systems to provide

GenAI capabilities that may be accessed via the internet. Edge-cloud systems utilize sophisticated AI models to produce content that imitates human creativity in various formats, such as text, graphics, audio, and video.

In mobile AIGC networks, **the cloud layer** is responsible for the pretraining and fine-tuning of AI models, requiring significant computational and storage resources. Initially, cloud-based data centers, exemplified by the Hugging Face Platform [47], are utilized to pretrain GenAI models such as GPT-3 and GPT-4. After that, these models are made available to users through cloud-based AIGC services. Despite the robust processing capabilities of cloud services, they are vulnerable to latency owing to their remote positioning. **The edge layer** acts as an intermediary, equipped with specialized hardware and software to manage data collection, inference, and service delivery closer to users. By deploying AIGC services on mobile edge networks, pretrained models can be enhanced and executed with significantly reduced transmission delays. This is particularly crucial for interactive services that require real-time responses [48]. At **the mobile device layer**, mobile units focus on efficient data acquisition and lightweight inference, employing techniques like FL and differential privacy to maintain user privacy while contributing to model improvement. The integration of these three layers enables personalized, secure, and efficient service delivery. The principal advantages of GAIaaMS are as follows:

- **Accessibility:** convenient access via open APIs or online interfaces.
- **Scalability:** capable of seamlessly managing large-scale initiatives.
- **Cost-effectiveness:** compensate for the actual utilization without significant upfront costs.
- **Wide-ranging Applications:** facilitates diverse generative functions across multiple sectors.
- **Continuous Improvement:** consistently employ AI model updates without the necessity of managing upgrades.

In summary, the cloud-edge-device stack has three complementary strengths: the cloud provides ample storage space, the edge provides low-latency contextual processing, and devices provide private, instant contact. When GenAI is delivered through this stack, it becomes a challenge in placement and adaptability. For each request, choose what model variant to use, where each stage executes (device/edge/cloud), and which runtime knobs (e.g., batching, caching, streaming) to activate so that latency, cost, and energy targets are met without sacrificing quality or privacy. The next section organizes concrete techniques at the model and system levels that make GenAI practical in mobile networks.

7.3 Network Resource Efficient GenAI Methods

In this section, we focus on optimizing GenAI models and services for deployment in resource-constrained mobile network environments. This section presents two complementary approaches: model-level optimizations that reduce computational and memory requirements of GenAI models themselves, and service-level optimizations that enhance the overall efficiency of GenAI service delivery within mobile networks.

7.3.1 Model Optimization Techniques

Model optimization techniques address the fundamental challenge of deploying bulky GenAI models within the constraints of mobile networks. These techniques aim to reduce model size, memory usage, and computational requirements while maintaining acceptable performance levels. Through methods like quantization, pruning, and knowledge distillation, these approaches make it possible to run sophisticated GenAI models on edge devices and servers with limited resources. The following sections detail key techniques that have proven effective in optimizing GenAI models for mobile network deployment.

7.3.1.1 Compression Techniques

- **Quantization:** This approach significantly reduces the memory usage of edge devices and the computational power needed to perform arithmetic operations by converting model weights and activations to a lower precision version (e.g., from 32-bit floating point to 8-bit). How to quantize LLMs effectively with minimal performance drop is an active field of research, and a multitude of articles has been published to solve this important issue [49–51]. For instance, the authors in Ref. [49] achieve up to 80% reduction in the memory consumption of OPT-175b (a LLM publicly available by Meta) by utilizing 3-bit activation and reordering channels in clusters. In another approach, the authors in Ref. [50] introduced SmoothQuant, a training-free, accuracy-preserving post-training quantization solution for large language models, enabling up to 1.56x speedup and 2x memory reduction with negligible loss in accuracy. Quantization is not only helpful for LLMs but can also be a game-changer in deploying diffusion models in mobile devices. Such models are known for their slow inference, high memory consumption, and computational intensity. Quantization can help deploy these models on resource-constrained edge devices. The authors in Ref. [52] introduced a novel method that can compress the noise estimation network to accelerate image generation while maintaining high image quality. In Ref. [53], the authors proposed a method that selectively performs computation for edited regions, accelerating various deep generative models, such as conditional GANs and diffusion models, on off-the-shelf hardware. Furthermore, in Ref. [54], the authors proposed a method to accelerate diffusion models by compressing the noise estimation network using post-training quantization, achieving effective performance with 8-bit models without retraining. These approaches make realistic data generation possible in edge devices by quantizing the cumbersome work of inferring from GenAI models.

- **Pruning and Sparsifying:** Several studies have shown that not all weights in neural networks contribute equally to producing the output. Consequently, one of the main ideas is to deploy only the parts of the network that have the most significant contribution to generating results during deployment. This concept has led to the practice of pruning, which has demonstrated a significant decrease in the memory and computation requirements of devices while maintaining much of the neural network's performance. For pruning and sparsifying neural networks, there are three strategies based on the timing of the pruning process. The most effective one is to prune the neural network after training. The most effective strategy is to prune the neural network after training. In this approach, a dense model is first trained to convergence, and then pruned and finally fine-tuned. Alternatively, pruning can occur during training, either gradually or iteratively, where elements are periodically pruned and re-added. Lastly, fully sparse training starts with a sparse model and maintains sparsity throughout the training process. The approaches can also vary in terms of which component of neural network we aim to prune. One common method is magnitude-based pruning, where weights with small absolute values are removed under the assumption that they contribute less to the output. Another approach is structured pruning, which involves removing entire neurons, filters, or channels. This method not only reduces the number of parameters but also simplifies the architecture, making it more efficient to implement on hardware. Additionally, there is dynamic sparsity, where the network structure is periodically pruned and regrown during training to maintain an optimal balance between sparsity and performance [55]. For our specific application in GenAI models, we can see that such approaches have been widely investigated [56–59]. For instance, the authors in Ref. [56] proposed an approach for pruning transformers that reduces inference cost and latency while maintaining accuracy, achieving up to a 2.0x reduction in FLOPs and a 1.56x speedup in inference latency in less than three minutes on a single GPU. Moreover, the authors in [58] proposed a method that allows large-scale GPT family models to be pruned to at least 50% sparsity in one shot without retraining, with minimal loss of accuracy.

7.3.1.2 Architecture Optimization

- **Low-rank Adaptation:** Low-rank adaptation (LoRA) is a method that authors proposed in Ref. [60] for fine-tuning the LLMs in an efficient manner. Based on the observation that the parameters of pretrained models are low rank, the authors hypothesized that fine-tuned parameters are also low rank. This led to the development of an approach where, instead of fine-tuning all parameters of a pretrained model for a downstream task, the pretrained weights are frozen, and a low-rank network is added in parallel, significantly reducing training time. Notably, this method only affects the attention module in transformers, which includes query, key, and value matrices. This method increases inference time minimally, as the low-rank parameters are integrated with the original parameters, maintaining the same latency. For more details, we encourage the reader to read the original paper. This low-rank adaptation not only expedites the fine-tuning process but also facilitates lightweight deployment on edge devices. Once the low-rank parameters have been trained, the low-rank model can be deployed on edge devices for downstream tasks, significantly reducing memory usage. There are numerous works inspired by LoRA that further help the deployment of LLMs on edge devices [57, 61–64]. For instance, Ref. [61] introduced a sparsifying scheduler to enable dynamic adjustments to the intrinsic rank during the adaptation process, resulting in better memory and generalization performance. Moreover, Ref. [62] introduced LoRA-FA, which reduces activation memory in LLMs without compromising fine-tuning accuracy. Additionally, Ref. [64] reduced the computational burden and memory consumption of LLMs by quantizing weights during fine-tuning while using an approach inspired from LoRA.
- **Neural Architecture Search:** Neural Architecture Search (NAS) automates the design of efficient neural network architectures tailored for specific tasks and resource constraints. NAS explores different architectural configurations to find the optimal model structure that meets the performance and resource requirements. This approach is particularly beneficial for deploying AI models on mobile and embedded devices, where resource limitations are a significant concern. By leveraging NAS, researchers can discover network architectures that balance performance and efficiency, ensuring that the models can run effectively on devices with limited computational power. This automation in architecture design helps in creating models that are both powerful and resource efficient.

7.3.1.3 Knowledge Transfer

- **Distillation:** The idea of knowledge distillation stems from the fact that the output of large models can be more informative for training than the labels themselves. To train a model with computational and memory constraints, we can first train a complex and bulky model (teacher) on a large dataset. Then, we train a smaller model (student) using the output of the teacher instead of the labels. This technique has proven to be highly effective for deploying small models on resource-constrained devices. This idea was originally presented in Ref. [65], where it was demonstrated that distillation can significantly reduce the computational burden without sacrificing accuracy. Following this work, we can see its application in generative models. The authors in Ref. [66] proposed a method that uses knowledge distillation to transform a multistep diffusion process into a single step, significantly improving the sampling speed while maintaining sample quality. Differently, other authors proposed miniLLM, which distills smaller language models from larger generative models. This method uses reverse Kullback–Leibler divergence to improve the distillation process, resulting in student models that perform well in generating precise and high-quality text responses [67].

7.3.2 Service Optimization Methods

In this section, we will focus on improving the overall efficiency of GenAI services in mobile networks beyond model-level optimizations. These methods address system-wide challenges such as distributed training, efficient inference, and resource allocation across the network infrastructure. By combining approaches like FL, intelligent prompt engineering, and adaptive task offloading, these techniques enable more efficient and scalable GenAI services while considering network constraints, user privacy, and quality of service requirements. The following subsections explore key methodologies that enhance the deployment and operation of GenAI services in mobile networks.

1. **FL:** FL is essential for users who lack high-quality data and need to collaborate to build robust models. Sharing personal data may not be feasible due to privacy concerns, making FL a valuable approach. In a standard FL setup, users collaboratively train high-quality models by sharing only their model weights with others [68]. In mobile networks, users typically have non-iid datasets due to diverse environments and varying computational capacities. To address such heterogeneity, various approaches have been developed [69–71].

   There are two main strategies for dealing with the computational heterogeneity of devices. The first involves sharing model parameters innovatively. For instance, HeteroFL [72] enables the training of heterogeneous local models by allowing each client to train a submodel that matches its computational resources. HeteroFL uses a static submodel extraction method where each submodel is a subset of the global model, ensuring that every part of the global model is trained on at least one client. This method allows clients with varying computational capacities to participate in the training process while still contributing to a unified global model. Another example is FedRolex [71], which employs a rolling submodel extraction scheme. This approach ensures different parts of the global server model are evenly trained, mitigating client drift caused by inconsistencies between individual client models and the server model architecture.

Alternatively, instead of sharing parameters, users can share information to others in completely different approach. Specifically, they can use knowledge distillation to share information within the network by sharing predictions on a public dataset instead of parameters. This approach offers greater flexibility, allowing models of arbitrary size to be trained in the network [73].

2. **Prompt Engineering:** The goal of engineering prompts is to create precise and optimized instructions to ensure generative models produce high-quality outputs efficiently. This optimization minimizes the need for multiple iterations and regenerations, conserving bandwidth and reducing latency, which ultimately enhances the overall QoE for users [74].

   One proposed solution for improving responses from GenAI models in mobile networks is netGPT [75]. This approach combines the strengths of both cloud and edge computing to optimize the deployment and operation of LLMs. By using edge LLMs for localized, personalized prompt processing, and more powerful cloud LLMs for comprehensive responses, netGPT significantly improves the efficiency and quality of GenAI services. This collaborative cloud-edge approach reduces latency and communication overhead by having edge servers handle initial prompt enhancements and data processing closer to the user, while the cloud provides robust computational resources for complex tasks. This synergy not only optimizes resource utilization across the network but also ensures that generative models deliver high-quality, context-aware outputs, meeting the stringent demands of modern mobile networks.

3. **Adaptive Task Offloading:** In a mobile network setting, task offloading involves transferring some or all computations from edge devices to edge or cloud servers. This process is particularly crucial for GenAI applications, which are computationally intensive and often beyond the capabilities of mobile devices (see Section 7.2.2.2). The decision to offload tasks is based on several factors, including the nature of the GenAI tasks, network conditions, and the availability of computational resources. Optimizing GenAI inference at the user end while minimizing total latency, power consumption of the edge device, and the use of network and computational resources is a key objective. This optimization can be achieved through a dynamic decision-making process that determines tasks should be offloaded, where they should be executed, and how resources should be allocated [76–79].

For solving such optimization problems adaptively—meaning the system continuously monitors device status, server load, and network conditions to make real-time decisions—various approaches have been proposed. Heuristic approaches and deep reinforcement learning techniques are common ways to tackle this problem [80–82]. These approaches allow the system to dynamically adjust its offloading strategies based on real-time data, thus optimizing performance and resource utilization.

Heuristic methods provide solutions that follow predefined criteria such as task priority, available bandwidth, and server load. These methods are generally not computationally expensive and can quickly offer near-optimal solutions, making them suitable for scenarios where quick decision-making is required or computational resources are limited. However, in highly complex environments, such approaches may not converge to a global optimum [81].

On the other hand, deep reinforcement learning methods offer solutions by adapting to the environment and gradually updating decision-making policies based on feedback from previous actions. This makes them ideal for scenarios where complex and highly variable environments must be managed. However, this higher performance comes at the cost of requiring more computational resources and more time to make decisions [80, 82].

By addressing these challenges and employing these techniques, mobile networks can effectively integrate GenAI, enhancing their capabilities and providing better services to users.

7.4 Security and Privacy at Application Layer

Security and privacy are paramount concerns in the implementation of AIGC models, which encompass threats such as **data poisoning**, **backdoor attacks during training**, and **intelligent theft post-deployment of the model**. Advanced models like ChatGPT are components particularly susceptible to emerging risks, such as **jailbreaks** and **prompt injections**. Privacy violations are significant due to AIGC models' reliance on extensive datasets containing users' confidential information, rendering them potential targets for data exploitation and criminal activity. The capacity of AIGC to disseminate disinformation and facilitate malicious activities underscores the imperative for implementing more robust protective measures and stringent regulations. For both business and consumer applications, ethical and privacy-preserving data management is crucial, along with protecting IP rights and ensuring the provenance of AI-produced content. Addressing these challenges is essential for fostering trust and responsible use of AIGC technologies [9].

This section delineates the classification of security and privacy within the AIGC paradigm regarding the issues alongside the present and prospective countermeasures.

7.4.1 Security Threat Models and Privacy Risks

AIGC systems expose a four-stage pipeline: (i) data supply and curation (web crawls, licensed corpora, user uploads), (ii) training and adaptation (pretraining, fine-tuning, RLHF), (iii) inference and orchestration (system prompts, user prompts, tools/plugins, RAG connectors, caches), and (iv) output and distribution (text, code, image, audio, video).

The following threats are organized along this pipeline and highlight unique risks for generative systems.

7.4.1.1 Data Supply and Training/Fine-tuning

- **Representational Bias in Generative Outputs:** Foundation and diffusion models inherit and sometimes amplify social biases present in training corpora, yielding skewed text and imagery across gender, race, and occupation. Empirical audits show such effects in both LLMs and text-to-image systems, with emerging mitigations (debiasing prompts, attention-level corrections, and fairness-aware fine-tuning) [83–85].
- **Training-time Data Poisoning (including creator-targeted poisoning):** Adversaries can seed web data or fine-tuning/RLHF datasets so models learn unsafe behaviors or embed "triggers." In vision, prompt-specific poisoning (e.g., Nightshade) demonstrates practical and targeted degradation of text-to-image models—a concern that is increasingly relevant as AIGC pipelines ingest creator content, underscoring the need for provenance, filtering, and robust training.
- **Memorization and Training-data Extraction:** Both LLMs and diffusion models can memorize and regurgitate training snippets or images, creating privacy and IP risk. Attacks have extracted exact passages from LLM APIs and original images from diffusion models; DP training and audit tooling help but do not eliminate risk [86, 87].
- **Model Poisoning in Federated/Participatory Settings:** When personalization or adaptation is performed across user devices (e.g., FL), malicious clients can submit adversarial updates. Robust aggregation and anomaly detection are required before applying cross-device training to AIGC [88].

7.4.1.2 Inference and Orchestration (Prompts, Tools, and RAG)

- **Jailbreaks (alignment bypass):** Carefully crafted prompts, role-play, and translation strategies can circumvent safety policies and elicit prohibited content. Continuous red-teaming, instruction-hierarchy hardening, and inference-time safety filters are necessary in production.
- **Prompt Injection (direct and indirect) in Tool-using/RAG Systems:** Untrusted documents, web pages, or tool outputs can carry hidden instructions that steer the model to exfiltrate secrets, issue unsafe tool calls, or override system prompts. Secure agent patterns—privilege separation, strict tool allow-lists, content sanitization, and output guards—are essential [89–91].
- **Availability/Efficiency Attacks ("sponge" prompts):** Inputs can be crafted to maximize compute (e.g., pathological decoding trajectories), inflating latency and cost and degrading SLOs—especially acute for edge/mobile deployments. Rate-limits, cost guards, and per-request budgeting help contain impact.
- **Covert Channels via Generative Media:** Diffusion-based steganography can embed payloads in generated images/audio, enabling silent data exfiltration through benign-seeming outputs. Providers should scan outbound media and apply policy constraints where appropriate [92].

7.4.1.3 Output, Distribution, and IP/Compliance

- **Content Provenance and Watermarking:** To trace and label AI-generated media (e.g., deepfakes), AIGC providers are adopting cryptographic provenance (C2PA Content Credentials) and watermarking tailored to diffusion models (e.g., Stable Signature; Tree-Ring). These improve attribution but are not yet tamper-proof, so multiple signals are advisable [93–95].
- **Intellectual Property and Licensing:** Risks include training on copyrighted material without authorization, unauthorized model/functionality cloning, and downstream plagiarism. Policy responses include dataset governance, license tracking, output filtering, and resilient watermarking [96].
- **Regulatory Obligations:** The EU AI Act introduces transparency and copyright-related duties for general-purpose and generative models (e.g., content labeling, training-data disclosures), complementing privacy regulations (e.g., GDPR, HIPAA). System design should anticipate disclosure, consent, and user-rights workflows [97–99].

7.4.1.4 Privacy Risks Across the Lifecycle

- **Interaction Leakage and Incident Risk:** Organizational incidents (e.g., internal code pasted into chatbots) and platform bugs have exposed sensitive data and payment metadata, underscoring the need for enterprise controls (no-log modes, DLP, SOC monitoring) [100].

- **Membership Inference and Auditing:** Adversaries may infer whether specific records were used in training, with elevated risk for fine-tuned models and certain diffusion setups. Providers should support red-team access and auditing, and consider DP/regularization where feasible [101–103].
- **Gradient-based Reconstruction in Federated Settings:** Without secure aggregation, homomorphic encryption, or local DP, uploaded gradients can leak local data. These protections trade off with model utility and latency and must be tuned to application risk.

7.4.2 Ethical Considerations in AIGC services

- **Scaled Phishing and Social Engineering:** Unlike template-based spam, LLMs generate context-aware spear-phishing by conditioning on public or leaked profiles (names, roles, recent projects) and by varying prompts to evade static filters. Attackers can also apply jailbreak phrasing to bypass safety rules and produce convincing lure text with credible tone and formatting [104].
- **Malicious Code Synthesis and Dual-use Assistance:** Code-capable AIGC systems can be steered to emit exploit scaffolds, obfuscated payloads, or step-by-step attack recipes (e.g., "port-scan, then privilege-escalate, then exfiltrate"), even when nominal guardrails exist. In practice, adversaries iterate prompts until they obtain runnable snippets or combine model output with simple tooling to assemble working malware [105].
- **Toxic, Violent, or Sexual Content:** Because text-to-image/video and chat models learn from web-scale corpora that include harmful material, prompt-conditioned generation can surface graphic or hateful outputs. Although providers employ input/output classifiers and safety-tuned models (e.g., refusal policies, NSFW filters), prompt variations and style transfers can still elicit problematic content, especially in long-tail or multilingual cases [106].
- **Cyber Fraud and Synthetic Impersonation:** Multimodal AIGC enables voice cloning from short samples, lip-synced or reenacted video, and dialog systems that sustain persuasive, real-time conversations. These capabilities raise the effectiveness of scams (e.g., "CEO voice" payment requests, romance scams with synthetic video), and allow cross-channel attacks that blend cloned audio, generated screenshots, and fluent chat.
- **Overreliance and Misuse:** The conversational style and confident tone of AIGC systems can induce automation bias, leading users to accept fabricated facts or outsource judgment. In education and assessment, students may submit generated work as if original, and anthropomorphic interfaces can further inflate user trust and expectations beyond what the models can reliably provide [107].

These concerns are inherently AIGC-specific: they stem from prompt-conditioned synthesis, few-shot imitation of style/voice, and multimodal generation—not from generic classification. Accordingly, mitigations should target the generative pipeline (prompt handling, safety tuning, moderation and provenance for outputs, and abuse-aware rate/cost controls) rather than rely on generic ML safeguards.

7.4.3 Enabling Secure AIGC-as-a-Service

Secure AIGC-as-a-Service (AIGCaaS) should close the loop between the threats previously discussed in the last two sections and deployment mechanisms that work in mobile/edge settings. Two complementary pillars cover the main risk surfaces: (i) provenance-aware, rights-preserving distribution to harden outputs and their downstream use, and (ii) privacy-preserving, robust adaptation to harden data, training, and personalization.

- **Provenance-aware, Rights-preserving Distribution:** This pillar addresses risks in production and distribution, such as fake abuse, fraud/impersonation, IP disputes, and accountability gaps. Each generated artifact carries a verifiable provenance manifest (model/version, prompt/context metadata, applied safety filters, license) and an associated content hash. The manifest is bound to the asset with Content Credentials/watermark claims and anchored on a ledger (permissioned chain or roll-up) to provide tamper-evident authorship and licensing. Smart contracts encode usage terms and royalty splits; decentralized identifiers sign publisher actions (issue, revoke, relabel). In practice, media stays off-chain (object storage/IPFS), while only pointers and hashes are recorded to avoid leaking personal data. This design lets high-risk channels (payments, KYC, voice/video intake) require signed media, enables automated down-ranking of unsigned or revoked content, and gives reliable evidence for takedown and copyright workflows. Provenance does not prevent generation; it labels and links it—so it should be paired with classifier-based detection and revocation lists [108–110].
- **Privacy-preserving and Robust Model Adaptation:** This pillar addresses risks at data supply/training and inference—interaction leakage, memorization/extraction, and model poisoning in cross-device personalization—while enabling responsiveness at the edge. Personalization proceeds via federated learning on devices/edge nodes: clients compute updates locally and share only model/adapter deltas. Secure aggregation hides individual updates;

differential privacy limits memorization of rare user data; robust aggregation (trust-weighted or outlier-resistant rules) reduces the impact of malicious clients. For mobile AIGC, adapting parameter-efficient heads/adapters (e.g., LoRA-style) keeps updates small, stabilizes convergence, and respects device energy/memory budgets. Combined with on-device retention limits and unlearning workflows, this approach localizes benefit while bounding privacy risk [111].

- **End-to-end Pattern:** At generation time, enforce safety prompts, injection guards, and SLO-aware rate/cost budgets to contain compute-inflating inputs. At output, attach provenance manifests and watermark claims, anchor hashes on a ledger, and gate critical channels to signed assets. For personalization/training, run FL with secure/robust aggregation and, where feasible, DP on adapter updates; keep only update summaries and manifest links. At ingest/consumption, verify provenance, auto-flag unsigned or revoked media, and route uncertain cases for review [112–114].

Together, these mechanisms map directly onto the problems identified earlier: provenance and rights management mitigate deepfakes, fraud, and IP/compliance at the distribution surface, while federated, privacy-preserving adaptation mitigates leakage, memorization, and poisoning at the data/training and inference surfaces—without sacrificing the latency, cost, and privacy constraints of mobile/edge AIGC deployments.

7.5 Use Cases of Mobile AIGC

This section examines use cases for effective mobile AIGC networks, such as using AIGC in social media, immersive streaming, and personalized AI services.

7.5.1 AI-generated Content in Social Media

AIGC has wholly transformed social media marketing, presenting advertisers with novel opportunities and obstacles. This technology enables enterprises to generate customized content on a significant level, construct captivating images, and establish complete marketing campaigns without the need for human involvement [115]. By 2025, almost 33% of the marketing messages distributed by big businesses will be generated by GenAI, prompting inquiries into consumer responses to material produced by GenAI.The COVID-19 epidemic has caused a paradigm shift in consumer-brand interactions toward online platforms, allowing marketers to effectively use GenAI to engage consumers through large-scale creative content [116].

This section discusses AIGC's social media applications and specific mechanisms, including incentive structures, privacy-preserving measures, and FL, guaranteeing effective, safe, and engaging user interactions.

- **Full-duplex Semantic Communications:**
One persistent challenge in social media is motivating users to create and share relevant content. Building on Ref. [117], the study considers full-duplex semantic communications (FDSC) not just as a physical-/link-layer innovation but as an application-layer enabler: instead of exchanging raw media, the client and platform continuously exchange semantic intents (prompts, topic vectors, style cues) in both directions while a post is being composed [118]. In practice, this supports concrete product flows—live co-creation with the audience, low-bandwidth "remix/duet" authoring on mobile connections, real-time safety/brand-suitability nudges before publishing, and on-the-fly personalization (e.g., language/style adaptation) without waiting for full uploads. Within this loop, a diffusion model supplies starter variants and contextual suggestions that evolve as the user edits, lowering time-to-first-post and creative friction while preserving user control. The incentive mechanism in Ref. [117] can then be surfaced in-session—for example, micro-rewards or badges for responding to community prompts, diversity/novelty bonuses for remixes, or credits for high-engagement variants—because the platform observes semantic progress in real-time rather than only final media. Empirically, this application of FDSC is expected to improve UGC quality and quantity (more on-topic posts, higher completion rates), strengthen participation loops during live or event-driven moments, and enhance QoE on constrained devices and networks by transmitting compact semantics rather than full-resolution assets. In short, FDSC provides the interaction substrate; the diffusion component supplies adaptive creative scaffolding; and the incentive scheme converts those gains into sustained, on-platform contribution.
- **Secure and Privacy-preserving Communication:**
This aspect is more critical since social media platforms now employ AI services to create content for their users, hence the need to protect their data. These issues are recapitulated in paper [34], where a GenAI-aided semantic communication (SemCom) system is proposed. The work also incorporates multimodal prompts—textual and visual descriptions—to decode prompts with high accuracy and low computational cost. In addition, it employs concealed modalities of information exchanges that involve the assistance of a friendly jammer to ensure that data passed between users cannot be accessed by known cyber threats. Not only does this new approach protect and safeguard the different processes conducted at AIGC, but it also ensures the sanctity and anonymity of the various communication that happens through various social media platforms.

- **FL for Improved Content Generation:**
  Organization-centric AI models face challenges in handling the diverse and ever-changing nature of social media content as user preferences vary and evolve. To address these issues, [119] highlights how FL can be leveraged within social media networks. By enabling decentralized AI model training without requiring users to share raw data, FL enhances user privacy, significantly reduces communication costs, and accelerates training time. This approach allows social media platforms to harness FL for AIGC, enabling the delivery of highly personalized content that adapts to individual user needs, ultimately boosting user engagement and satisfaction.
- **Content Transmission with Multimodal Prompts:**
  The SemCom system introduced in Ref. [120] is an AI-assisted, generative deep learning model that heavily improves the security and reliability of content delivery in social media. The extension of multimodal prompts comprises visual and textual information, whereby the system can translate messages end-to-end without the need to train both encoder and decoder in the system in the traditional manner. Due to the way it addresses concerns about the integrity and privacy of user-generated content, this strategy is particularly appropriate for social media platforms. Including a covert communication method also protects information when transmitted so that social media platforms can disseminate content without compromising the content's security, especially where there could be a security risk. However, this innovation in secure semantic communications provides a direct solution to sharing AIGC through social media networks.

7.5.2 Immersive Streaming (AR/VR)

Immersive streaming is an advanced use of AIGC that gives users a near-real-time environment they can interact with in a lifelike way. Beyond simply streaming fixed 360° video or pre-authored 3D scenes, AIGC enables on-demand generation of scenes, dialog, overlays, and viewpoints that adapt to user intent, context, and device capabilities. This makes AR/VR experiences more interactive and scalable: providers no longer need to preproduce every branch or localization because content can be personalized per viewer—language, difficulty, accessibility cues, and cultural context—at generation time. Distribution can also be more bandwidth-efficient: instead of transporting high-bitrate video or bulky meshes end-to-end, systems can transmit compact semantic descriptors (prompts, scene graphs, seeds) and render near the user (edge or device), reducing backbone load and enabling rapid scene updates. These properties broaden utility across education, entertainment, and social media services, where users benefit from live features such as custom camera angles, explanatory overlays, and responsive tutors or NPCs. At the same time, integrating AIGC with immersive streaming raises risks: algorithms may introduce bias into generated video, and the modality's high data rates and strict motion-to-photon budgets can create latency that degrades the experience. Future work should prioritize mitigating content bias and managing bandwidth/latency—for example, with safety-tuned generation, provenance controls, edge co-inference, and predictive caching—to ensure consistently high quality of experience.

This section investigates how AIGC can support the improvement of immersive streaming through AI models, blockchain technologies, and network optimization. These technologies enhance the progressive and secure delivery of content that ensures the touch point, or the experience the user gets in the immersive environment, is high quality [2].

- **Addressing Latency and Accessibility for Immersive Services:** The world of immersive streaming with AIGC services needs to adopt suitable edge computing solutions to address the demanding computational tasks tied to the content creation in real time. For optimizing the AIGC Service Providers (ASPs) selection process, diffusion-based reinforcement learning is presented in Ref. [121]. This approach takes advantage of the fact that AIGC models can be run on edge servers, thus minimizing impairments such as latency and inaccessibility to immersive streaming services in conditions where such devices may be primary. The proposed AI-generated optimal decision (AGOD) algorithm takes selections over the ASPs to improve the quality of the service further, satisfying all user requirements and ensuring smooth streaming without interruption.
- **Blockchain-enabled Content Management in Web3:** To solve content management problems in immersive streaming, Ref. [122] proposes a blockchain-semantic framework based on wireless edge intelligence. There is a remarkable convergence here, combining semantic communication with a blockchain solution to enhance the security and efficiency of delivering AIGC. Regarding immersive streaming, blockchain guarantees the work's authenticity and its components' immunity to alterations, while semantic communication supports fast content transmission, making it highly relevant to viewers' preferences. The outcome is a far more secure streaming process and, quite correspondingly, a higher level of trust toward the origin and quality of the content.
- **Lifecycle Management for Seamless User Experience:** The process of AIGC in immersive environments needs to be well maintained throughout its usage and lifecycle to prevent numerous issues with user experience. The authors in Ref. [123] discuss contingency for the more decentralized execution of the lifecycle management of AIGC products using blockchain across edge networks. The proposed solutions are beneficial in meeting characteristics such as service latency, security, and trust, which are significant

in immersive streaming. By building a blockchain's distributed ledger with added intelligent contracts, the framework protects content ownership and copyright and guarantees the authenticity and safety of the AIGC product exchange. More than that, this approach improves the ease of use for end users while at the same time protecting the rights of content creators and giving them their due for their creativity.

- **Integrated Sensing and Communication:**
  Using integrated sensing and communication technology is another significant component in realizing the success of immersive streaming. Ref. [124] explains how GenAI models help optimize the ISAC system's physical layer for communication and perception. In immersive streaming, these technologies assist in real-time content control based on the user's environment. Thus, the immersive experience of streaming continues to be smooth. Some advanced methods include using diffusion models for tasks like signal direction estimation and channel state information (CSI) compression, which add to streaming and decrease bit error rates.
- **Wireless Perception to Generate Immersive Content:**
  Another example of enhancement in immersive streaming is using wireless perception to direct AI-created content. In Ref. [125], authors describe a wireless sensing platform to maximize the performance of AIGC services with multi-scale perception. Besides, by employing a signal transmitter and multiple receivers, the system can acquire and understand the user environment to produce content that will be contextually meaningful and realistic. This approach minimizes individual devices' load, making streaming more complex and real-life scenes possible. The wireless perception system improves the overall immersion by ensuring the content created is relevant and adapts to the user's environment.
- **Collaborative AIGC Framework:**
  Ref. [126] provides a detailed understanding of the development of collaborative distributed diffusion-based AIGC in the wireless network system to optimize the allocation of resources. This enables AIGC tasks to be executed using a framework that decentralizes computationally heavy tasks across devices more efficiently, especially in resource-constrained devices like mobile phones. For example, the collaborative approach in creating content entails passing intermediate outcomes to complete the content generation process after several denoising steps. This approach dramatically optimizes energy usage and guarantees that only high-quality texts are used despite real-time content production and distribution. Within immersive streaming, where content continuity and quality are paramount, this distributed architecture provides a means to deliver immersion while minimizing resource overhead.

- **Deep Generative Network Management:**
  Ref. [127] has given a study of deep generative models (DGMs) for efficient realization of wireless network management, especially in resource allocation. Thus, the paper highlights the benefits of these models in capturing intricate patterns and sampling realistic data, thus showing their discipline applicability for enhancing the effectiveness and stability of wireless networks. In particular, DGMs are employed in immersive streaming, which relies on the timely and efficient delivery of high-quality content. The paper focuses on specific areas in which DGMs can contribute to improving decision-making related to network routing, allocating particular resources, and network economics. With the aid of the generative characteristics of these models, the proposed arithmetic framework can adjust to dynamic changing network circumstances to guarantee seamless delivery of the immersive streaming content. This is especially important when the available network resources are limited, and the quality of service provided needs to be controlled.

In conclusion, Table 7.1 summarizes the key applications of AR/VR and their respective benefits and drawbacks.

Table 7.1 Use cases of AR/VR, their benefits, and drawbacks.

Application	Benefits	Challenges
Metaverse and Beyond [121, 125, 126]	• Resource efficiency • Enhanced decision-making • Real-time performance	• High computational needs • Environmental uncertainty • Human-aware utility modeling
Digital Asset Management [122]	• Resource efficiency • Secure, authentic content • Fast, efficient delivery	• Scalability limits • On/off chain integration • Semantic verification
Digital Art [123]	• Reduced service latency • Reliable economic incentives • Trust-based ESP selection	• Security threats • Integration complexity • Complex economic dynamics
Smart Cities and IoT [124, 127]	• Efficient resource utilization • Improved sensing capabilities • Lower error rates	• Real-time constraints • Conflicting design requirements • Robustness in environments

7.5.3 Personalized AI Services

AIGC provides several benefits in tailored services, particularly compared to PGC and UGC. GenAI models demonstrate rapid production of customized, high-quality content that caters to the specific requirements of consumers [2]. A vital function of local edge servers is to tailor services to the particular needs of local users and deliver customized services according to user choices. With DALL-E 2, users can modify inputs to personalize outputs, such as producing a wide range of graphics. ChatGPT can generate contextually suitable text and provide personalized replies by analyzing input, dialog history, and user profiles. Within mobile AIGC networks, edge servers can optimize AI models by utilizing extensive datasets such as LAION-400M [128]. This allows for providing dynamic and location-based personalized services with little delay. Notwithstanding the advantages, hurdles such as intricacy, expenses, and privacy issues persist, requiring additional investigation into efficient caching techniques and customized service provision.

- **Mobile AIGC for Personalized Healthcare:** In the healthcare field, the idea of Human Digital Twins (HDTs) has created significant progress in implementing AI services. Ref. [129] discusses how mobile AIGC redefines personalized healthcare solutions by creating high-fidelity digital twins. Moreover, these HDTs can represent the human body in detail, which means that synthetic data for such pathological conditions can be generated, surgical operations can be effectively planned according to individual incidences, and patients' curative tables can be developed. Consequently, the framework described in the paper employs mobile edge networks to carry out interaction-intensive healthcare services with low latency, enabling patients to receive individualized, real-time medical services. By often updating the HDT with the physical twin's status, the system offers a progressive healthcare system customized to the patient's needs.
- **Edge Intelligence for Personalized AI Services:** The study [130] looks at how hard it is to manage and keep pretrained foundation models (PFMs) up to date for GenAI services in mobile edge networks while also looking at the performance of inference latency, accuracy, and resource use. As such, this paper presents a joint model caching and inference model that could enhance the usage of PFMs in delivering customized services. It uses the AoC metric and the LC algorithm to ensure the best models for each situation are stored and used in the inference process. This lowers the system's cost and improves the services' accuracy. This is especially valuable where the content is generated based on the user's preferences, interests, and prior actions while online; this is common in applications like advertising, content recommendation, and customized interfaces.

- **AIGC for Autonomous Driving:** AI is also being personalized in the automobile industry by applying GenAI to improve self-driving vehicles' safety and reliability. Ref. [131] puts forward a GenAI simulation framework for autonomous driving within vehicular mixed reality metaverses. This framework also incorporates digital twin (DT) technologies to generate simulations to create physical-digital environments that enable autonomous vehicles (AVs) to rehearse and optimize their behaviors in diverse DT settings. Using a multitask digital twin offloading model and a diffusion model-based traffic sign generator, the model helps the AVs incorporate the different driving techniques required by various users, promoting safer roads and better traffic management.
- **Personalized Services in Intelligent Vehicular Networks:** This paper [132] describes the implementation of GenAI in vehicular networks in the context of improving the quality of services provided in intelligent transportation systems (ITS). This study aims to outline a multimodality semantic-aware framework that relies on GenAI and is deemed to create and manage multimodal content critical for adapting and delivering personalized and contextualized services to vehicles on the road. This approach provides more reliable guidance toward ensuring better usability on vehicular networks by addressing the needs of individual consumers. It also introduces a DRL-based approach for efficient resource management in vehicle-to-vehicle (V2V) communication to ensure the effectiveness of the proposed system under dynamic conditions. As the communication and content generation process is aligned with each vehicle's specific contexts, the framework supports personalized AI services that collectively improve driving experience and safety and optimize traffic conditions.
- **Personalized Digital Twin in Agriculture Using YOLO-based Semantic Communication and GenAI:** The research in Ref. [133] provides a semantic communication framework based on the YOLO object detection approach combined with GenAI-assisted resource management for creating digital twins, especially in the agri-food chain. This model extracts and emphasizes semantically essential information for interaction from images received from the edge devices, minimizing the amount of data to send and improving the quality of communication. Due to the use of AIGC, the framework provides context-sensitive and highly accurate copies of entities or processes, known as digital twins, such as a virtual apple orchard. These digital twins may be applied to such personalized uses as precision farming, where particular activities may be optimized depending on the unique conditions of each farm. Incorporating GenAI for resource management guarantees that these digital replicas are precise, detailed, and optimally utilized, thus making them effective in delivering personalized AI services in agriculture and other sectors.

- **Optimizing Personalized Experiences with GPT-based Caching and Inference:** Ref. [134] considers using GPTs in edge intelligence frameworks that aim at improving personalization of experiences in the Metaverse. This paper presents a caching and inference approach that addresses the computational burden of GPTs to deliver real-time and user-specific content in a virtual environment. By optimizing the caching of contextually relevant models, the framework minimizes the latency and enhances the accuracy of personalized content delivery. Meeting the need for highly realistic and personalized content using AI while dealing with large amounts of data and computational resources is even more critical in the Metaverse. In this context, the emphasis on using GPTs for time-sensitive real-time predictions and content personalization aligns with the work of the broader field of personalized AI services, which will be essential for continuously delivering quality and related services within constantly growing immersive environments.

Finally, Table 7.2 summarizes the key personalized AI services and their respective benefits and drawbacks.

Table 7.2 A summary of personalized AI services in mobile networks, their benefits, and drawbacks.

Application	Benefits	Challenges
Healthcare [129]	• Real-time monitoring • Personalized care • Resource efficiency	• Data updates • Privacy concerns • Integration of multi-modal data
Semantic Content Generation [130, 134]	• Resource optimization • Higher inference accuracy • Lower latency	• Caching complexity • Dynamic network condition • Scalability and integration
Autonomous Driving [131, 132]	• Reduced inference latency • Safer driving • Cost efficiency	• Trade-off management • Dynamic context handling • Resource limitations
Agriculture [133]	• Precision farming • Real-time monitoring • Improved object detection	• Robust semantic extraction • Data management challenges • Edge device limits in rural setting

7.6 Conclusion and Research Directions

Mobile AIGC networks improve cloud-based AIGC services by letting people interact quickly and easily by processing data locally on mobile edge networks. This cuts down on the wait times that come from servers that are far away. Their services include localization and mobility, enabling base stations to customize models according to regional and user requirements. Enhanced customization and personalization are achieved by restricting data exposure to local servers, thereby improving privacy and security. Nevertheless, tackling obstacles such as resource allocation, latency, energy consumption, mobility management, and privacy concerns is important to get the best possible performance.

- **Resource Management:** The AIGC services necessitate substantial processing and storage resources, which present difficulties for the current mobile edge computing architecture. The key performance indicators (KPIs) used to measure edge resource distribution are **accuracy**, **bandwidth utilization**, and **resource consumption**. Future research on resource allocation should focus on creating adaptive and context-aware strategies that balance the demands of high-quality, generative models with limited edge resources. This entails investigating lightweight model architectures, transfer learning, and model compression to fine-tune model complexity while optimizing bandwidth usage and storage dynamically [135]. Moreover, novel algorithms that consider user mobility, evolving demand patterns, and energy constraints must be developed to determine how best to schedule, distribute, and reconfigure workloads across edge nodes [136]. Coordination mechanisms leveraging distributed optimization tools like the alternate direction method of multiplier (ADMM) along with the advanced auction-based and game-theoretic frameworks can further ensure that resource allocation remains efficient, equitable, and nearly optimal in dynamic environments [137, 138].
- **Edge Caching:** Innovations in edge caching must go beyond traditional content storage strategies to accommodate the complexity of GenAI models and their associated inference tasks [139]. Future work should focus on intelligent cache management policies that leverage predictive analytics to store and retrieve models or model fragments at opportune moments, thereby minimizing delays and backhaul traffic. Further, dividing large generative models into smaller functional units and exploring federated caching architectures can help maintain privacy and resilience in mobile AIGC networks [140]. By integrating caching solutions like EdgeServe and DeepCache with adaptive policies and QoS-driven optimization techniques, future systems can ensure prompt, reliable, and energy-efficient access to AIGC services, even under fluctuating mobility and user density conditions.

- **Mobility Management:** Managing mobility will require the development of robust frameworks capable of predicting user and node trajectories, dynamically adjusting routing decisions, and selectively offloading computations in real-time [141]. Predictive models can integrate machine learning and reinforcement learning to forecast movement patterns and reconfigure edge infrastructure resources accordingly. UAV-assisted solutions and IoV-based frameworks will need innovative strategies to ensure uninterrupted AIGC service, including seamless handovers, efficient coverage extension, and reduced latency, even in remote or high-mobility scenarios [142]. As AIGC services become integral to transportation, navigation, and vehicular edge inference systems, continuously updated optimization techniques must navigate the complexities of nonstationary network conditions, ensuring that mobility management [143] evolves alongside the demands of future applications.
- **Incentive Mechanisms and Fairness:** Promoting fairness involves reconciling the interests of diverse stakeholders—edge nodes, service providers, and users—and ensuring that resource allocation and incentive mechanisms are both transparent and equitable [144]. Future research should design incentive structures that motivate greater participation from edge nodes, possibly leveraging blockchain-based solutions for verifiable fairness, trust, and safe transactions. Equitable FL frameworks and hierarchical reinforcement learning methods must guarantee that no single entity's poor-quality updates degrade the global model's performance. By incorporating market-based approaches such as Stackelberg games and multiagent reinforcement learning, future systems can balance social welfare, revenue generation, and risk aversion, ultimately fostering a more inclusive and sustainable AIGC ecosystem [145].

7.7 Summary

This chapter explores the integration of GenAI into mobile networks, focusing on AIGC applications such as text, image, video, and 3D content creation. It addresses challenges in resource-constrained mobile environments, including computational scalability, end-to-end latency, and security-privacy risks, by explaining a collaborative edge-cloud-user device infrastructure that optimizes resource allocation, leverages edge computing for low-latency inference, and balances privacy via on-device processing. Model optimization techniques (quantization, pruning, knowledge distillation) and service-level methods (federated learning, adaptive offloading) are detailed to enhance efficiency. Security threats like data poisoning, adversarial attacks, and privacy risks are mitigated through blockchain-based decentralization, federated learning, and ethical frameworks.

Use cases span social media, AR/VR streaming, and personalized AI services in healthcare, autonomous driving [146–149], and agriculture. Future research directions emphasize resource management, edge caching, mobility handling, and fairness mechanisms to ensure scalable, secure, and user-centric AIGC deployment in mobile networks.

References

1 Jinsong Wu, Weiming Gan, Zhongming Chen, Shuang Wan, and Hui Lin. Ai-generated content (AIGC): A survey. *arXiv preprint arXiv:2304.06632*, 2023.

2 Ming Xu, Haijun Du, Dusit Niyato, Jiawen Kang, Zehui Xiong, Shiwen Mao, and H. Vincent Poor. Unleashing the power of edge-cloud generative AI in mobile networks: A survey of AIGC services. *IEEE Communications Surveys & Tutorials*, 2024.

3 Yiheng Cao, Shouyuan Li, Yuhui Liu, Zhiqiang Yan, Yu Dai, Philip S. Yu, and Le Sun. A comprehensive survey of AI-generated content (AIGC): A history of generative AI from GAN to ChatGPT *arXiv preprint arXiv:2303.04226*, 2023.

4 Tom B. Brown, Benjamin Mann, Nick Ryder, Melanie Subbiah, Jared Kaplan, Prafulla Dhariwal, Arvind Neelakantan, Pranav Shyam, Girish Sastry, Amanda Askell, Sandhini Agarwal, Ariel Herbert-Voss, Gretchen Krueger, Tom Henighan, Rewon Child, Aditya Ramesh, Daniel M. Ziegler, Jeffrey Wu, Clemens Winter, Christopher Hesse, Mark Chen, Eric Sigler, Mateusz Litwin, Scott Gray, Benjamin Chess, Jack Clark, Christopher Berner, Sam McCandlish, Alec Radford, Ilya Sutskever, and Dario Amodei. Language models are few-shot learners, arXiv:2005.14165v4, 2020. Available at https://arxiv.org/pdf/2005.14165.pdf.

5 OpenAI. GPT-4 technical report, 2023. Available at https://arxiv.org/abs/2303.08774.

6 H. Brendan McMahan, Eider Moore, Daniel Ramage, Seth Hampson, and Blaise Aguera y Arcas. Communication-efficient learning of deep networks from decentralized data. In *Proceedings of the 20th International Conference on Artificial Intelligence and Statistics*, volume 54 of *Proceedings of Machine Learning Research*, pages 1273–1282. PMLR, 2017. https://proceedings.mlr.press/v54/mcmahan17a.html.

7 Tim Dettmers, Artidoro Pagnoni, Ari Holtzman, and Luke Zettlemoyer. Qlora: Efficient finetuning of quantized LLMs. *arXiv preprint arXiv:2305.14314*, 2023. Available at https://arxiv.org/pdf/2305.14314.pdf.

8 Evan Crothers, Nathalie Japkowicz, and Herna Viktor. Machine generated text: A comprehensive survey of threat models and detection methods. *arXiv preprint arXiv:2210.07321*, 2022.

9 Yen-Cheng Wang, Jiaqi Xue, Cheng Wei, and C.-C. Jay Kuo. An overview on generative AI at scale with edge-cloud computing. *IEEE Open Journal of the Communications Society*, 4: 2952–2971, 2023.

10 Xiaoming Wang, Yu Hong, and Xiang He. Exploring artificial intelligence generated content (AIGC) applications in the metaverse: Challenges, solutions, and future directions. *IET Blockchain*, 2024.

11 Zhenzhong Lan, Mingda Chen, Sebastian Goodman, Kevin Gimpel, Piyush Sharma, and Radu Soricut. Albert: A lite BERT for self-supervised learning of language representations. In *Proceedings of the International Conference on Learning Representations (ICLR)*, pages 1–17, Addis Ababa, Ethiopia, 2019.

12 Zhiqing Sun, Hongbin Yu, Xiaodan Song, Renjie Liu, Yiming Yang, and Denny Zhou. MobileBERT: A compact task-agnostic BERT for resource-limited devices. In *Proceedings of the 58th Annual Meeting of the Association for Computational Linguistics (ACL)*, pages 2158–2170, 2020.

13 Marat Makhmutov, Sawsan Varouqa, and John A. Brow. Survey on copyright laws about music generated by artificial intelligence. In *Proceedings of the IEEE Symposium Series on Computational Intelligence*, pages 3003–3009, 2020.

14 Po-Han Chi, Po-Han Chung, Tsung-Han Wu, Cheng-Chieh Hsieh, Yu-Huai Chen, Shang-Wen Li, and Hung-yi Lee. Audio albert: A lite BERT for self-supervised learning of audio representation. In *Proceedings of the IEEE Spoken Language Technology Workshop*, pages 344–350, 2021.

15 Andreas Blattmann, Robin Rombach, Kutay Oktay, Jonas Müller, and Björn Ommer. Semi-parametric neural image synthesis. In *Advances in Neural Information Processing Systems*, 2022.

16 Gary Marcus, Ernest Davis, and Scott Aaronson. A very preliminary analysis of dall-e 2. *arXiv preprint arXiv:2204.13807*, 2022.

17 Zehui Liang and Fei Tao. Research on the application of artificial intelligence in e-commerce design. In *2020 International Conference on Innovation Design and Digital Technology (ICIDDT)*, pages 455–458. IEEE, 2020.

18 Jonathan Ho et al. Imagen video: High definition video generation with diffusion models. *arXiv preprint arXiv:2210.02303*, 2022.

19 Hao Zhang, Haibo Zhang, Bohan Di, Marco Di Renzo, Zhu Han, H Vincent Poor, and Lingyang Song. Holographic integrated sensing and communication. *IEEE Journal on Selected Areas in Communications*, 40(7): 2114–2130, 2022.

20 Borna Barahimi, Hina Tabassum, Mohammad Omer, and Omer Waqar. Context-aware predictive coding: A representation learning framework for WiFi sensing. *IEEE Open Journal of the Communications Society*, 5: 6119–6134, 2024. https://doi.org/10.1109/OJCOMS.2024.3465216.

21 Xianjun Deng, Yajun Jiang, Laurence T. Yang, Min Lin, Lin Yi, and Meiqin Wang. Data fusion based coverage optimization in heterogeneous sensor networks: A survey. *Information Fusion*, 52: 90–105, 2019.

22 Pedro Cruz, Nadjib Achir, and Aline Carneiro Viana. On the edge of the deployment: A survey on multi-access edge computing. *ACM Comput. Surv.*, 55(5), dec 2022. ISSN 0360–0300. https://doi.org/10.1145/3529758.

23 Abderrahime Filali, Amine Abouaomar, Soumaya Cherkaoui, Abdellatif Kobbane, and Mohsen Guizani. Multi-access edge computing: A survey. *IEEE Access*, 8: 197017–197046, 2020. https://doi.org/10.1109/ACCESS.2020.3034136.

24 Xu Chen, Lei Jiao, Wenzhong Li, and Xiaoming Fu. Efficient multi-user computation offloading for mobile-edge cloud computing. *IEEE/ACM Transactions on Networking*, 24: 2795–2808, 2015. https://doi.org/10.1109/TNET.2015.2487344.

25 Panagiotis Apostolopoulos, Eirini-Eleni Tsiropoulou, and Symeon Papavassiliou. Risk-aware data offloading in multi-server multi-access edge computing environment. *IEEE/ACM Transactions on Networking*, 28: 1405–1418, 2020. https://doi.org/10.1109/TNET.2020.2983119.

26 Michael P. J. Mahenge, Chunlin Li, and C. Sanga. Collaborative mobile edge and cloud computing: Tasks unloading for improving users' quality of experience in resource-intensive mobile applications. *2019 IEEE 4th International Conference on Computer and Communication Systems (ICCCS)*, pages 322–326, 2019. https://doi.org/10.1109/CCOMS.2019.8821787.

27 Xiaoqian Chen, Tieliang Gao, Hui Gao, Baoju Liu, Ming Chen, and Bo Wang. A multi-stage heuristic method for service caching and task offloading to improve the cooperation between edge and cloud computing. *PeerJ Computer Science*, 8:e1012, 2022. https://doi.org/10.7717/peerj-cs.1012.

28 Huaming Wu, Ziru Zhang, Chang Guan, K. Wolter, and Minxian Xu. Collaborate edge and cloud computing with distributed deep learning for smart city internet of things. *IEEE Internet of Things Journal*, 7: 8099–8110, 2020. https://doi.org/10.1109/JIOT.2020.2996784.

29 Yun-Cheng Wang, Jintang Xue, Chengwei Wei, and C.-C Kuo. An overview on generative AI at scale with edge-cloud computing. *IEEE Open Journal of the Communications Society*, 4:2952–2971, 2023. https://doi.org/10.1109/OJCOMS.2023.3320646.

30 Hengchang Du, Zheng Li, Dusit Niyato, Junshan Kang, Zehui Xiong, and Dong In Kim. Enabling AI-generated content (AIGC) services in wireless edge networks. *arXiv preprint arXiv:2301.03220*, 2023.

31 Mehdi Monemi and Hina Tabassum. Performance of UAV-assisted d2d networks in the finite block-length regime. *IEEE Transactions on Communications*, 68(11): 7270–7285, 2020. https://doi.org/10.1109/TCOMM.2020.3017623.

32 Batyr Charyyev, Engin Arslan, and Mehmet Hadi Gunes. Latency comparison of cloud datacenters and edge servers. In *GLOBECOM 2020 - 2020 IEEE Global Communications Conference*, pages 1–6, 2020. https://doi.org/10.1109/GLOBECOM42002.2020.9322406.

33 Shalin Parikh, Dharmin Dave, Reema Patel, and Nishant Doshi. Security and privacy issues in cloud, fog and edge computing. *Procedia Computer Science*, 160: 734–739, 2019. https://doi.org/10.1016/j.procs.2019.11.018.

34 Hongyang Du, Dusit Niyato, Jiawen Kang, Zehui Xiong, Kwok-Yan Lam, Yuguang Fang, and Yonghui Li. Spear or shield: Leveraging generative AI to tackle security threats of intelligent network services, arXiv:2306.02384, 2023. https://arxiv.org/abs/2306.02384.

35 Mohammad Shoeybi, Mostofa Patwary, Raul Puri, Patrick LeGresley, Jared Casper, and Bryan Catanzaro. Megatron-LM: Training multi-billion parameter language models using model parallelism, 2019. 2020 (v4 version) — Available at https://arxiv.org/abs/1909.08053.

36 Google Cloud. Enabling next-generation AI workloads: Announcing tpu v5p and AI hypercomputer. Blog post, 2023. Available at https://cloud.google.com/blog/products/ai-machine-learning/introducing-cloud-tpu-v5p-and-ai-hypercomputer.

37 NVIDIA. Wide open: NVIDIA accelerates inference on meta llama3. Blog post, 2024. Available at https://blogs.nvidia.com/blog/meta-llama3-inference-acceleration/.

38 Meta Llama 3 Researchers. Meta llama 3, 2024. https://ai.meta.com/blog/meta-llama-3/.

39 David Patterson, Joseph Gonzalez, Quoc Le, Chen Liang, Lluis-Miquel Munguia, Daniel Rothchild, David So, Maud Texier, and Jeff Dean. Carbon emissions and large neural network training, 2021. https://arxiv.org/abs/2104.10350.

40 Satyam Kumar, Dayima Musharaf, Seerat Musharaf, and Anil Kumar Sagar. A comprehensive review of the latest advancements in large generative AI models. In Rabindra Nath Shaw, Marcin Paprzycki, and Ankush Ghosh, editors, *Advanced Communication and Intelligent Systems*, pages 90–103, Cham, 2023. Springer Nature Switzerland. ISBN 978-3-031-45121-8.

41 Jilei Hou Ziad Asghar. World's first on-device demonstration of stable diffusion on an android phone, 2023. https://www.qualcomm.com/news/onq/2023/02/worlds-first-on-device-demonstration-of-stable-diffusion-on-android. Accessed: August, 28, 2024.

42 Google Cloud. Memory-optimized machine family for compute engine. Documentation, 2025. Last updated July 30, 2025; Available at https://cloud.google.com/compute/docs/memory-optimized-machines.

43 Microsoft. Microsoft releases SAP HANA standard sizing certified VM with close to 24TB memory. Tech Community blog post, 2022. Published Oct 12, 2022; Available at techcommunity.microsoft.com.

44 NVIDIA. H100 tensor core GPU — data center. Webpage, 2025. Accessed July 30, 2025; Available at https://www.nvidia.com/en-us/data-center/h100/.

45 Andrew Hard, Kanishka Rao, Rajiv Mathews, Swaroop Ramaswamy, Françoise Beaufays, Sean Augenstein, Hubert Eichner, Chloé Kiddon, and Daniel Ramage. Federated learning for mobile keyboard prediction. Google Research page, 2018. Accessed Jul 30, 2025; Available at https://research.google/pubs/federated-learning-for-mobile-keyboard-prediction-2/.

46 Yuyi Mao, Changsheng You, Jun Zhang, Kaibin Huang, and Khaled B. Letaief. Mobile Edge Computing: Survey and Research Outlook, 2017. Available at https://ar5iv.labs.arxiv.org/html/1701.01090.

47 Shashank Mohan Jain. *Introduction to Transformers for NLP: With the Hugging Face Library and Models to Solve Problems*. Apress, 2022. ISBN 9781484288443.

48 Yuyi Mao, Changsheng You, Jun Zhang, Kaibin Huang, and Khaled B. Letaief. A survey on mobile edge computing: The communication perspective. *IEEE Communications Surveys & Tutorials*, 19(4): 2322–2358, 2017.

49 Zhihang Yuan, Lin Niu, Jia-Wen Liu, Wenyu Liu, Xinggang Wang, Yuzhang Shang, Guangyu Sun, Qiang Wu, Jiaxiang Wu, and Bingzhe Wu. RPTQ: Reorder-based post-training quantization for large language models. *ArXiv, abs/2304.01089,* 2023. https://doi.org/10.48550/arXiv.2304.01089.

50 Guangxuan Xiao, Ji Lin, Mickael Seznec, Julien Demouth, and Song Han. Smoothquant: Accurate and efficient post-training quantization for large language models. *ArXiv, abs/2211.10438,* 2022. https://doi.org/10.48550/arXiv.2211.10438.

51 Zechun Liu, Barlas Og˘uz, Changsheng Zhao, Ernie Chang, Pierre Stock, Yashar Mehdad, Yangyang Shi, Raghuraman Krishnamoorthi, and Vikas Chandra. LLM-QAT: Data-free quantization aware training for large language models. *ArXiv*, abs/2305.17888, 2023. https://doi.org/10.48550/arXiv.2305.17888.

52 Xiuyu Li, Long Lian, Yijia Liu, Hua Yang, Zhen Dong, Daniel Kang, Shanghang Zhang, and Kurt Keutzer. Q-diffusion: Quantizing diffusion models. *ArXiv*, abs/2302.04304, 2023. https://doi.org/10.48550/arXiv.2302.04304.

53 Muyang Li, Ji Lin, Chenlin Meng, Stefano Ermon, Song Han, and Jun-Yan Zhu. Efficient spatially sparse inference for conditional GANs and diffusion models. *IEEE Transactions on Pattern Analysis and Machine Intelligence*, 45: 14465–14480, 2022. https://doi.org/10.1109/TPAMI.2023.3316020.

54 Yuzhang Shang, Zhihang Yuan, Bin Xie, Bingzhe Wu, and Yan Yan. Post-training quantization on diffusion models. *2023 IEEE/CVF Conference on Computer Vision and Pattern Recognition* (*CVPR*), pages 1972–1981, 2022. https://doi.org/10.1109/CVPR52729.2023.00196.

55 Torsten Hoefler, Dan Alistarh, Tal Ben-Nun, Nikoli Dryden, and Alexandra Peste. Sparsity in deep learning: pruning and growth for efficient inference and training in neural networks. *Journal of Machine Learning Research*, 22(1): 1–124, 2021. ISSN 1532-4435.

56 Woosuk Kwon, Sehoon Kim, Michael W. Mahoney, Joseph Hassoun, Kurt Keutzer, and Amir Gholami. A fast post-training pruning framework for transformers. *ArXiv*, abs/2204.09656, 2022. https://doi.org/10.48550/arXiv.2204.09656.

57 Xinyin Ma, Gongfan Fang, and Xinchao Wang. LLM-pruner: On the structural pruning of large language models. In *Thirty-seventh Conference on Neural Information Processing Systems*, 2023. https://openreview.net/forum?id=J8Ajf9WfXP.

58 Elias Frantar and Dan Alistarh. Sparsegpt: Massive language models can be accurately pruned in one-shot. *ArXiv*, abs/2301.00774, 2023. https://doi.org/10.48550/arXiv.2301.00774.

59 Xiaolong Ma, Minghai Qin, Fei Sun, Zejiang Hou, Kun Yuan, Yi Xu, Yanzhi Wang, Yen kuang Chen, Rong Jin, and Yuan Xie. Effective model sparsification by scheduled grow-and-prune methods. *ArXiv, abs/2106.09857*, 2021.

60 Edward J. Hu, Yelong Shen, Phillip Wallis, Zeyuan Allen-Zhu, Yuanzhi Li, Shean Wang, Lu Wang, and Weizhu Chen. Lora: Low-rank adaptation of large language models, arXiv:2106.09685, 2021. https://arxiv.org/abs/2106.09685.

61 Ning Ding, Xingtai Lv, Qiaosen Wang, Yulin Chen, Bowen Zhou, Zhiyuan Liu, and Maosong Sun. Sparse low-rank adaptation of pretrained language models. In *Conference on Empirical Methods in Natural Language Processing*, 2023. https://api.semanticscholar.org/CorpusID:265294736.

62 Longteng Zhang, Lin Zhang, S. Shi, X. Chu, and Bo Li. LoRA-FA: Memory-efficient low-rank adaptation for large language models fine-tuning. *ArXiv*, abs/2308.03303, 2023. https://doi.org/10.48550/arXiv.2308.03303.

63 Zhengmao Ye, Dengchun Li, Jingqi Tian, Tingfeng Lan, Jie Zuo, Lei Duan, Hui Lu, Yexi Jiang, Jian Sha, Ke Zhang, and Mingjie Tang. Aspen: High-throughput lora fine-tuning of large language models with a single GPU. *ArXiv, abs/2312.02515*, 2023. https://doi.org/10.48550/arXiv.2312.02515.

64 Yuhui Xu, Lingxi Xie, Xiaotao Gu, Xin Chen, Heng Chang, Hengheng Zhang, Zhensu Chen, Xiaopeng Zhang, and Qi Tian. QA-LoRA: Quantization-aware low-rank adaptation of large language models. *ArXiv, abs/2309.14717*, 2023. https://doi.org/10.48550/arXiv.2309.14717.

65 Geoffrey Hinton, Oriol Vinyals, and Jeff Dean. Distilling the knowledge in a neural network, arXiv:1503.02531, 2015. https://arxiv.org/abs/1503.02531.

66 Eric Luhman and Troy Luhman. Knowledge distillation in iterative generative models for improved sampling speed. *ArXiv, abs/2101.02388*, 2021.

67 Yuxian Gu, Li Dong, Furu Wei, and Minlie Huang. MiniLLM: Knowledge distillation of large language models, arXiv:2306.08543, 2024. https://arxiv.org/abs/2306.08543.

68 H. Brendan McMahan, Eider Moore, Daniel Ramage, and Blaise Agüera y Arcas. Federated learning of deep networks using model averaging. *CoRR, abs/1602.05629*, 2016. http://arxiv.org/abs/1602.05629.

69 Mariel Werner, Lie He, Michael Jordan, Martin Jaggi, and Sai Praneeth Karimireddy. Provably personalized and robust federated learning. *Transactions on Machine Learning Research*, 2023. ISSN 2835–8856. https://openreview.net/forum?id=B0uBSSUy0G.

70 Alexander Jung, Yasmin SarcheshmehPour, and Amirhossein Mohammadi. Personalized federated learning via active sampling, arXiv:2409.02064, 2024. https://arxiv.org/abs/2409.02064.

71 Samiul Alam, Luyang Liu, Ming Yan, and Mi Zhang. Fedrolex: Model-heterogeneous federated learning with rolling sub-model extraction. *arXiv preprint arXiv:2212.01548*, 2022.

72 Enmao Diao, Jie Ding, and Vahid Tarokh. HeteroFL: Computation and communication efficient federated learning for heterogeneous clients. *arXiv preprint arXiv:2010.01264*, 2020.

73 Alexander Jung, S Abdurakhmanova, and Y Sarcheshmehpour. Towards model-agnostic federated learning over networks. In *2023 31st European Signal Processing Conference (EUSIPCO)*, pages 1614–1618. IEEE, 2023.

74 Yinqiu Liu, Hongyang Du, Dusit Niyato, Jiawen Kang, Shuguang Cui, Xuemin Shen, and Ping Zhang. Optimizing mobile-edge AI-generated everything (AIGX) services by prompt engineering: Fundamental, framework, and case study, arXiv:2309.01065, 2023. https://arxiv.org/abs/2309.01065.

75 Yuxuan Chen, Rongpeng Li, Zhifeng Zhao, Chenghui Peng, Jianjun Wu, Ekram Hossain, and Honggang Zhang. NetGPT: An AI-native network architecture for provisioning beyond personalized generative services. *IEEE Network*, pages 1–1, 2024. https://doi.org/10.1109/MNET.2024.3376419.

76 Tuyen X. Tran and Danilo Pompili. Joint task offloading and resource allocation for multi-server mobile-edge computing networks. *IEEE Transactions on Vehicular Technology*, 68: 856–868, 2017. https://doi.org/10.1109/TVT.2018.2881191.

77 Chuan Sun, Hui Li, Xiuhua Li, Junhao Wen, Qingyu Xiong, Xiaofei Wang, and V. Leung. Task offloading for end-edge-cloud orchestrated computing in mobile networks. *2020 IEEE Wireless Communications and Networking Conference (WCNC)*, pages 1–6, 2020. https://doi.org/10.1109/WCNC45663.2020.9120496.

78 Umber Saleem, Yu Liu, Sobia Jangsher, Xiaoming Tao, and Yong Li. Latency minimization for d2d-enabled partial computation offloading in mobile edge computing. *IEEE Transactions on Vehicular Technology*, 69: 4472–4486, 2020. https://doi.org/10.1109/TVT.2020.2978027.

79 Wei Feng, Hao Liu, Yingbiao Yao, Diqiu Cao, and Mingxiong Zhao. Latency-aware offloading for mobile edge computing networks. *IEEE*

Communications Letters, 25(8): 2673–2677, 2021. ISSN 1558–2558. https://doi.org/10.1109/LCOMM.2021.3074621.

80 Samrat Nath and Jingxian Wu. Dynamic computation offloading and resource allocation for multi-user mobile edge computing. *GLOBECOM 2020 - 2020 IEEE Global Communications Conference*, pages 1–6, 2020. https://doi.org/10.1109/GLOBECOM42002.2020.9348161.

81 Jing Mei, Zhao Tong, Kenli Li, Lianming Zhang, and Keqin Li. Energy-efficient heuristic computation offloading with delay constraints in mobile edge computing. *IEEE Transactions on Services Computing*, 16(6): 4404–4417, Nov 2023. ISSN 1939–1374. https://doi.org/10.1109/TSC.2023.3324604.

82 Liang Huang, Suzhi Bi, and Y. Zhang. Deep reinforcement learning for online computation offloading in wireless powered mobile-edge computing networks. *IEEE Transactions on Mobile Computing*, 19: 2581–2593, 2018. https://doi.org/10.1109/TMC.2019.2928811.

83 Emily M. Bender, Timnit Gebru, Angelina McMillan-Major, and Shmargaret Shmitchell. On the dangers of stochastic parrots: Can language models be too big? In *Proceedings of the 2021 ACM Conference on Fairness, Accountability, and Transparency (FAccT '21)*, pages 610–623, 2021. https://doi.org/10.1145/3442188.3445922.

84 NeurIPS Datasets and Benchmarks Track Authors. Datasets and benchmarks track paper. In *Advances in Neural Information Processing Systems 36 (Datasets and Benchmarks Track)*, 2023. Accessed via PDF link: https://proceedings.neurips.cc/paper_files/paper/2023/file/b01153e7112b347d8ed54f317840d8af-Paper-Datasets_and_Benchmarks.pdf.

85 Felix Friedrich, Manuel Brack, Lukas Struppek, Dominik Hintersdorf, Patrick Schramowski, Sasha Luccioni, and Kristian Kersting. Auditing and instructing text-to-image generation models on fairness. *AI Ethics*, 5(3): 2103–2123, 2025. https://doi.org/10.1007/s43681-024-00531-5. (PMCID: PMC12103484).

86 Milad Nasr, Nicholas Carlini, Jonathan Hayase, Matthew Jagielski, A. Feder Cooper, Daphne Ippolito, Christopher A. Choquette-Choo, Eric Wallace, Florian Tramèr, and Katherine Lee. Extracting training data from ChatGPT. Not Just Memorization blog, 2023. Available at https://not-just-memorization.github.io/extracting-training-data-from-chatgpt.html.

87 Nicolas Carlini, Jamie Hayes, Milad Nasr, Matthew Jagielski, Vikash Sehwag, Florian Tramèr, Borja Balle, Daphne Ippolito, and Eric Wallace. Extracting training data from diffusion models. In *32nd USENIX Security Symposium (USENIX Security 23)*, pages 5253–5270, Anaheim, CA, 2023. ISBN 978-1-939133-37-3.

88 Xuefei Cao, Minghong Fang, Jinyuan Liu, and Neil Zhenqiang Gong. Fltrust: Byzantine-robust federated learning via trust bootstrapping. In *Proceedings of NDSS*, 2021.

89 Microsoft. Architecting secure generative AI applications: Preventing indirect prompt injection attacks. Microsoft Security Blog post, 2024. Available at https://techcommunity.microsoft.com/blog/microsoft-security-blog/architecting-secure-gen-ai-applications-preventing-indirect-prompt-injection-att/.

90 Microsoft Security Response Center. How Microsoft defends against indirect prompt injection attacks. MSRC blog post, 2025. Available at https://msrc.microsoft.com/blog/2025/07/how-microsoft-defends-against-indirect-prompt-injection-attacks/.

91 Luca Beurer-Kellner, Beat Buesser, Ana-Maria Crețu, Edoardo Debenedetti, Daniel Dobos, Daniel Fabian, Marc Fischer, David Froelicher, Kathrin Grosse, Daniel Naeff, Ezinwanne Ozoani, Andrew Paverd, Florian Tramèr, and Václav Volhejn. Design patterns for securing LLM agents against prompt injections. *arXiv preprint arXiv:2506.08837v3,* 2025.

92 Jiwen Yu, Xuanyu Zhang, Youmin Xu, and Jian Zhang. Cross: Diffusion model makes controllable, robust and secure image steganography. Advances in Neural Information Processing Systems, 36 (2023):80730–80743. 2023.

93 Coalition for Content Provenance and Authenticity. C2pa specification version 2.2. Technical Specification, 2025. Version 2.2, Available at https://spec.c2pa.org/specifications/specifications/2.2/specs/C2PA_Specification.html.

94 OpenAI. C2pa in ChatGPT images. OpenAI Help Center article, 2024. Updated slightly in 2025; accessed Jul 30, 2025 via Help Center (ID 8912793); Available at https://help.openai.com/en/articles/8912793-c2pa-in-chatgpt-images.

95 Pierre Fernandez, Guillaume Couairon, Hervé Jégou, Matthijs Douze, and Teddy Furon. The stable signature: Rooting watermarks in latent diffusion models, arXiv:2303.15435, 2023. Available at https://arxiv.org/abs/2303.15435.

96 Yuxin Wen, John Kirchenbauer, Jonas Geiping, and Tom Goldstein. Tree-ring watermarks: Invisible fingerprints for diffusion images. In *Advances in Neural Information Processing Systems 36* (*NeurIPS 2023*), 2023. Main conference track.

97 European Commission. Artificial intelligence act. EU regulatory framework page, 2024. Regulation (EU) 2024/1689; first comprehensive AI law in force; Available at https://digital-strategy.ec.europa.eu/en/policies/regulatory-framework-ai.

98 European Parliament & Council of the European Union. Article 50: Transparency obligations for providers and deployers of certain AI systems. Regulation (EU) 2024/1689, 2024. Official transparency provisions for Article 50, Available via AI Act Explorer or EUR-Lex.

99 Amanda K. Sarata. The HIPAA privacy rule: Overview and issues. CRS In Focus (#IF12759), Library of Congress / Congress.gov, 2024. Provides federal standards for the protection of personal health information (HIPAA Privacy Rule).

100 OpenAI. March 20 ChatGPT outage: Here's what happened. OpenAI blog post, 2023. Explains data leakage and service disruption; Available at https://openai.com/index/march-20-chatgpt-outage/.

101 Michael Duan, Anshuman Suri, Niloofar Mireshghallah, Sewon Min, Weijia Shi, Luke Zettlemoyer, Yulia Tsvetkov, Yejin Choi, David Evans, and Hannaneh Hajishirzi. Do membership inference attacks work on large language models?, arXiv:2402.07841, 2024. Accepted at the Conference on Language Modeling (COLM) 2024; Available at https://arxiv.org/abs/2402.07841v2.

102 Wenjie Fu, Huandong Wang, Chen Gao, Guanghua Liu, Yong Li, and Tao Jiang. Membership inference attacks against fine-tuned large language models via self-prompt calibration. In *NeurIPS 2024 Workshop on Privacy (poster)*, 2024. Available on Open Review at https://openreview.net/forum?id=PAWQvrForJ.

103 Hailong Hu and Jun Pang. Membership inference of diffusion models, 2023. Available as arXiv preprint.

104 Mike Caulfield. ChatGPT is changing the phishing game, 2023. https://www.securityinfowatch.com/cybersecurity/information-security/breach-detection/article/53057705/chatgpt-is-changing-the-phishing-game. Accessed: May. 17, 2023.

105 Jaikumar Vijayan. Attackers are already exploiting ChatGPT to write malicious code, 2023. https://www.darkreading.com/attacks-breaches/attackers-are-already-exploiting-chatgpt-to-write-malicious-code. Accessed: May. 17, 2023.

106 Lingjuan Lyu. A pathway towards responsible AI-generated content, 2023. https://towardsdatascience.com/a-pathway-towards-responsible-ai-generated-content-6c915e8155f9. Accessed: May. 17, 2023.

107 Matthew Leung and Sahar Niazi. Universities on alert over ChatGPT and other AI-assisted tools, 2023. https://www.universityworldnews.com/post.php?story=20230222132357841. Accessed: May. 17, 2023.

108 Amjad Aldweesh. The impact of blockchain on digital content distribution: a systematic review. *Wireless Networks*, pages 763–779, 2023. https://link.springer.com/article/10.1007/s11276-023-03524-0.

109 Aki J. Serada, Tanja Sihvonen, and J. Tuomas Harviainen. Cryptokitties and the new ludic economy: How blockchain introduces value, ownership, and scarcity in digital gaming. *Games and Culture*, 16(4): 457–480, 2021. https://journals.sagepub.com/doi/abs/10.1177/1555412019877302.

110 Yutao Li, Chunhua Chen, Ning Liu, Huajie Huang, Zibin Zheng, and Qiben Yan. A blockchain-based decentralized federated learning framework with committee consensus. *IEEE Network*, 35(1): 234–241, 2020. https://ieeexplore.ieee.org/document/9357544.

111 Xiaofei Liu, Yansha Deng, and Tao Mahmoodi. Wireless distributed learning: A new hybrid split and federated learning approach. *IEEE Transactions on Wireless Communications*, 22(4): 2650–2665, 2023.

112 Jiawen Kang, Hongyang Du, Dinh C. Nguyen, Zehui Xiong, Dusit Niyato, and Chong Zhang. Blockchain-based federated learning for industrial metaverses: Incentive scheme with optimal AOI. In *Proc. IEEE Int. Conf. Blockchain (Blockchain)*, pages 71–78, Espoo, Finland, 2022.

113 Jiawen Kang, Hongyang Du, Dinh C. Nguyen, Zehui Xiong, Dusit Niyato, and Chong Zhang. Communication-efficient and cross-chain empowered federated learning for artificial intelligence of things. *IEEE Transactions on Network Science and Engineering*, 9(5): 2966–2977, 2022.

114 Lingjuan Cui, Zhihao Shi, Zhuo Zou, and Mikael Gidlund. Security and privacy-enhanced federated learning for anomaly detection in IoT infrastructures. *IEEE Transactions on Industrial Informatics*, 18(5): 3492–3500, 2022.

115 Yuxin Wei and Gareth Tyson. Understanding the impact of AI generated content on social media: The pixiv case. *arXiv preprint arXiv:2402.18463*, 2024. https://arxiv.org/abs/2402.18463.

116 Johannes D. Brüns and Martin Meißner. Do you create your content yourself? using generative artificial intelligence for social media content creation diminishes perceived brand authenticity. *Journal of Retailing and Consumer Services*, 79: 103790, 2024. https://doi.org/10.1016/j.jretconser.2024.103790.

117 Hongyang Du, Jiacheng Wang, Dusit Niyato, Jiawen Kang, Zehui Xiong, and Dong In Kim. AI-generated incentive mechanism and full-duplex semantic communications for information sharing. *techrxiv*, 2023. https://doi.org/10.36227/techrxiv.22209178.

118 Sheyda Zarandi, Ata Khalili, Mehdi Rasti, and Hina Tabassum. Multi-objective energy efficient resource allocation and user association for in-band full duplex small-cells. *IEEE Transactions on Green Communications and Networking*, 4(4): 1048–1060, 2020. https://doi.org/10.1109/TGCN.2020.3006794.

119 Xumin Huang, Peichun Li, Hongyang Du, Jiawen Kang, Dusit Niyato, Dong In Kim, and Yuan Wu. Federated learning-empowered AI-generated content in wireless networks. *IEEE Network*, pages 1–1, 2024. ISSN 1558-156X. https://doi.org/10.1109/MNET.2024.3353377.

120 Hongyang Du, Guangyuan Liu, Dusit Niyato, Jiayi Zhang, Jiawen Kang, Zehui Xiong, Bo Ai, and Dong In Kim. Generative AI-aided joint training-free secure semantic communications via multi-modal prompts, 09 2023.

121 Hongyang Du, Zonghang Li, Dusit Niyato, Jiawen Kang, Zehui Xiong, Huawei Huang, and Shiwen Mao. Diffusion-based reinforcement learning for edge-enabled AI-generated content services. *IEEE Transactions on Mobile Computing*, pages 1–16, 2024. ISSN 1558-0660. https://doi.org/10.1109/TMC.2024.3356178.

122 Yijing Lin, Zhipeng Gao, Hongyang Du, Dusit Niyato, Jiawen Kang, Ruilong Deng, and Xuemin Sherman Shen. A unified blockchain-semantic framework for wireless edge intelligence enabled web 3.0. *IEEE Wireless Communications*,

31(2): 126–133, 2024. ISSN 1558–0687. https://doi.org/10.1109/MWC.018.2200568.

123 Yinqiu Liu, Hongyang Du, Dusit Niyato, Jiawen Kang, Zehui Xiong, Chunyan Miao, Xuemin Shen, and Abbas Jamalipour. Blockchain-empowered lifecycle management for AI-generated content products in edge networks. *IEEE Wireless Communications*, 31(3): 286–294, 2024. ISSN 1558–0687. https://doi.org/10.1109/MWC.003.2300053.

124 Jiacheng Wang, Hongyang Du, Dusit Niyato, Jiawen Kang, Shuguang Cui, Xuemin Shen, and Ping Zhang. Generative AI for integrated sensing and communication: Insights from the physical layer perspective, 2023.

125 Jiacheng Wang, Hongyang Du, Dusit Niyato, Zehui Xiong, Jiawen Kang, Shiwen Mao, and Xuemin Sherman Shen. Guiding AI-generated digital content with wireless perception. *IEEE Wireless Communications*, pages 1–8, 2024. ISSN 1558–0687. https://doi.org/10.1109/MWC.008.2300162.

126 Hongyang Du, Ruichen Zhang, Dusit Niyato, Jiawen Kang, Zehui Xiong, Dong In Kim, Xuemin Shen, and H. Vincent Poor. Exploring collaborative distributed diffusion-based AI-generated content (AIGC) in wireless networks. *IEEE Network*, 38(3): 178–186, 2024. ISSN 1558–156X. https://doi.org/10.1109/MNET.006.2300223.

127 Yinqiu Liu, Hongyang Du, Dusit Niyato, Jiawen Kang, Zehui Xiong, Dong In Kim, and Abbas Jamalipour. Deep generative model and its applications in efficient wireless network management: A tutorial and case study, *IEEE Wireless Communications*, 31(4): 199–207, 2023.

128 Christoph Schuhmann, Richard Vencu, Romain Beaumont, Robert Kaczmarczyk, Clayton Mullis, Aarush Katta, Theo Coombes, Jenia Jitsev, and Aran Komatsuzaki. LAION-400M: Open dataset of CLIP-filtered 400 million image-text pairs. *arXiv preprint arXiv:2111.02114*, 2021. https://arxiv.org/abs/2111.02114.

129 Jiayuan Chen, Changyan Yi, Hongyang Du, Dusit Niyato, Jiawen Kang, Jun Cai, Xuemin, and Shen. A revolution of personalized healthcare: Enabling human digital twin with mobile AIGC, *IEEE Network*, 38(6):234–242. 2023.

130 Minrui Xu, Dusit Niyato, Hongliang Zhang, Jiawen Kang, Zehui Xiong, Shiwen Mao, and Zhu Han. Joint foundation model caching and inference of generative AI services for edge intelligence. In *GLOBECOM 2023 - 2023 IEEE Global Communications Conference*, pp. 3548–3553. IEEE, 2023.

131 Minrui Xu, Dusit Niyato, Junlong Chen, Hongliang Zhang, Jiawen Kang, Zehui Xiong, Shiwen Mao, and Zhu Han. Generative AI-empowered simulation for autonomous driving in vehicular mixed reality metaverses. *IEEE Journal of Selected Topics in Signal Processing*, 17(5): 1064–1079, 2023. ISSN 1941–0484. https://doi.org/10.1109/JSTSP.2023.3293650.

132 Ruichen Zhang, Ke Xiong, Hongyang Du, Dusit Niyato, Jiawen Kang, Xuemin Shen, and H. Vincent Poor. Generative AI-enabled vehicular networks: Fundamentals, framework, and case study, 2023.

133 Baoxia Du, Hongyang Du, Haifeng Liu, Dusit Niyato, Peng Xin, Jun Yu, Mingyang Qi, and You Tang. Yolo-based semantic communication with generative AI-aided resource allocation for digital twins construction, 2023.

134 Minrui Xu, Dusit Niyato, Hongliang Zhang, Jiawen Kang, Zehui Xiong, Shiwen Mao, and Zhu Han. Sparks of GPTS in edge intelligence for metaverse: Caching and inference for mobile AIGC services, arXiv preprint arXiv:2304.08782, 2023.

135 Yue Yang, Zihan Zhang, Yunan Tian, Zhihua Yang, Chao Huang, Chen Zhong, and Kai Kit Wong. Over-the-air split machine learning in wireless MIMO networks. *IEEE Journal on Selected Areas in Communications*, 41(4): 1007–1022, 2023.

136 Hina Tabassum, Ekram Hossain, Md. Jahangir Hossain, and Dong In Kim. On the spectral efficiency of multiuser scheduling in RF-powered uplink cellular networks. *IEEE Transactions on Wireless Communications*, 14(7): 3586–3600, 2015. https://doi.org/10.1109/TWC.2015.2408328.

137 Zhaohui Lin, Shenghui Bi, and Ying Jun Angela Zhang. Optimizing AI service placement and resource allocation in mobile edge intelligence systems. *IEEE Transactions on Wireless Communications*, 20(11): 7257–7271, 2021.

138 Saba Asaad, Hina Tabassum, Chongjun Ouyang, and Ping Wang. Joint antenna selection and beamforming for massive MIMO-enabled over-the-air federated learning. *IEEE Transactions on Wireless Communications*, 23(8): 8603–8618, 2024. https://doi.org/10.1109/TWC.2024.3352504.

139 Mengyuan Yao, Lixing Chen, Jun Zhang, Jie Huang, and Jun Wu. Loading cost-aware model caching and request routing for cooperative edge inference. In *ICC 2022 - IEEE International Conference on Communications*, pages 2327–2332. IEEE, 2022.

140 Minrui Xu, Dusit Niyato, Zehui Yang, Zehui Xiong, Jiawen Kang, Dong In Kim, and Xuemin (Sherman) Shen. Privacy-preserving intelligent resource allocation for federated edge learning in quantum internet. *IEEE Journal of Selected Topics in Signal Processing*, 17(1): 142–157, 2022.

141 Hongyang Du, Dusit Niyato, Yanjie A. Xie, Yu Cheng, Jiawen Kang, and Dong In Kim. Performance analysis and optimization for jammer-aided multiantenna UAV covert communication. *IEEE Journal on Selected Areas in Communications*, 40(10): 2962–2979, 2022.

142 J. Wang, K. Zhu, and E. Hossain. Green internet of vehicles (IoV) in the 6G era: Toward sustainable vehicular communications and networking. *IEEE Transactions on Green Communications and Networking*, 6(1): 391–423, 2021.

143 Md Tanvir Hossan and Hina Tabassum. Mobility-aware performance in hybrid RF and terahertz wireless networks. *IEEE Transactions on Communications*, 70(2): 1376–1390, 2021.

144 Hongyang Du, Jiawen Kang, Dusit Niyato, Jie Zhang, and Dong In Kim. Reconfigurable intelligent surface-aided joint radar and covert communications: Fundamentals, optimization, and challenges. *IEEE Vehicular Technology Magazine*, 17(3): 54–64, 2022.

145 Jie Ren, Guanding Yu, Yuchen He, and Geoffrey Ye Li. Collaborative cloud and edge computing for latency minimization. *IEEE Transactions on Vehicular Technology*, 68(5): 5031–5044, 2019.

146 Haider Shoaib and Hina Tabassum. Optimization of speed and network deployment for reliable v2i communication in the presence of handoffs and interference. *IEEE Wireless Communications Letters*, 12(6): 1051–1055, 2023. https://doi.org/10.1109/LWC.2023.3259719.

147 Mohammad Amin Saeidi, Haider Shoaib, and Hina Tabassum. A tractable handoff-aware rate outage approximation with applications to thz-enabled vehicular network optimization. In *GLOBECOM 2023 - 2023 IEEE Global Communications Conference*, pages 5092–5097, 2023. https://doi.org/10.1109/GLOBECOM54140.2023.10436904.

148 Zijiang Yan, Wael Jaafar, Bassant Selim, and Hina Tabassum. Multi-UAV speed control with collision avoidance and handover-aware cell association: DRL with action branching. In *GLOBECOM 2023 - 2023 IEEE Global Communications Conference*, pages 5067–5072, 2023. https://doi.org/10.1109/GLOBECOM54140.2023.10436730.

149 Zijiang Yan and Hina Tabassum. Reinforcement learning for joint v2i network selection and autonomous driving policies. In *GLOBECOM 2022 - 2022 IEEE Global Communications Conference*, pages 1241–1246, 2022. https://doi.org/10.1109/GLOBECOM48099.2022.10001396.

8

Applications of GenAI on Wireless and Cybersecurity

Brian Kim[1,], Yalin E. Sagduyu[2], Tugba Erpek[2], Yi Shi[3], and Sennur Ulukus[4]*

[1] *DEVCOM U.S. Army Research Laboratory, Adelphi, MD, USA*
[2] *Nexcepta, Inc., Gaithersburg, MD, USA*
[3] *Commonwealth Cyber Initiative, Virginia Tech, Arlington, VA, USA*
[4] *Department of Electrical and Computer Engineering, University of Maryland, College Park, MD, USA*

*Corresponding author: brian.kim628@gmail.com

8.1 Introduction to GenAI in Wireless and Cybersecurity

Recent advances in deep learning (DL) have revolutionized how wireless communication systems operate, enabling them to handle increasingly complex challenges in dynamic environments. The integration of DL within wireless communication systems harnesses its ability to learn rich data representations from large-scale datasets, enhancing the efficiency and adaptability of these systems. The specific use cases, such as spectrum sensing [1], channel estimation [2–4], and signal classification [5], are prime examples of how DL transforms communication technologies.

The application of generative AI (GenAI) techniques, such as generative adversarial networks (GANs) and variational encoders (VAEs), is a newer yet promising approach in wireless communication. GenAI refers to models that generate new data, solutions, or insights by learning from existing data, which can significantly benefit wireless communication systems where generating new data is inherently challenging due to the time-varying channel environments. Key applications of GenAI used for wireless communications are network design/optimization [6], resource allocation [7, 8], and channel estimation [9, 10]. An example of using GenAI for network optimization was studied in Ref. [11], which estimated the load

Generative AI for Communications Systems: Fundamentals, Applications, and Prospects,
First Edition. Edited by Diep N. Nguyen et al.

at the cell using a GAN-based algorithm for 5G mobile communication networks to enhance performance by making correct real-time decisions. Further, estimating the channel using synthetic training data generated by GenAI had been studied in Ref. [12], where GAN was used to generate channel samples according to its distribution and use this knowledge as prior information to estimate the current channel. The topics covered in this chapter are shown in Figure 8.1.

As these GenAI applications enhance the performance of wireless communication systems, cybersecurity becomes increasingly important as cyber threats continue to evolve and perform more sophisticated attacks, mainly due to the inherent openness of wireless networks. There are various types of cyber attacks in general, as seen in Figure 8.2, such as phishing, malware, denial of service (DoS), distributed denial of service (DDoS), man-in-the-middle (MitM), and spoofing attacks. Similarly, the wireless domain is susceptible to physical layer security attacks such as eavesdropping, jamming, Trojan horse attacks, and signal spoofing. These attacks aim to exploit vulnerabilities in systems for malicious purposes where the attack undermines the confidentiality, integrity, and availability of communication systems.

Specifically, phishing is a threat where attackers seek to obtain private information, such as personal details and passwords, by sending deceptive messages that appear to come from trustworthy sources. Another threat, known as a DoS attack, is designed to disrupt the normal operation of a computer, network, or service by overwhelming it with an excessive amount of requests or traffic loads. A variation of this, the DDoS attack, involves using multiple compromised devices to flood a target system with excessive traffic loads, overwhelming its resources and causing service disruption. In an MitM attack, the attacker secretly intercepts, alters, or

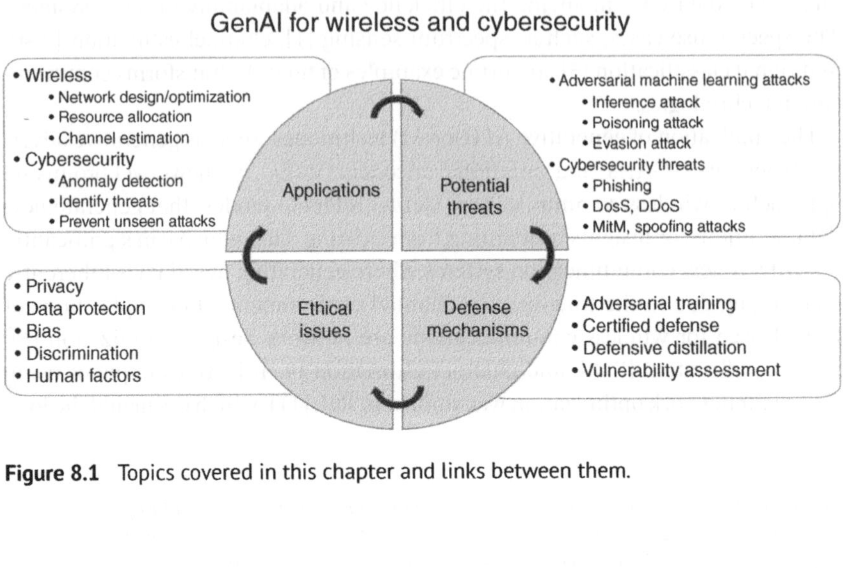

Figure 8.1 Topics covered in this chapter and links between them.

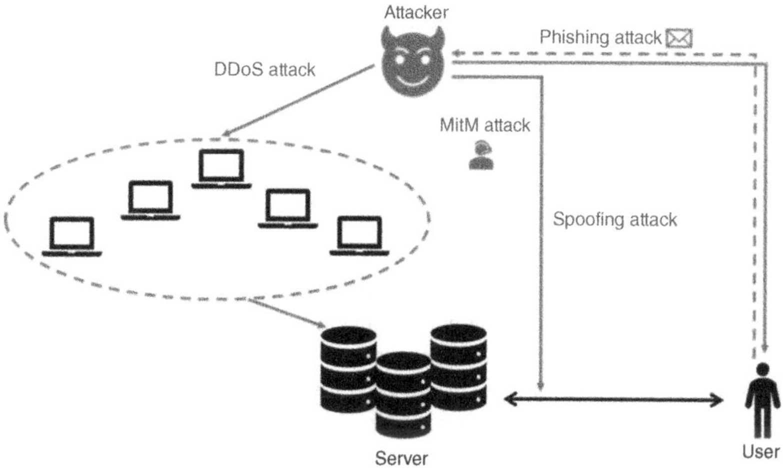

Figure 8.2 Different types of cyber threats.

monitors communications between two parties without their awareness. A spoofing attack involves the attacker impersonating a legitimate entity or falsifying their identity to deceive the target, often to gain unauthorized access to systems and steal sensitive information.

As these cybersecurity threats emerge, not only does GenAI play an important role in wireless communications, but it also plays an increasingly vital role in cybersecurity. As previously mentioned, GenAI can effectively learn insights by learning from existing data, which also can be effectively applied to detect anomalies, identify emerging threats, and predict potential attacks. Additionally, GenAI significantly strengthens cybersecurity defenses by creating new, synthetic attack scenarios based on known attack patterns that can be used to prevent unseen attacks and enhance overall security measures [13–16].

However, these GenAI applications themselves are susceptible to various types of attacks, such as evasion attacks (adversarial attacks), inference attacks, poisoning attacks, and backdoor (Trojan) attacks, where the details of which will be discussed in the following sections. Also, GenAI applications can be used by adversaries for adversarial purposes such as generating convincing phishing content, deepfake images, and analyzing traffic patterns and vulnerabilities. Therefore, it is essential to understand the vulnerabilities of the GenAI applications that we are using and how adversaries can take advantage of them. To this end, we highlight defense mechanisms for different types of attacks and discuss ethical issues related to GenAI applications for wireless communications and cybersecurity.

8.2 Adversarial Machine Learning in Wireless Communications

GenAI algorithms used to address the challenges in wireless communications are inherently vulnerable to different machine learning (ML) attacks due to their reliance on patterns learned from data, which can be subtly manipulated. These vulnerabilities stem from the high-dimensional nature of input spaces, where even small perturbations, which are hard to recognize, can cause a model to make incorrect predictions with high confidence. As adversaries can manipulate the inputs during both the training and testing phases of GenAI, the field of *adversarial ML* has emerged to examine how ML models behave in the presence of adversaries and to ensure the safe integration of ML in emerging applications [17]. In particular, DNNs are highly vulnerable to even minor adversarial attacks, as shown in the computer vision domain [18]. In wireless systems, adversarial attacks can manifest as carefully crafted perturbations are easily added to the input signals, such as introducing noise or modifying transmitted waveforms to mislead GenAI models. For instance, an adversarial signal could make a spectrum sensing model fail to detect a legitimate signal, leading to spectrum interference or inefficient resource usage. Similarly, attacks on models used for signal estimation could degrade signal quality, resulting in communication outages or reduced system performance. Also, the shared and broadcast nature of the wireless medium further increases the risk of adversaries tampering with GenAI applications used for wireless communication over the air. However, transferring attack and defense strategies from other domains, like computer vision, to wireless communications is not straightforward due to their distinct characteristics, such as channel effects and the broadcast nature of the wireless medium, as explained below.

- **Propagation Characteristics:** In wireless systems, signals propagate through complex and often unpredictable environments, encountering multipath fading and Doppler shifts. Adversarial perturbations applied to transmitted signals must survive these distortions to remain effective at the receiver, adding complexity compared to the relatively controlled domain of image processing.
- **Signal-to-Noise Ratio (SNR) Constraints:** Wireless communication systems operate within a defined SNR regime. Adversarial perturbations must be crafted carefully to remain within noise floor and avoid detection, which is a more stringent requirement than in computer vision, where pixel-level perturbations do not face such physical constraints.
- **Attack Objectives:** In computer vision, adversarial attacks often aim to cause misclassification or misidentification, but in wireless communications, the objectives may include jamming, eavesdropping, or interference in

resource allocation. These goals require a deeper understanding of communication protocols and physical layer parameters.

- **Real-time and Distributed Nature:** Wireless communication systems operate in real-time and are distributed across multiple users and devices, making it more challenging to coordinate attacks. Adversarial attacks must be synchronized and computationally efficient to disrupt communication effectively.

In this section, we study different types of attacks that aim to manipulate GenAI algorithms used for wireless communications and corresponding defense mechanisms.

8.2.1 Different Types of Attacks Against GenAI-driven Wireless Applications

Recent studies [19, 20] have explored various types of attacks, highlighting the vulnerabilities of DL-based models when applied to wireless networks. These attacks, which can manipulate the behavior of GenAI models during both training and inference phases, pose a considerable threat to essential wireless communication tasks such as signal classification, spectrum sensing, and power allocation. As illustrated in Figure 8.3, the landscape of adversarial threats is diverse, with different types of attacks targeting specific components and phase of wireless systems and GenAI model. Understanding and mitigating these attacks is critical for ensuring the robustness and security of future wireless communication networks, making it an active area of research in the field.

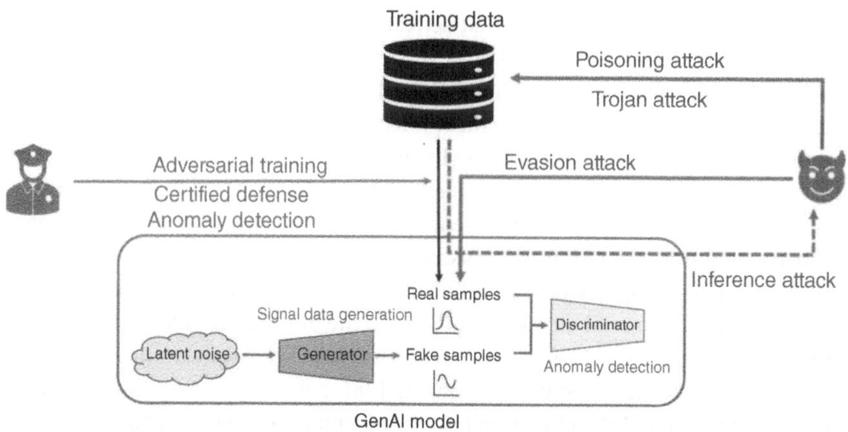

Figure 8.3 Different types of attacks against GenAI application and defense mechanisms.

Exploratory (inference) Attacks: One category of attacks, known as exploratory (inference) attacks against generative models, was first examined in Ref. [21]. In these attacks, an adversary attempts to recognize previously trained inputs without direct access to the training data. Specifically, the adversary queries a crafted input to the model to observe the model's response where the model tends to behave differently for training samples, such as giving a higher confidence score. Inference attacks targeting wireless ML applications were explored in Refs. [22, 23], where adversaries seek to determine whether a specific data sample was used during the model's training phase. This knowledge can reveal key waveform, channel, and radio hardware characteristics, which could later be exploited for adversarial purposes. Additionally, model inference attacks, which focus on constructing a deep neural network (DNN) that replicates the behavior of a legitimate user's wireless signal classifier, were studied in Refs. [24–26]. Furthermore, an inference attack against GenAI model used for RF fingerprinting was analyzed in Ref. [27].

Poisoning (causative) Attacks: Another type of attack, called poisoning (causative) attacks, aims to manipulate the training data to cause a malfunction of the GenAI algorithm, such as a decrease in overall accuracy during the test phase. An example of poisoning attacks was explored in Refs. [28, 29], where the adversary manipulates the transmitter's spectrum sensing data (training data) by transmitting during the spectrum sensing period, thereby poisoning the data. These poisoning attacks have also been extended to cooperative spectrum sensing [30, 31] and IoT data aggregation [32]. Further, Jin and Li [33] studied the effect of poisoning attacks on GenAI applications in a federated learning framework. A similar type of attack, called backdoor or Trojan attacks, has been studied in Ref. [34] involving adversaries subtly modifying training data by embedding Trojans (in the form of phase shifts) that activate during testing.

Adversarial (evasion) Attacks: Adversarial (evasion) attacks introduce small modifications to the input data that go into the DNNs such that the DL algorithm cannot reliably classify the input data or intentionally change the class of the input from the true class to another class. In the computer vision domain, Kos et al. [35] has shown that GenAI models, such as VAE and VAE-GAN, are susceptible to adversarial attacks where the latent space of the generative model gets perturbed. For wireless communications, Sadeghi and Larsson [36] has shown that the end-to-end autoencoder communication systems, proposed in Ref. [37], are vulnerable to adversarial attacks that increase the block-error-rate at the receiver. Adversarial attacks were also considered in various ML applications for wireless communications such as spectrum sensing [38], power allocation [39, 40], beam selection [41], and MIMO channel state information (CSI) feedback [42]. We summarize these attacks in Table 8.1.

Table 8.1 Different types of attacks and their impacts.

Attacks	Time	Impact
Exploratory attacks	Inference	Compromise data privacy, Undermines model integrity
Poisoning attacks	Training	Performance degradation, Biased predictions
Adversarial attacks	Inference	Bypass detection, Manipulate classification

8.2.2 Defense Against Adversarial Attacks for GenAI-driven Wireless Applications

To counter attacks, various defense strategies have been explored to enhance the robustness of GenAI applications used in wireless communications. One such defense is *adversarial training* [43, 44], which involves incorporating adversarial examples, artificially generated attacks, into the training data to make the GenAI model more resilient to adversarial manipulation. However, this kind of defense is only helpful if we know the attack type in advance to create an adversarial training dataset. Thus, another defense, called *certifiable defense*, was discussed in Ref. [45], which provides guarantees of robustness against adversarial attacks within a specific perturbation range. This approach involves adding random noise to the training data to smooth the model's decision boundaries, making it more difficult for attackers to cause misclassifications through small input changes. Additionally, defensive distillation, explored in Refs. [46, 47], leverages a smaller "student" model trained by a larger "teacher" model, which results in a more robust model due to its smoother decision boundaries.

8.3 GenAI for Wireless Security and Cybersecurity

GenAI holds significant promise for transforming the security of wireless communications and cybersecurity by addressing data scarcity, adapting to dynamic environments, and enhancing both offensive and defensive capabilities. The ability of GenAI to synthesize data and simulate real-world conditions has unlocked new opportunities to address challenges inherent to dynamic and complex wireless and cyber environments. To this end, we explore how GenAI can be exploited for wireless security and cybersecurity in this section.

8.3.1 GenAI for Wireless Security

GenAI enables systems to adapt to dynamic conditions, augment limited datasets, and improve the reliability and security of wireless operations. Specifically, GenAI

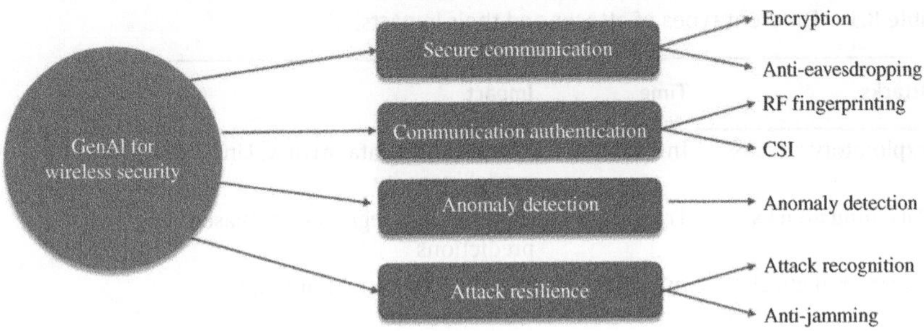

Figure 8.4 Different implementations of GenAI for wireless security.

is being employed in anomaly detection, authentication, and attack defense by leveraging the ability to learn the underlying features of legitimate input data. More details about different applications of GenAI for wireless security are presented in Figure 8.4. The key highlighted use cases of GenAI for wireless security are as follows.

GenAI for Secure Communication: Recently, GenAI has gained significant attention for enhancing confidentiality in wireless communications, particularly at the physical and semantic layers. Unlike traditional methods that encrypt raw bits, GenAI enables obfuscation at the semantic level, rendering messages unintelligible to adversaries without the appropriate semantic keys. Semantic communication techniques leveraging GenAI have been explored in Refs. [48, 49], demonstrating improved reliability through semantic-aware data encoding and compression. The use of GenAI to mitigate eavesdropping threats has also been investigated in recent literature. In these approaches, GenAI is employed to reconstruct or alter signals in ways that prevent unauthorized decoding. For example, Wang et al. [50] utilizes diffusion models to generate protective signals that inhibit unauthorized devices from extracting true CSI during wireless sensing.

GenAI for Communication Authentication: Roy et al. [51] studied an adversarial learning technique to identify rogue RF transmitters and classify legitimate ones by implementing GANs. To detect such adversarial activities, GAN exploits in-phase and quadrature imbalances that are present in all transmitters to learn the unique high-dimensional features that can be used as fingerprints for transmitter authentication. Similar work has been investigated in Ref. [52], where they use GAN to authenticate a transmitting device by measuring CSI.

GenAI for Anomaly Detection: Ayanoglu et al. [53] investigated GAN applications for spectrum sharing, anomaly detection, and security attack mitigation within NextG networks. Here, GANs' ability to synthesize high-resolution field data and anomaly detection in signal classification has been leveraged to enhance the competitive resource allocation problem. The research underscores GANs' comparative advantages over autoencoders and explores their potential to enable efficient and robust NextG communications. Further, Ref. [54] has exploited GAN to detect abnormal signals in the spectrum, which can help the system to apply abnormality mitigation strategies.

GenAI for Attack Resilience: Davaslioglu et al. [55] explored the application of GANs within the context of end-to-end autoencoder communications (AECs) to enhance 6G systems by mitigating interference and improving reliability under adversarial conditions. AECs, comprising DNNs for encoding and decoding, outperformed traditional communication methods by achieving lower bit-error rates under interference, limited training data, and practical implementation constraints. The GAN-augmented framework aided in the training phase of the encoder and decoder networks to adapt to unknown and evolving jamming attacks, enhancing the system's robustness. This approach was further fortified with interference training and randomized smoothing, allowing the system to suppress interference by up to 36 dB, even across multiple OFDM symbols and in MIMO setups. As another approach to being resilient to attacks, GAN-based preprocessing techniques were introduced in Ref. [56] to defend against adversarial attacks in wireless modulation classification.

The studies above illustrate the versatility of GenAI in the security of wireless communications, but they also reveal the future direction of research regarding the security of wireless communications. Extensions could include:

- **Explainable Generative Models:** Developing explainable GANs tailored for wireless applications could improve trust in AI-driven decisions, particularly for security-critical systems such as military networks and emergency communications.
- **Multidomain Integration:** Future research could explore combining GANs with other generative models like VAEs to synthesize data across diverse wireless domains, such as mmWave, THz, and satellite communications.
- **Real-time Adaptation:** Leveraging GenAI for real-time adaptation in highly dynamic environments, such as vehicular and drone networks, could enhance system reliability and security under varying conditions.
- **Federated GenAI:** Distributed GenAI models, integrated with federated learning, could allow collaborative data synthesis across nodes while preserving privacy and reducing reliance on centralized datasets.
- **Cross-layer Optimization:** GenAI could extend beyond the physical layer to jointly optimize MAC, network, and application-layer performance, facilitating end-to-end wireless network efficiency and security.
- **Adversarial Robustness:** Extending GenAI-based frameworks to build adversarially robust applications could mitigate emerging threats like adversarial spoofing and jamming in a broader range of scenarios.

8.3.2 GenAI for Cybersecurity

GenAI models play a crucial role in enhancing cybersecurity by learning normal data distributions and detecting anomalies that deviate from expected patterns. These models have been widely explored in fields such as computer vision, where GANs have proven effective in identifying irregularities through adversarial learning. In cybersecurity, GenAI has been employed for anomaly detection, adversarial defense, and vulnerability assessment, significantly improving the ability to counter sophisticated cyber threats. The key applications of GenAI in cybersecurity are as follows:

Anomaly Detection in Cybersecurity: GenAI has been leveraged to detect malicious behavior by identifying deviations from normal network traffic and system activities. For instance, Toma et al. [57] explored the use of GAN-based anomaly detection in cognitive radio networks to identify malicious users attempting to manipulate the radio spectrum. Additionally, Siddiqui et al. [58] demonstrated how GenAI could be applied to cyber attack detection, flagging anomalies and providing explanations for suspicious activities.

Defending Against Cybersecurity Attacks: GenAI has been employed to mitigate various cybersecurity threats, including DDoS attacks and Deepfake-based social engineering. In Ref. [59], GAN-based adversarial training was integrated into intrusion detection systems to enhance resilience against DDoS attacks. Similarly, GANs were utilized in Ref. [60] to counter poisoning attacks by mimicking the target model and augmenting data, thereby improving model resilience. The ability of GANs to generate adversarial samples and simulate attack patterns has significantly improved cybersecurity defenses, enabling more adaptive and proactive security mechanisms.

Vulnerability Assessment and Penetration Testing: Another critical application of GenAI in cybersecurity is vulnerability assessment and penetration testing. GenAI-driven models can create realistic attack vectors and simulate complex penetration testing scenarios, uncovering security weaknesses that conventional methods might overlook [61, 62]. By generating diverse attack patterns, these models provide a more comprehensive evaluation of system vulnerabilities, allowing organizations to strengthen their defenses preemptively. Additionally, GenAI can assist in generating synthetic exploit code to test security measures in both wireless and traditional cybersecurity frameworks.

Phishing and Spoofing Attack Detection: GenAI also plays a vital role in detecting phishing and spoofing attacks by leveraging anomaly detection techniques. By continuously learning from normal user behavior, generative models can identify subtle deviations indicative of malicious activities, such as phishing attempts or impersonation. This proactive approach enhances email security, user authentication, and fraud detection, making GenAI a valuable tool in combating social engineering attacks.

The applications of GenAI in cybersecurity illustrate its potential to revolutionize cyber defense mechanisms. Future research should focus on expanding these capabilities to create more robust and explainable cybersecurity solutions. Areas for further exploration include:

- **Explainable GenAI Models:** Enhancing interpretability in GenAI-driven cybersecurity solutions can improve trust and accountability, especially for critical infrastructures and enterprise security.
- **Multilayered Security:** Integrating GenAI with traditional security frameworks can provide a more holistic defense strategy, combining anomaly detection, encryption, and real-time response mechanisms.
- **Federated Learning for Cybersecurity:** Implementing federated GenAI models can enable decentralized threat intelligence sharing while preserving data privacy and minimizing centralized vulnerabilities.
- **Automated Cyber Threat Simulation:** Using GenAI to generate evolving cyber threat scenarios can improve the effectiveness of incident response planning and real-time mitigation strategies.
- **Robust Adversarial Defenses:** Expanding adversarial training techniques for cybersecurity applications can enhance resilience against sophisticated attacks, including AI-generated exploits and adversarial malware.

By advancing research in these areas, GenAI has the potential to redefine cybersecurity strategies, offering more intelligent, adaptive, and proactive defense mechanisms against emerging cyber threats.

8.3.3 GenAI-driven Attacks Against Wireless and Cybersecurity Applications

GenAI introduces new threats to wireless security and cybersecurity by enabling highly sophisticated and evolving attack strategies. Generative models can be exploited to craft adversarial inputs that deceive machine learning-based security systems, such as intrusion detection and anomaly detection tools [63–65]. Additionally, GenAI can be used to automate phishing, deepfake generation, and adversarial attacks against wireless communication networks, significantly enhancing the capabilities of cyber adversaries. The key applications of GenAI-driven attacks are as follows:

Adversarial Attacks Against ML-based Security Systems: GenAI-driven adversarial attacks can generate deceptive inputs that compromise the effectiveness of mechanisms. For example, in Ref. [66], GAN-generated adversarial perturbations were used to manipulate modulation classifiers in wireless networks, significantly reducing their accuracy by leveraging multi-objective loss functions. Various applications for wireless communication, such as signal classification and traffic, were considered against adversarial attacks generated using GenAI [67–69]. Similarly, poisoning attacks for wireless communications were explored in Refs. [70, 71], where GANs were used to execute label flipping and backdoor attacks. These adversarial tactics highlight the ability of GenAI to subvert ML-based security models by dynamically adapting attack strategies.

Deepfake and Phishing Attacks: GenAI has been employed to generate hyper-realistic phishing emails, fake websites, and deepfake-based identity theft attacks. In Ref. [72], deepfake attacks were analyzed using GenAI, where perturbed facial images were synthesized to bypass biometric authentication systems. These attacks pose a growing challenge to cybersecurity as they become more realistic and harder to detect, allowing adversaries to impersonate trusted entities and steal sensitive information.

GAN-based Wireless Signal Spoofing: Spoofing attacks in wireless security involve generating signals that deceive authentication systems, allowing adversaries to impersonate legitimate transmitters. In Refs. [73, 74], GANs were used to create synthetic wireless signals that are indistinguishable from authorized transmissions. Attackers use trained generators to mimic legitimate signals while defenders attempt to counteract the deception. Additionally, statistical methods for rogue signal generation were explored in Ref. [75], showing the potential for combining GenAI techniques with statistical models to enhance signal spoofing capabilities.

The use of GenAI in offensive security highlights its dual nature, where it enhances security capabilities; it also introduces new vulnerabilities that require advanced countermeasures. To mitigate these risks, future research should focus on:

- **Adversarial Defense Mechanisms:** Developing robust adversarial training techniques to counteract GenAI-generated attacks in wireless and cybersecurity applications.
- **Explainable AI for Threat Detection:** Enhancing interpretability in security models to detect adversarial examples and provide transparency in threat assessment.
- **Automated Cyber Threat Intelligence:** Leveraging GenAI for real-time adversarial behavior analysis and rapid adaptation to emerging cyber threats.
- **Defensive GenAI Models:** Exploring ways to use GenAI for proactive threat detection, enabling AI-driven models to anticipate and neutralize adversarial attacks.
- **Cross-layer Security Strategies:** Integrating GenAI defenses across multiple layers of network security, from physical-layer signal authentication to application-layer anomaly detection.

By addressing these challenges, the security community can develop resilient countermeasures against the rapidly evolving landscape of GenAI-driven cyber and wireless threats.

8.4 Ethical Issues Related to GenAI for Wireless Communications and Cybersecurity

As mentioned in the previous sections, GenAI has been used for various applications for wireless communications and cybersecurity to improve performance and robustness. However, there are a number of ethical issues [76] that come out with using GenAI, especially as these technologies are integrated into critical parts of the system. One of the key issues related to GenAI is privacy and data protection. In wireless communications, GenAI models often need access to large datasets, including user behavior and communication patterns, for those used for network optimization and traffic prediction. This poses privacy risks if sensitive data is not properly protected or anonymized. Also, in cybersecurity, GenAI tools designed to detect threats using personal data could also breach user privacy where the personal data can be abused, especially if the data is not handled with transparency or accountability.

Also, GenAI has the potential to reflect and amplify biases and discrimination because the vast amounts of data used for training might contain inherent prejudices. For example, GenAI models are trained on large datasets sourced from the internet, books, social media, and other digital content where these datasets may contain biased language, stereotypes, or discriminatory views present in society. If the training data reflects certain historical or societal inequalities, the GenAI model might learn and reproduce these biases. Similarly, in wireless communications, if the GenAI model is trained on historical data, which could include inherent biases related to network access, user behavior, or location-based performance, it can lead to AI systems that unfairly prioritize certain areas or users while neglecting others, promoting inequality in network access and performance.

Cybersecurity using GenAI may exhibit similar bias when detecting potential threats. For example, if the training for the anomaly detection model reflects a biased decision, favoring certain types of attacks from specific regions, the GenAI will fail to perform well in real-world scenarios involving unseen threats. To address these challenges, research is needed on methods to reduce bias and discrimination in GenAI systems by regularly testing GenAI methods for biased outcomes and adjusting models accordingly, ensuring training datasets are diverse, inclusive, and representative of various groups, developing algorithms that aim to reduce bias and improve fairness, and implementing human-in-the-loop processes to monitor and correct the model when necessary.

As GenAI raises ethical concerns due to its characteristics, there is another aspect of ethical issue related to GenAI, which is human factors. One key question about using GenAI or general ML is who should be responsible for the failure of the ML applications, which results in bad consequences. There are numerous factors caused by humans, such as programming codes, training data, improper operation, and many others, that may cause failure. Thus, accountability is an ethical issue that concerns the human factors included in designing, implementing, and deploying GenAI. Further, to mitigate such undesirable consequences from human factors, people related to AI system design should take training courses that teach human rights laws.

8.5 Summary

This chapter explores the role of GenAI in wireless and cybersecurity, highlighting both its benefits and potential risks. GenAI has significantly transformed wireless communications, improving tasks like spectrum sensing, channel estimation, and signal classification. Also, the integration of GenAI has further enhanced network optimization, resource allocation, and synthetic data generation, addressing challenges in dynamic wireless environments. However, the openness of wireless networks exposes them to cybersecurity threats such as phishing, DoS/DDoS, MitM, and spoofing attacks. While GenAI can strengthen cybersecurity through anomaly detection, attack prediction, and vulnerability assessment, it also introduces new risks as adversaries can exploit it to develop more sophisticated attacks.

One major concern is adversarial ML, where GenAI-driven wireless systems are vulnerable to adversarial ML attacks that manipulate input data, leading to misclassifications and degraded performance. Unlike in computer vision, these attacks face unique challenges in wireless networks, such as signal propagation effects and real-time constraints. Also, there exist other threats, including exploratory attacks that extract model information and poisoning attacks that corrupt training data. Adversaries can also leverage GenAI to craft sophisticated attacks against ML applications. Thus, addressing these vulnerabilities is crucial for ensuring the security and reliability of ML-powered wireless communications.

GenAI also plays a crucial role in enhancing wireless security and cybersecurity by improving anomaly detection, authentication, and defensive strategies. Its ability to learn from data distributions makes it effective in detecting cyber threats, securing wireless transmissions, and mitigating attacks. On the other hand, malicious adversaries can exploit GenAI to craft adversarial inputs, automate phishing, generate deepfakes, and conduct data poisoning attacks. These capabilities highlight the dual-use nature of GenAI, which can be used for both defensive and offensive purposes in cybersecurity and wireless networks.

Beyond security concerns, ethical issues also arise with GenAI applications, particularly regarding privacy risks and data protection. Since GenAI models require access to large-scale datasets, there is a risk of exposing sensitive user information in wireless communications and cybersecurity applications. Additionally, GenAI-driven threat detection tools could compromise user privacy if personal data is mishandled. As the adoption of GenAI continues to grow, it is crucial to develop robust defense mechanisms and establish ethical guidelines to ensure responsible use.

References

1 Kemal Davaslioglu and Yalin E. Sagduyu. Generative adversarial learning for spectrum sensing. In *IEEE International Conference on Communications (ICC)*, 2018.

2 Xuan Song, Jun Wang, Jie Wang, Guan Gui, Tomoaki Ohtsuki, Haris Gacanin, and Hikmet Sari. Saldr: Joint self-attention learning and dense refine for massive MIMO CSI feedback with multiple compression ratio. *IEEE Wireless Communications Letters*, 2021.

3 Jie Wang, Guan Gui, Tomoaki Ohtsuki, Bamidele Adebisi, Haris Gacanin, and Hikmet Sari. Compressive sampled CSI feedback method based on deep learning for FDD massive MIMO systems. *IEEE Transactions on Communications*, 2021.

4 Yu Jin, Jiayi Zhang, Shi Jin, and Bo Ai. Channel estimation for cell-free mmWave massive MIMO through deep learning. *IEEE Transactions on Vehicular Technology*, 68(10): 10325–10329, 2019.

5 Timothy J. O'Shea, Johnathan Corgan, and T. Charles Clancy. Convolutional radio modulation recognition networks. In *International Conference on Engineering Applications of Neural Networks*, 2016.

6 Yinqiu Liu, Hongyang Du, Dusit Niyato, Jiawen Kang, Zehui Xiong, Dong In Kim, and Abbas Jamalipour. Deep generative model and its applications in efficient wireless network management: A tutorial and case study. *IEEE Wireless Communications*, 2024.

7 Baoxia Du, Hongyang Du, Haifeng Liu, Dusit Niyato, Peng Xin, Jun Yu, Mingyang Qi, and You Tang. YOLO-based semantic communication with generative AI-aided resource allocation for digital twins construction. *IEEE Internet of Things Journal*, 11(5): 7664–7678, 2023.

8 Jie Zheng, Baoxia Du, Hongyang Du, Jiawen Kang, Dusit Niyato, and Haijun Zhang. Energy-efficient resource allocation in generative AI-aided secure semantic mobile networks. *IEEE Transactions on Mobile Computing*, 2024.

9 Tianyu Hu, Yang Huang, Qiuming Zhu, and Qihui Wu. Channel estimation enhancement with generative adversarial networks. *IEEE Transactions on Cognitive Communications and Networking*, 7(1): 145–156, 2020.

10 Hao Jiang, Mingyao Cui, Derrick Wing Kwan Ng, and Linglong Dai. Accurate channel prediction based on transformer: Making mobility negligible. *IEEE Journal on Selected Areas in Communications*, 40(9): 2717–2732, 2022.

11 Changfa Leng, Chungang Yang, Sifan Chen, Qing Wu, and Yao Peng. GAN for load estimation and traffic-aware network selection for 5G terminals. *IEEE Internet of Things Journal*, 9(17): 16353–16362, 2022.

12 Eren Balevi and Jeffrey G Andrews. Wideband channel estimation with a generative adversarial network. *IEEE Transactions on Wireless Communications*, 20(5): 3049–3060, 2021.

13 Ravindra Changala, S Kayalvili, Mansoor Farooq, L Malleswara Rao, Vuda Sreenivasa Rao, and S Muthuperumal. Using generative adversarial networks for anomaly detection in network traffic: Advancements in AI cybersecurity. In *2024 International Conference on Data Science and Network Security (ICDSNS)*, pages 1–6. IEEE, 2024.

14 Aeryn Dunmore, Julian Jang-Jaccard, Fariza Sabrina, and Jin Kwak. A comprehensive survey of generative adversarial networks (GANS) in cybersecurity intrusion detection. *IEEE Access*, 11: 76071–76094, 2023.

15 Andrei Kucharavy, Zachary Schillaci, Loïc Maréchal, Maxime Würsch, Ljiljana Dolamic, Remi Sabonnadiere, Dimitri Percia David, Alain Mermoud, and Vincent Lenders. Fundamentals of generative large language models and perspectives in cyber-defense. *arXiv preprint arXiv:2303.12132*, 2023.

16 Siva Sai, Utkarsh Yashvardhan, Vinay Chamola, and Biplab Sikdar. Generative AI for cyber security: Analyzing the potential of ChatGPT, DALL-E and other models for enhancing the security space. *IEEE Access*, 2024.

17 Y Vorobeychik and M Kantarcioglu. Adversarial machine learning. *Synthesis Lectures on Artificial Intelligence and Machine Learning*, 12(3): 1–169, 2017.

18 Christian Szegedy, Wojciech Zaremba, Ilya Sutskever, Joan Bruna, Dumitru Erhan, Ian Goodfellow, and Rob Fergus. Intriguing properties of neural networks. *arXiv preprint arXiv: 1312.6199*, 2013.

19 Yalin E Sagduyu, Yi Shi, Tugba Erpek, W. Headley, B. Flowers, G. Stantchev, and Z. Lu. When wireless security meets machine learning: Motivation, challenges, and research directions. *arXiv preprint arXiv:2001.08883*, 2020.

20 Damilola Adesina, Chung-Chu Hsieh, Yalin E Sagduyu, and Lijun Qian. Adversarial machine learning in wireless communications using RF data: A review. *arXiv preprint arXiv:2012.14392*, 2020.

21 Jamie Hayes, Luca Melis, George Danezis, and Emiliano De Cristofaro. Logan: Membership inference attacks against generative models. *arXiv preprint arXiv:1705.07663*, 2017.

22 Yi Shi, Kemal Davaslioglu, and Yalin E Sagduyu. Over-the-air membership inference attacks as privacy threats for deep learning-based wireless signal classifiers. In *ACM WiSec Workshop on Wireless Security and Machine Learning (WiseML)*, 2020.

23 Yi Shi and Yalin E Sagduyu. Membership inference attack and defense for wireless signal classifiers with deep learning. *arXiv preprint arXiv:2107.12173*, 2021.

24 Yi Shi, Yalin E Sagduyu, Tugba Erpek, Kemal Davaslioglu, Zhuo Lu, and Jason H. Li. Adversarial deep learning for cognitive radio security: Jamming attack and defense strategies. In *IEEE International Communications Conference (ICC) Workshop on Machine Learning in Wireless Communications*, 2018.

25 Tugba Erpek, Yalin E. Sagduyu, and Yi Shi. Deep learning for launching and mitigating wireless jamming attacks. *IEEE Transactions on Cognitive Communications and Networking*, 5(1): 2–14, 2019.

26 Yi Shi and Yalin E Sagduyu. Membership inference attack and defense for wireless signal classifiers with deep learning. *IEEE Transactions on Mobile Computing*, 22(7): 4032–4043, 2022.

27 Sunder Ali Khowaja, Parus Khuwaja, Kapal Dev, Angelos Antonopoulos, and Maurizio Magarini. DASTAN-CNN: RF fingerprinting for the mitigation of membership inference attacks in 5G. In *GLOBECOM 2023-2023 IEEE Global Communications Conference*, pages 5524–5529. IEEE, 2023.

28 Yi Shi, Tugba Erpek, Yalin E Sagduyu, and Jason H. Li. Spectrum data poisoning with adversarial deep learning. In *IEEE Military Communications Conference (MILCOM)*, 2018.

29 Yalin E Sagduyu, Tugba Erpek, and Yi Shi. Adversarial deep learning for over-the-air spectrum poisoning attacks. *IEEE Transactions on Mobile Computing*, (1):2–14, 2021.

30 Zhengping Luo, Shangqing Zhao, Zhuo Lu, Jie Xu, and Yalin Sagduyu. When attackers meet AI: Learning-empowered attacks in cooperative spectrum sensing. *IEEE Transactions on Mobile Computing*, 2020.

31 Zhengping Luo, Shangqing Zhao, Rui Duan, Zhuo Lu, Yalin E. Sagduyu, and Jie Xu. Low-cost influence-limiting defense against adversarial machine learning attacks in cooperative spectrum sensing. In *ACM WiSec Workshop on Wireless Security and Machine Learning (WiseML)*, 2021.

32 Zhengping Luo, Shangqing Zhao, Zhuo Lu, Yalin E Sagduyu, and Jie Xu. Adversarial machine learning based partial-model attack in IoT. In *ACM WiSec Workshop on Wireless Security and Machine Learning (WiseML)*, 2020.

33 Ruinan Jin and Xiaoxiao Li. Backdoor attack is a devil in federated GAN-based medical image synthesis. In *International Workshop on Simulation and Synthesis in Medical Imaging*, pages 154–165. Springer, 2022.

34 K. Davaslioglu and Y. E. Sagduyu. Trojan attacks on wireless signal classification with adversarial machine learning. In *IEEE International Symposium on Dynamic Spectrum Access Networks (DySPAN) Workshop on Data-Driven Dynamic Spectrum Sharing*, 2019.

35 Jernej Kos, Ian Fischer, and Dawn Song. Adversarial examples for generative models. In *2018 IEEE Security and Privacy Workshops (SPW)*, pages 36–42. IEEE, 2018.

36 Meysam Sadeghi and Erik G. Larsson. Physical adversarial attacks against end-to-end autoencoder communication systems. *IEEE Communications Letters*, 23(5): 847–850, 2019.

37 Timothy O'Shea and Jakob Hoydis. An introduction to deep learning for the physical layer. *IEEE Transactions on Cognitive Communications and Networking*, 3(4): 563–575, 2017.

38 Yalin E. Sagduyu, Yi Shi, Tugba Erpek, William Headley, Bryse Flowers, George Stantchev, and Zhuo Lu. IoT network security from the perspective of adversarial deep learning. In *IEEE International Conference on Sensing, Communication and Networking (SECON) Workshop Machine Learning for Communication and Networking in IoT*, 2019.

39 BR Manoj, Meysam Sadeghi, and Erik G Larsson. Adversarial attacks on deep learning-based power allocation in a massive MIMO network. *arXiv preprint arXiv:2101.12090*, 2021.

40 Brian Kim, Yi Shi, Yalin E. Sagduyu, Tugba Erpek, and Sennur Ulukus. Adversarial attacks against deep learning-based power control in wireless communications. In *IEEE Global Communications Conference (Globecom)*, 2021.

41 Brian Kim, Yalin E Sagduyu, Tugba Erpek, and Sennur Ulukus. Adversarial attacks on deep learning-based mmWave beam prediction in 5G and beyond. In *IEEE Statistical Signal Processing Workshop*, 2021.

42 Qing Liu, Jiajia Guo, Chao-Kai Wen, and Shi Jin. Adversarial attack on DL-based massive MIMO CSI feedback. *arXiv preprint arXiv:2002.09896*, 2020.

43 Alexey Kurakin, Ian Goodfellow, and Samy Bengio. Adversarial machine learning at scale. *arXiv preprint arXiv:1611.01236*, 2016.

44 Javier Maroto, Gérôme Bovet, and Pascal Frossard. SafeAMC: Adversarial training for robust modulation recognition models. *arXiv preprint arXiv:2105.13746*, 2021.

45 Brian Kim, Yalin E Sagduyu, Kemal Davaslioglu, Tugba Erpek, and Sennur Ulukus. Channel-aware adversarial attacks against deep learning-based wireless signal classifiers. *IEEE Transactions on Wireless Communications*, 21(6): 3868–3880, 2021.

46 Nicolas Papernot, Patrick McDaniel, Xi Wu, Somesh Jha, and Ananthram Swami. Distillation as a defense to adversarial perturbations against deep neural networks. In *2016 IEEE Symposium on Security and Privacy (SP)*, pages 582–597. IEEE, 2016.

47 Azuka Chiejina, Brian Kim, Kaushik Chowhdury, and Vijay K Shah. System-level analysis of adversarial attacks and defenses on intelligence in O-RAN-based cellular networks. In *Proceedings of the 17th ACM Conference on Security and Privacy in Wireless and Mobile Networks*, pages 237–247, 2024.

48 Chengsi Liang, Hongyang Du, Yao Sun, Dusit Niyato, Jiawen Kang, Dezong Zhao, and Muhammad Ali Imran. Generative AI-driven semantic communication networks: Architecture, technologies and applications. *IEEE Transactions on Cognitive Communications and Networking*, 2024.

49 Bin Han, Ye Yuan, and Hans D Schotten. Venena: A deceptive visual encryption framework for wireless semantic secrecy. *arXiv preprint arXiv:2501.10699*, 2025.

50 Jiacheng Wang, Hongyang Du, Yinqiu Liu, Geng Sun, Dusit Niyato, Shiwen Mao, Dong In Kim, and Xuemin Shen. Generative AI-based secure wireless sensing for ISAC networks. *IEEE Transactions on Information Forensics and Security*, 2025.

51 Debashri Roy, Tathagata Mukherjee, Mainak Chatterjee, and Eduardo Pasiliao. Detection of rogue RF transmitters using generative adversarial nets. In *2019 IEEE Wireless Communications and Networking Conference (WCNC)*, pages 1–7. IEEE, 2019.

52 Ken St Germain and Frank Kragh. Physical-layer authentication using channel state information and machine learning. In *2020 14th International Conference on Signal Processing and Communication Systems (ICSPCS)*, pages 1–8. IEEE, 2020.

53 Ender Ayanoglu, Kemal Davaslioglu, and Yalin E Sagduyu. Machine learning in nextG networks via generative adversarial networks. *IEEE Transactions on Cognitive Communications and Networking*, 8(2): 480–501, 2022.

54 Nguyen Van Huynh, Jiacheng Wang, Hongyang Du, Dinh Thai Hoang, Dusit Niyato, Diep N Nguyen, Dong In Kim, and Khaled B Letaief. Generative AI for physical layer communications: A survey. *IEEE Transactions on Cognitive Communications and Networking*, 10(3): 706–728, 2024.

55 Kemal Davaslioglu, Tugba Erpek, and Yalin Sagduyu. End-to-end autoencoder communications with optimized interference suppression. *Physical-Layer Security for 6G*, pages 153–184, 2024.

56 Zhaowei Wang, Weicheng Liu, and Hui-Ming Wang. GAN against adversarial attacks in radio signal classification. *IEEE Communications Letters*, 26(12): 2851–2854, 2022.

57 Andrea Toma, Ali Krayani, Muhammad Farrukh, Haoran Qi, Lucio Marcenaro, Yue Gao, and Carlo S Regazzoni. AI-based abnormality detection at the PHY-layer of cognitive radio by learning generative models. *IEEE Transactions on Cognitive Communications and Networking*, 6(1): 21–34, 2020.

58 Md Amran Siddiqui, Jack W Stokes, Christian Seifert, Evan Argyle, Robert McCann, Joshua Neil, and Justin Carroll. Detecting cyber attacks using anomaly detection with explanations and expert feedback. In *ICASSP 2019-2019 IEEE International Conference on Acoustics, Speech and Signal Processing (ICASSP)*, pages 2872–2876. IEEE, 2019.

59 Maged Abdelaty, Sandra Scott-Hayward, Roberto Doriguzzi-Corin, and Domenico Siracusa. GADoT: GAN-based adversarial training for robust DDoS attack detection. In *2021 IEEE Conference on Communications and Network Security (CNS)*, pages 119–127. IEEE, 2021.

60 Jian Chen, Xuxin Zhang, Rui Zhang, Chen Wang, and Ling Liu. De-Pois: An attack-agnostic defense against data poisoning attacks. *IEEE Transactions on Information Forensics and Security*, 16: 3412–3425, 2021.

61 Jeremy Charlier, Aman Singh, Gaston Ormazabal, Radu State, and Henning Schulzrinne. Syngan: Towards generating synthetic network attacks using GANs. *arXiv preprint arXiv:1908.09899*, 2019.

62 Jinyin Chen, Haibin Zheng, Hui Xiong, Shijing Shen, and Mengmeng Su. MAG-GAN: Massive attack generator via GAN. *Information Sciences*, 536: 67–90, 2020.

63 Zhipeng Liu, Junyi Hu, Yang Liu, Kaushik Roy, Xiaohong Yuan, and Jinsheng Xu. Anomaly-based intrusion on IOT networks using AIGAN-a generative adversarial network. *IEEE Access*, 11: 91116–91132, 2023.

64 Xiaolin Chen, Daoguang Zan, Wei Li, Bei Guan, and Yongji Wang. A GAN-based data poisoning framework against anomaly detection in vertical federated learning. *arXiv preprint arXiv:2401.08984*, 2024.

65 Zilong Lin, Yong Shi, and Zhi Xue. IDSGAN: Generative adversarial networks for attack generation against intrusion detection. In *Pacific-Asia Conference on Knowledge Discovery and Data Mining*, pages 79–91. Springer, 2022.

66 Paulo Freitas de Araujo-Filho, Georges Kaddoum, Mohamed Naili, Emmanuel Thepie Fapi, and Zhongwen Zhu. Multi-objective GAN-based adversarial attack technique for modulation classifiers. *IEEE Communications Letters*, 26(7): 1583–1587, 2022.

67 Kawtar Zerhouni, Gurjot Singh Gaba, Mustapha Hedabou, Taras Maksymyuk, Andrei Gurtov, and El Mehdi Amhoud. GAN-based evasion attack in filtered multicarrier waveforms systems. *IEEE Transactions on Machine Learning in Communications and Networking*, 2024.

68 Fenghui Liu. Target attack GAN: Black box target attack in specific emitter identification. In *Proceedings of the 2024 11th International Conference on Wireless Communication and Sensor Networks*, pages 62–67, 2024.

69 Muhammad Usama, Adnan Qayyum, Junaid Qadir, and Ala Al-Fuqaha. Black-box adversarial machine learning attack on network traffic classification. In *2019 15th International Wireless Communications & Mobile Computing Conference (IWCMC)*, pages 84–89. IEEE, 2019.

70 Jiale Zhang, Bing Chen, Xiang Cheng, Huynh Thi Thanh Binh, and Shui Yu. PoisonGAN: Generative poisoning attacks against federated learning in edge computing systems. *IEEE Internet of Things Journal*, 8(5): 3310–3322, 2020.

71 Dongwei Xu, Ruochen Fang, Qi Xuan, Weiguo Shen, Shilian Zheng, and Xiaoniu Yang. AT-GAN: A backdoor attack against radio signals modulation based on adaptive trigger. *IEEE Transactions on Circuits and Systems II: Express Briefs*, 2024.

72 Chaofei Yang, Leah Ding, Yiran Chen, and Hai Li. Defending against GAN-based deepfake attacks via transformation-aware adversarial faces. In *2021 International Joint Conference on Neural Networks (IJCNN)*, pages 1–8. IEEE, 2021.

73 Yi Shi, Kemal Davaslioglu, and Yalin E Sagduyu. Generative adversarial network for wireless signal spoofing. In *ACM WiSec Workshop on Wireless Security and Machine Learning (WiseML)*, 2019.

74 Yi Shi, Kemal Davaslioglu, and Yalin E Sagduyu. Generative adversarial network in the air: Deep adversarial learning for wireless signal spoofing. *IEEE Transactions on Cognitive Communications and Networking*, 7(1): 294–303, 2021.

75 Yalin E Sagduyu and Tugba Erpek. Adversarial attacks on LoRA device identification and rogue signal detection with deep learning. In *IEEE Military Communications Conference (MILCOM)*, pages 385–390, 2023.

76 Changwu Huang, Zeqi Zhang, Bifei Mao, and Xin Yao. An overview of artificial intelligence ethics. *IEEE Transactions on Artificial Intelligence*, 4(4): 799–819, 2022.

9

Challenges and Opportunities for Generative AI in Wireless Communications and Networking

Songyang Zhang[1,*] *and Zhi Ding*[2]

[1] *Department of Electrical and Computer Engineering, University of Louisiana at Lafayette, LA, USA*
[2] *Department of Electrical and Computer Engineering, University of California at Davis, CA, USA*

*Corresponding author: songyang.zhang@louisiana.edu

9.1 Introduction

The rapid development of information technologies has stimulated many novel concepts in next-generation wireless communications, such as beyond-fifth-generation (B5G) and sixth-generation (6G) cellular networks [1], leading to a dramatic increase in the deployment of edge devices and Internet-of-Things (IoT). The widespread applications of these novel techniques generate gigantic amounts of data to be processed in edges, resulting in the increasing demand for efficient processing of large-size data under limited bandwidth and constrained computation resource. A typical example is the drone-based emergency communication system, where drones collect multimodal data from carry-on sensors and communicate with remote working base stations [2] with very limited communication capacity. Meanwhile, with the expansion of sensor networks and edge devices, spectrum environments become increasingly dynamic and complex, resulting in urgent needs for panoramic and efficient spectrum sensing of spectrum distribution, quick and accurate network management, distributed data storage and processing in cooperative environments, and assurance of privacy and security [3]. All these applications motivate a more intelligent and efficient wireless communication system in the era of big data, where emerging artificial intelligence (AI) technologies could play important roles.

Generative AI for Communications Systems: Fundamentals, Applications, and Prospects,
First Edition. Edited by Diep N. Nguyen et al.

With the development of machine learning (ML) techniques, AI-powered wireless networks have become important parts in next-generation communications. Different communication systems generate large amounts of data, while AI techniques offer intelligent solutions to more efficient network design and optimization. Most of the existing successes have been achieved by the utilization of neural networks (NNs) inspired by the structure of the human brain [4]. In these applications, the structure of the NNs, represented by a set of parameters denoted by a function $g(\cdot)$, is trained from data for certain objectives. Generally, existing ML methodologies can be categorized as supervised learning, unsupervised learning, self-supervised learning, and reinforcement learning [5]. Among all these AI infrastructures, generative learning has recently emerged as a fashionable tool for data processing in many applications, such as natural language processing (NLP), computer vision, biomedical data analysis, and autonomous driving. Interested readers could refer to the previous chapters for a detailed introduction. Despite its successes in conventional multimedia applications, the utilization of generative AI(GenAI) in wireless communications remains a newly developed field, where more explorations are required to fulfill the new life of wireless communication intelligence.

In this chapter, we will provide a tutorial on the application of GenAI for next-generation wireless communications. To avoid repeated presentation, instead of providing a detailed introduction of GenAI models, we will focus mainly on the challenges and opportunities in GenAI-enabled wireless communications. To offer practical guidance and stimulate innovative technologies for overcoming these challenges, we will explore potential solutions and present case studies in two exemplary scenarios:

- Generative Learning in Spectrum Cartography and Management for Physical Wireless Networking;
- Generative Learning for Privacy-preserving and Bandwidth Efficient Distributed Wireless Communications.

9.2 Challenges of Applying Generative AI in Wireless Communications

As an emerging field in intelligent wireless communications, GenAI faces multiple challenges in various aspects of technologies, ethics, and regulation, as shown in Figure 9.1. In this section, we will first discuss multifaceted open research challenges and point out potential solutions to handle them.

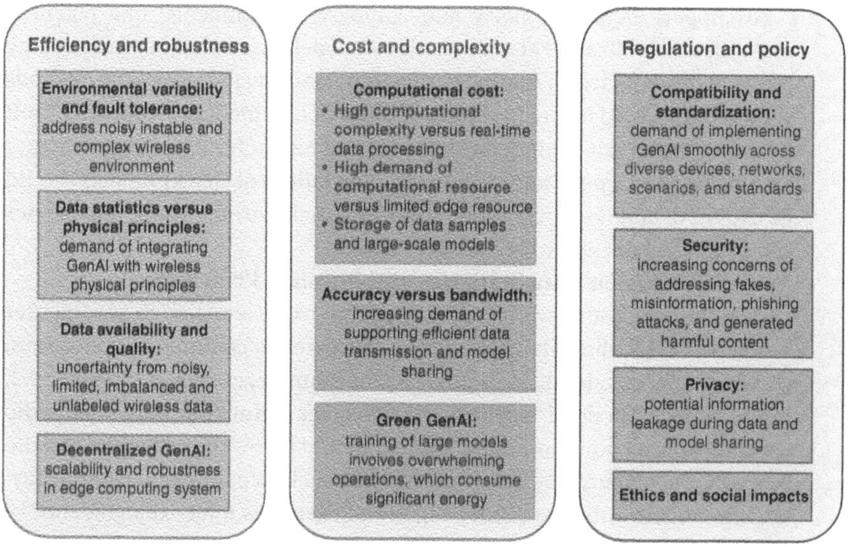

Figure 9.1 Challenges and concerns of applying GenAI in wireless communications.

9.2.1 Efficiency and Robustness

As aforementioned in the previous chapters, GenAI models are able to learn the underlying distributions from data samples and then generate synthetic data accordingly. In conventional applications, GenAI models focus more on the data statics while ignoring their underlying nature. However, in wireless communications, the data shall follow some strict physical models, which motivates more efficient integration of GenAI with the domain knowledge of wireless communications. Moreover, to ensure the effectiveness of GenAI in future wireless communications, they should be robust against complex environments and various conditions of noise/errors. Thus, the demand for efficiency and robustness of GenAI in the background of wireless communications contributes to the first category of challenges, discussed as follows.

- **Environmental Variability and Fault Tolerance:**
  In modern wireless communication systems, there is a divergence between system configurations, the mixture of various mobile devices, and dynamic spectrum environments, which raises the demand for robust GenAI to address noisy, instable, and complex wireless data. Moreover, in many wireless communication applications, such as emergency communications or underwater communications, generative learning should be able to process data even with the failure of a communication link or data streams.

To support GenAI in noisy and complex environments, one potential method is the utilization of a conventional model-based method to correct the errors in the data-driven results. Various efficient data preprocessing methods should be explored before inputting into the GenAI models. To enhance fault tolerance, novel self-healing mechanisms and hierarchical learning structure are investigated. For example, the cascading failure of GenAI models should be explored, and the integration of model-based/learning-based approaches can be applied to ensure the success of tasks.

- **Tradeoff Between Data Statistics and Physical Principles:**
Different from existing well-studied applications of GenAI, such as computer vision and NLP, the data generated in the wireless communication system should follow certain physical models. For example, as shown in Figure 9.2, in the application of AI-enabled radio coverage estimation, generated radio coverages (also known as radiomaps [6]) should follow the principles of radio propagation, captured by empirical models, such as path loss models or ray-tracing models. Efficient processing of data samples under the guidance of wireless models is in urgent demand.

  Most existing approaches for wireless data analysis can be categorized into model-based or learning-based methods. Model-based methods usually assume specific functions of physical models and attempt to estimate the parameters of the given model. However, inappropriate utilization of physical models may limit the accuracy of model-based methodologies. Moreover, with the use of various wireless standards and different types of edge devices, even a single wireless application can be too complex to be captured or formulated by an empirical physical model. On the other hand, learning-based

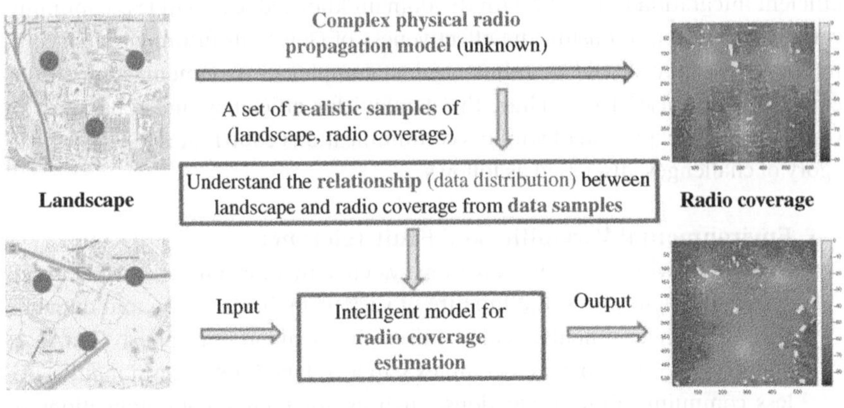

Figure 9.2 Examples of AI-enabled radio coverage estimation: the generated radio coverage should follow the physical principles of radio propagation. The color version of this figure is available on Wiley.com.

methods, including GenAI, usually do not assume specific models but leverage the power of ML techniques to characterize the data statistics. However, learning-based methods usually rely on sufficient and high-quality training samples, which are usually inaccessible in practical applications. Thus, how to integrate model-based and learning-based methods to address the tradeoff between data statistics and physical principles remains an open question.

To tackle the aforementioned challenges, recent interest has been attracted by the physics-inspired GenAI. The ideas of physics-inspired ML could originate from physics-informed neural networks, which seamlessly integrate data and mathematical physics models [7] to solve partial differential equations (PDEs). To generalize the applications of physics-inspired ML, different categories of integration between AI models and physical principles are introduced in Ref. [8] for the imaging of electromagnetic data. Inspired by these ideas, three physics-inspired AI approaches are introduced for radiomap estimation (RME) in Ref. [3], which can be illustrated in Figure 9.3. Similarly, we can develop physics-inspired GenAI to address the wireless communication problem. One of the potential methods is to first process the data via a model-based approach and then input the processed data into the neural network for the final decision. Another method is to customize the loss function of generative learning models, where additional regularization terms are designed based on physical principles. A case study of this method will be discussed in Section 9.3. Moreover, one can also change the neural network architecture or learning strategies based on physical models. An intuitive way is to use the features extracted from physics as the condition of conditional GenAI models.

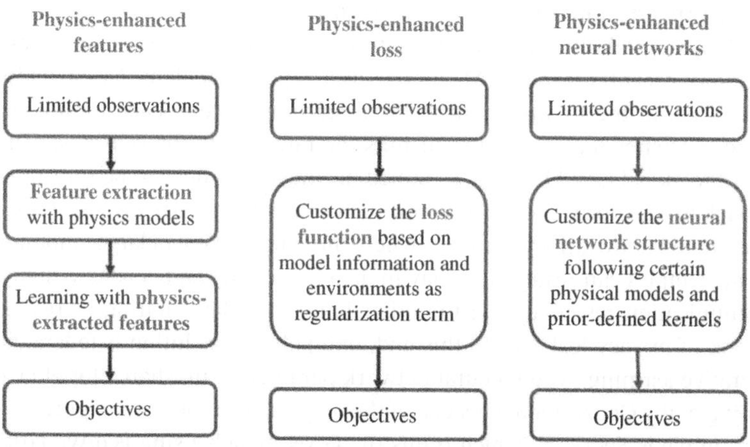

Figure 9.3 Three types of physics-inspired GenAI.

- **Data Availability and Quality:**
  The power of neural networks is highly dependent on the quality and quantity of training samples. Unlike other multimedia applications, even though modern wireless communication systems generate large amounts of data, these datasets are usually noisy, which could lead to large uncertainty in the training of ML approaches. In addition, many wireless communications applications are only accessible to simulated data, while the lack of realistic data results in an inaccurate interpretation of real-world communication behavior. Moreover, many wireless datasets miss human labels, where conventional supervised training strategies cannot work efficiently. Hence, the lack of high-fidelity, training-friendly, and interpretable data is a critical concern for applying GenAI in wireless communications.

  To take advantage of GenAI power in wireless communications, optimizing both simulator and hardware can lead to better data collection platforms. The communication community should encourage more efforts to expand the available testbed and open datasets. In addition to data collections, additional values can be introduced to GenAI with other data-augmentation techniques. One potential method is integration with self-supervised learning, where the model is trained on a task using the data itself to generate supervisory signals rather than external labels. Another method is that GenAI can be utilized to generate synthetic data conditioned on physical models to increase the data sets. Beyond the single modality of data usage, GenAI can benefit from combining different modality of data from multiple types of sensors. A typical example is integrated sensing and communications (ISAC) [9], where high-quality sensing images can aid in the processing of wireless data.
- **Decentralized GenAI for Edge Computing:**
  To enhance scalability and robustness in edge computing systems, the investigation of decentralized GenAI emerges as a popular problem. As shown in Figure 9.4, in the traditional distributed learning system, each local edge device, also shown as a client, transmits the raw data to the central server, while the global GenAI model is trained on the server sides to be transmitted back to each local client. However, such direct data sharing leads to concerns about bandwidth inefficiency, where the transmission of large amounts of data samples can result in a delay in data processing and consume many communication resources. On the other hand, raw data sharing increases the risk of privacy leakage. To this end, one potential solution is federated generative learning, where neural network parameters are shared for global training of the GenAI model instead of the original data shown in Figure 9.5. An example is FedGAN, which trains the GAN models using FedAvg [10]. Despite some successes, gradient-based federated learning (FL) suffers from its ability to process heterogeneous data and reduce communication overhead. More efficient methods shall be explored. We will offer one exemplary

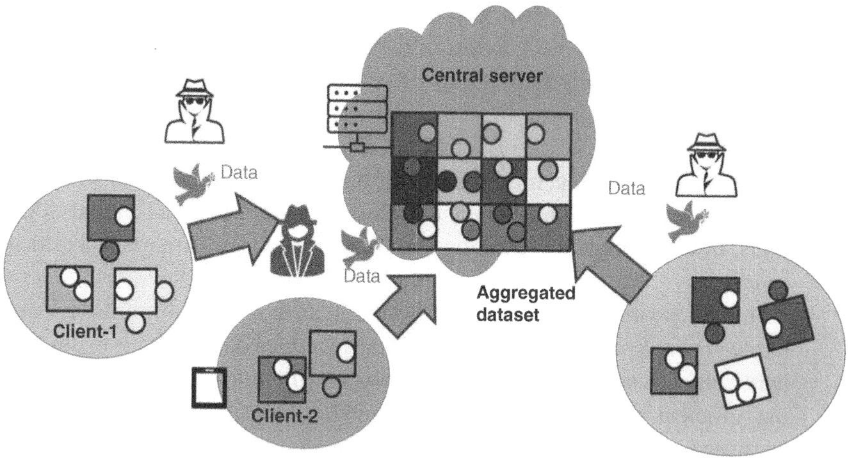

Figure 9.4 Conventional decentralized training: local data is transmitted to the central server.

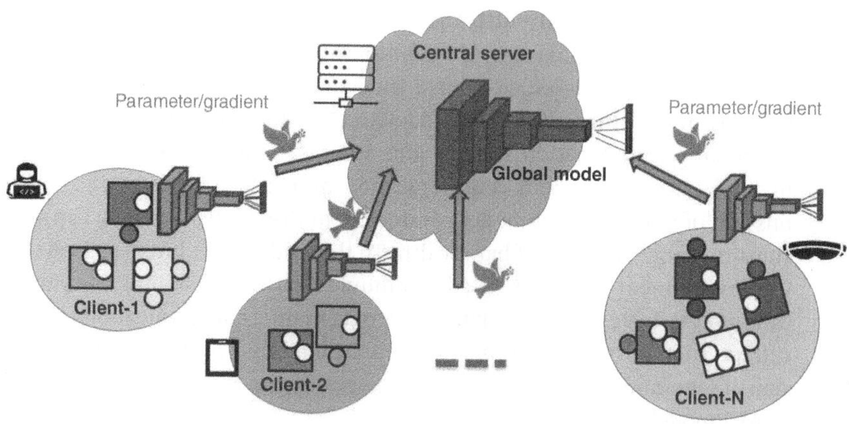

Figure 9.5 Federated learning: global model is trained without direct data sharing.

solution in Section 9.3. Beyond the efficiency and privacy of model training, additional concerns about decentralized GenAI include resource allocation, routing, and data storage.

9.2.2 Cost and Complexity

The computation and communication overhead presents another major challenge in deploying GenAI for wireless communications. In this part, we will discuss the challenges of complexity issues for GenAI.

- **High Computational Cost Versus Demand of Real-time Processing:**
  Next-generation wireless communications are expected to support real-time applications, such as autonomous driving, UAV-based emergency communications, online disease diagnosis, and virtual reality. However, the high computational complexity of GenAI can restrict its ability in wireless communications. For example, training a medium size diffusion model on an image dataset with around 1 million images may cost 6–14 days on RTX 4090 GPUs. Thus, how to enable fast training and inference of GenAI remains a critical concern.

  To address these concerns, improvements in both algorithm design and hardware can contribute. An intuitive way to reduce complexity is to design a lightweight GenAI. Some promising directions include neural network pruning, efficient sampling, knowledge distillation, parallel training, and split learning [11–13]. Meanwhile, the design of GenAI-specialized chips could further enhance the efficiency. Typical research directions include high-performance GPU design, memory optimization, precision improvement, and parallelization.
- **Accuracy Versus Bandwidth Usage:**
  The ever-evolving landscape of communications and networking shall be capable of supporting efficient data transmission and model sharing, particularly in a multiagent distributed edge computing system. However, conventional communication systems aim at bit-wise message delivery, which may raise concerns about communication overhead. Moreover, the transmission of gradients and model parameters in distributed learning can also consume large amounts of bandwidth. To this end, recent interests focus on the development of task-oriented communications and semantic communications [14]. Instead of exact data recovery, semantic communications aim to extract the task-oriented "semantic" representation. Through the transmission of semantic representation, the receiver side can reconstruct the data via GenAI models.
- **Green GenAI:**
  The high computational demands of GenAI usually lead to increased energy consumption. The training of large models involves overwhelming amounts of floating-point operations, which consume significant energy. In addition, the training of GenAI models can sometimes take weeks or months on high-power GPUs/TPUs, leading to sustained energy use. Moreover, during inference, the repeated computations also cost considerable energy.

  To enable green GenAI in wireless communications, energy-efficient algorithms and hardware acceleration techniques play an important role. Similarly to reducing computational cost, one promising research direction is to develop lightweight GenAI for model compression, where pruning, quantization, and knowledge distillation are important techniques. In addition, adaptive training strategies, such as dynamic batch sizes and early stopping, can further enhance energy efficiency. From hardware efficiency, it is crucial to use energy-efficient GPUs/TPUs and optimize hardware utilization. Other related areas also include the efficient management of next-generation data centers and optimization of workflow, such as scheduling and recycling.

9.2.3 Standardization, Regulation, and Policy

Beyond technical challenges, we next introduce some potential concerns on ethics and regulation.

- **Compatibility and Standardization:**
  With the broad impact of AI techniques in our daily life, it is crucial to ensure that GenAI can work smoothly across diverse devices, networks, scenarios, and standards. Since the design and architecture of GenAI can vary widely in different applications to meet their specialized demand, achieving interoperability is highly challenging. Sometimes, even a slight bias in model initialization can lead to totally different performances. Moreover, the GenAI models are usually trained in a specific dataset, where shifting to another application could lead to many efforts in model fine-tuning. Additional concerns also come from the format of data storage. Due to the different computing processes and the management of computational resources, the implementation of the same algorithm in various coding languages can also result in different performance and efficiency. Thus, it is crucial to ensure algorithm reproducibility across diverse datasets, devices, and programming languages to support practical applications. One potential solution is to provide large foundation models for different wireless communication applications, similar to NLP and computer vision.
- **Security and Privacy:**
  GenAI poses significant security and privacy challenges due to its ability to generate synthetic content and process large amounts of data. On the security front, it can be exploited to create fakes, spread misinformation, conduct phishing attacks, and generate harmful content, raising concerns about malicious use. Additionally, vulnerabilities in generative models, such as adversarial attacks [15] or data poisoning [16], can compromise their integrity. Since GenAI is capable of capturing the underlying data distribution, it can expose certain sensitive information. For example, anyone with the generator can generate the synthetic data with the same categories as the original data, leading to a leakage of membership information. Moreover, unintended memorization and the unauthorized use of user interactions for fine-tuning

further increase the risks of the exposure of personal information, particularly for decentralized learning. To address these issues, robust safeguards, including adversarial testing, differential privacy techniques, transparency in data usage, and strict governance policies, are essential to ensure responsible and ethical deployment of GenAI. Machine unlearning [17] is another potential technique for addressing data privacy concerns, which can unlearn sensitive data from trained models.

- **Ethics and Social Impacts:**
  The key ethical concerns of GenAI include the potential for misuse, which can create false data and spread misinformation. Additionally, issues of bias and fairness arise from the utilization of GenAI in wireless communications, which can lead to unfair resource allocation and inappropriate service delivery, leading to direct social impacts. General guidelines of ethics shall be defined to regularize the utilization of GenAI. Ethical use also demands transparency, ensuring users understand the origins and limitations of AI-generated content. Developers and organizations must prioritize fairness, accountability, and inclusivity in their designs while adhering to clear regulatory frameworks to prevent misuse. The use of GenAI in wireless communications should ensure compliance with government policies and data protection regulations, such as the General Data Protection Regulation (GDPR).

9.3 Adopting Generative AI in NextG Communications: Case Studies

To offer practical guidance and stimulate innovative technologies, we next provide several application examples of GenAI for wireless communications, which shall provide hints for addressing some of the aforementioned challenges.

9.3.1 Integration of Generative AI and Physical Communications Models

As illustrated in Section 9.2.1, one of the most critical challenges is to address the tradeoff between data statistics and physical principles. In the first case, we will introduce the development of physics-inspired GenAI for wireless communications, with RME as an exemplary application based on conditional generative adversarial networks (cGAN).

9.3.1.1 Introduction of RME

Before we discuss the potential design of physics-inspired GenAI (PIGL), we first introduce the background of RME.

Wireless networks and technologies are evolving to a new era with increasingly dynamic and complex spectrum environments, which has urgent needs for efficient characterization of spectrum environment. A vital tool for gaining panoramic awareness of the RF environment and spectrum coverage is radiomap, which entails mapping radio strength distribution in the spatial domain across multiple frequency bands and over certain time periods [18], as shown in Figure 9.6. Enriching the spectrum distribution, radiomap has broad applications in spectrum management and network optimization, including UAV path planning and system diagnosis. Usually, a dense radiomap is reconstructed from sparse samples collected from deployed sensors and mobile devices [20]. To leverage the power of radiomap in next-generation spectrum management, the RME from sparse samples is a critical problem.

RME is usually formulated as a sparse sensing problem [3] as shown in Figure 9.7. Suppose that $\mathbf{c}$ denotes two-dimensional (2D) coordinates, whose PSD is a function of location $\mathbf{c}$, frequency f, and sometimes with time t, denoted by $r(\mathbf{c}, f, t)$. In this case, we mainly focus on the 2D spatial RME of a region $\mathcal{C}$, that is, $R(\mathcal{C})$ consisting of all the PSD values $r(\mathbf{c}_i)$, where $\mathbf{c}_i \in \mathcal{C}$. Given a set of sparse observations $\{r(\mathbf{c}_{oi})\}_{i=1}^{M}$, whose locations are denoted by $\mathcal{C}_o$ consisting of $\{\mathbf{c}_{o1}, \mathbf{c}_{o2}, \dots, \mathbf{c}_{oM}\}$. The general formulation of RME is to find a function $g(\cdot)$, that is,

$$\hat{R}(\mathcal{C}) = g(\mathcal{X}, R(\mathcal{C}_o)), \tag{9.1}$$

where $\mathcal{X}$ is the environment information, such as the locations of transmitters and building distribution. In this example, the function $g(\cdot)$ is the generative learning models to generate the dense radiomap from sparse observations.

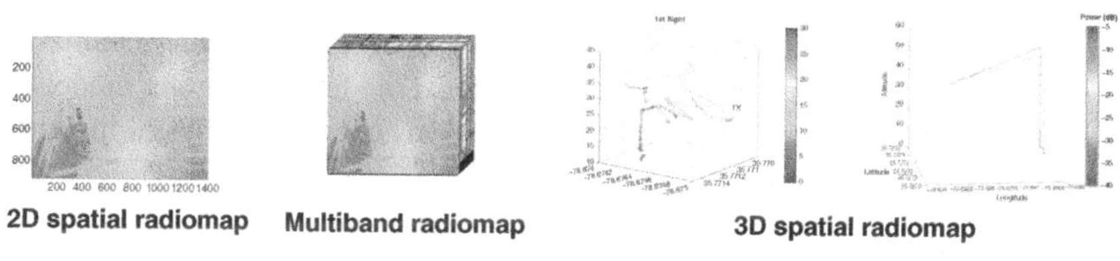

Figure 9.6 Examples of different types of radiomaps (simulated by WinProp [19]).

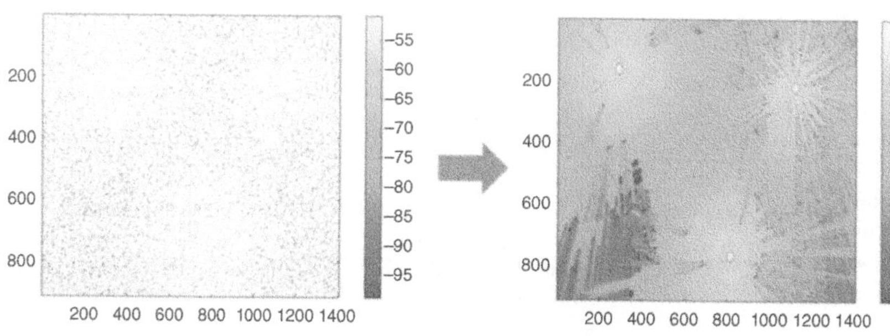

Figure 9.7 Illustration of 2D radiomap estimation. The color version of this figure is available on Wiley.com.

9.3.1.2 Conditional Generative Learning Models

In this example, we introduce the RME based on cGAN. cGAN is one of the emerging generative learning models, which study a set of training examples to acquire sufficient information on the sample distribution to generate synthetic data with a similar distribution conditioned on certain additional information [21]. Compared to other GenAI models, such as diffusion models and transformers, GANs are known for their speed in generating high-fidelity samples, which could benefit the timely data processing and reduce the computational cost in practical wireless communications. Thus, we utilize cGAN as an exemplary GenAI for RME. In cGAN-based RME, two neural networks interact in the form of a zero-sum game, in which the generator G aims to generate a dense radiomap with the same statistics as the training dataset with sparse observations as condition, while the discriminator D focuses on the identification of the real and fake radiomaps. Suppose that the generator learns the distribution p_g on the data $\mathbf{x}$ by mapping a prior noise distribution $p_z(\mathbf{z})$ to the data space. The cGAN model can be trained via

$$\min_{G}\max_{D} \quad V(D,G) = \mathbb{E}_{x\sim p_{data}(\mathbf{x})}[\log\ D(\mathbf{x}|\mathbf{I})] + \mathbb{E}_{z\sim p_z(\mathbf{z})}[\log(1 - D(G(\mathbf{z}|\mathbf{I})))], \tag{9.2}$$

where $\mathbb{E}[\cdot]$ refers to the expectation operation and $\mathbf{I}$ is the additional information, as shown in Figure 9.8. The ideas of cGAN match the formulation of the RME problem very well, where transmitter locations, sparse observations, building distributions, and radio propagation models can serve as conditions $\mathbf{I}$ to restrict the generation of dense radiomaps.

Next, we will introduce a physics-enhanced cGAN model for RME. As shown in Figure 9.3, one of the approaches to integrate GenAI models and physics models is to customize the loss functions of cGAN training. In general, the physics-inspired loss function can be formulated as

$$\min_{\Theta} L_1(\hat{R}(C_o), R(C_o)) + \alpha L_2(\hat{R}(C)) + \lambda L_3(\hat{R}(C), \mathcal{X}) \tag{9.3}$$

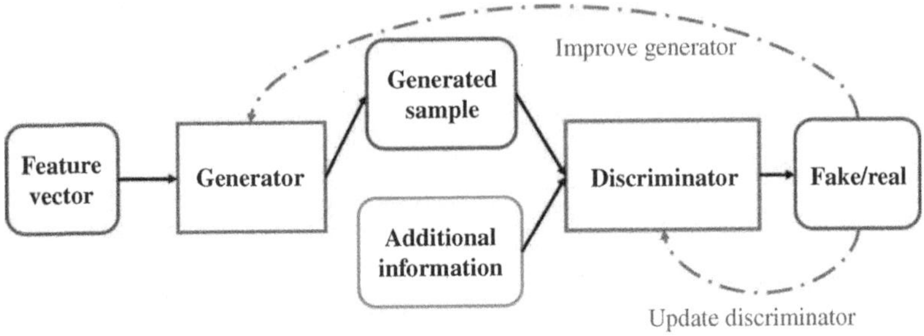

Figure 9.8 Structure of cGAN model.

where $\hat{R}(\mathcal{C}) = G_\Theta(\mathcal{X}, R(\mathcal{C}_o))$ is the generator with parameters Θ, and $\mathcal{X}$ is the additional physics-based inputs.

Here, L_1 is the empirical loss in measuring the quality of the generated radiomaps. It can refer to the traditional cGAN loss to optimize G as

$$L_G = \mathbb{E}_{x \sim p_{\text{data}}(x)}[\log(1 - D(G(\mathbf{x}), \mathbf{I}))]. \tag{9.4}$$

As a pair, the loss function to optimize D is denoted by

$$L_D = -V(D, G). \tag{9.5}$$

L_2 is a regularization term, which could help optimize to capture better geometric intuition and possess certain properties. Typical examples include the Laplacian regularizer or total variation to ensure the smoothness of radiomaps. L_3 aims to guide the training of the generator following certain models, such as the path-loss model. Other potential design of L_3 may also consider downstream tasks, such as segmentation of the outage area, the intensity of radiomap patterns, and the preselected observations based on the model [6].

In the following part, we will introduce an example of physics-inspired GenAI for RME with customized loss, namely RME-GAN [22].

9.3.1.3 Framework of RME-GAN

The general framework of the generator in RME-GAN is shown in Figure 9.9. Instead of inputting random noise into the generator, the RME-GAN takes the sparse observations, together with the environmental information, as the conditions of cGAN model. In particular, given a region $\mathcal{C} \in \mathbb{R}^{N \times N}$, with size $N \times N$, we first interpolate the sparse observations $R(\mathcal{C}_o)$ to the same size as the region size by zero-padding, that is, $\mathbf{R}_{obs} \in \mathbb{R}^{N \times N}$. Since we consider a convolutional neural network (CNN) structure for Unet, the interpolated zeros would not influence other features. In addition to sparse observations, we further input the corresponding urban map $\mathbf{m} \in \mathbb{R}^{N \times N}$ and the transmitter locations to provide guidelines for radio propagation. Note that, to ensure all the input features to have the same dimension, we rearrange the size of transmitter positions as $\mathbf{p} \in \mathbb{R}^{N \times N}$, whose values at transmitter locations are ones; otherwise, zeros. Now, our objective is to train an efficient generator $G(\cdot)$ so that the dense radiomaps can be generated, that is,

$$\tilde{R}(\mathcal{C}) = \tilde{\mathbf{y}} = G(\mathbf{m}, \mathbf{p}, \mathbf{R}_{obs}). \tag{9.6}$$

For the discriminator, we consider a four-layer CNN with generated radiomap and prior knowledge of model information as input, which generates a one-dimensional probability to identify the uncertainty of the generated radiomap.

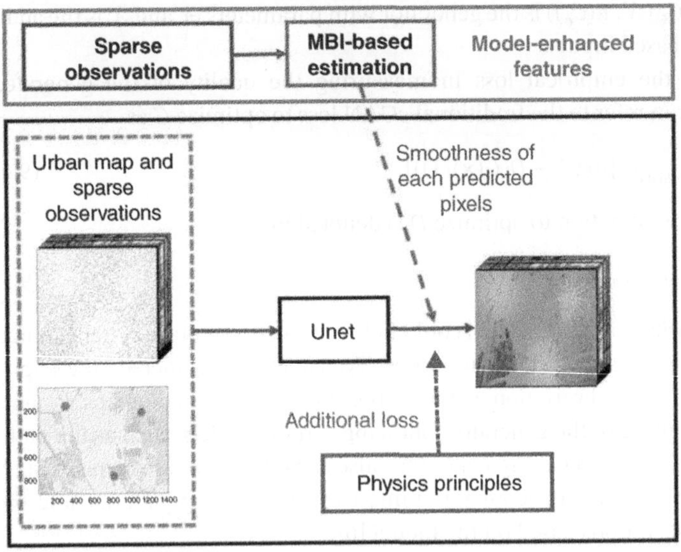

Figure 9.9 Structure of the generator for RME-GAN [22] / IEEE: Input $\mathcal{X} = \{\mathbf{m}, \mathbf{p}, \mathbf{R}_{obs}\}$; Output $\tilde{R}(\mathcal{C}) = \tilde{\mathbf{y}}$. The color version of this figure is available on Wiley.com.

9.3.1.4 Two-phase Training in RME-GAN

To capture the features of the overall radio propagation behavior and the details influenced by surrounding environments. We introduce a two-phase training strategy and customize the loss function of cGAN following Eq. (9.3). More specifically, we first introduce two sampling strategies for feature extraction in RME.

- **Upsampling for Global Information**:
  Large-scale radiomaps are usually supposed to follow certain path loss models, such as the log-distance path loss model (LDPL) [23]. To capture the global patterns in the radiomap, we first up-sample the sparse observations to a template $\mathbf{R}_{MBI} \in \mathbb{R}^{N \times N}$ using model-based interpolation (MBI).

  In MBI-based RME, the PSD value at position $\mathbf{c} = (a, b)$ in a system with N_t transmitters can be calculated as

$$P(a, b) = \sum_{k=1}^{N_t} (\alpha_k - 10\theta_k \log_{10} \left(\sqrt{(a_k - a)^2 + (b_k - b)^2)}\right), \qquad (9.7)$$

  where α_k is the power from kth transmitter with location (a_k, b_k) at the reference distance. With the sparse observations, we could optimize the parameters of the LDPL model by

$$\min_{\alpha_k,\theta_k} \sum_{j=1}^{K_i} (r(\mathbf{c}_o) - P(\mathbf{c}_o))^2. \tag{9.8}$$

With the estimated LDPL parameters, a dense model-interpolated radiomap can be constructed as $\mathbf{R}_{MBI}$. Although MBI results cannot capture the shadow impacts, they characterize the pathloss patterns, which could provide guidelines for the first phase of global feature extraction.

- **Downsampling for Detailed Features:**
In the second phase of detailed feature extraction, we consider two downsampling strategies: (i) geometry-downsampling and (ii) frequency-downsampling. Since the sparse observations may be nonuniformly distributed over the whole space, we can downsample the most important observations in different regions to balance the training samples and avoid model overfitting. Moreover, since the details and outliers are usually related to high frequencies in Fourier space, we could also transform the estimated radiomap into the frequency domain and extract the high-frequency features for the second phase, which favor the details in the radiomap.
With the extracted features under different sampling strategies, we propose a two-phase design on the loss functions for the training of the generator, that is,

$$L'_G = \begin{cases} L_G + L_{Global} & \text{if} \quad \epsilon < \tau \,, \\ L_G + L_{Local} & \text{if} \quad \epsilon \geq \tau \end{cases} \tag{9.9}$$

where L_{Global} consist of loss functions capturing global model information and L_{Local} are the loss functions focusing on the local details. ϵ is a pre defined criteria for phase changing.

9.3.1.5 Loss Functions

Next, we introduce the design of loss functions.

- **Phase 1 - Global Model Prediction:**
In the first phase, we prefer to utilize the MBI-based radiomap as the template. In Eq. (9.3), the L_1 is the empirical loss, where the squared error (MSE) can be applied, that is,

$$L_{\text{MSE}} = \frac{1}{N \times N} \cdot ||\mathbf{y} - \tilde{\mathbf{y}}||_F^2, \tag{9.10}$$

where $|| \cdot ||_F$ is the Frobenius norm, $\mathbf{y}$ is the groundtruth, and $\tilde{\mathbf{y}}$ is the generated radiomap.

To ensure the smoothness of the predicted radiomap, we use the spatial total variation as L_2, which is calculated by

$$L_{\mathrm{TV}} = \sum_{i,j \in \mathcal{N}} ||\tilde{\mathbf{y}}(i) - \tilde{\mathbf{y}}(j)||_2^2, \tag{9.11}$$

where $\mathcal{N}$ indicates pixel neighborhood and $\tilde{\mathbf{y}}(i)$ is the ith element of the estimated radio map $\tilde{\mathbf{y}}$.

To enhance the RME by the pathloss model, we also investigate the upsampled radiomap $\mathbf{R}_{MBI}$ in L_3, where gradient patterns are used to define a gradient-based loss function. In particular, we first calculate the gradient of each pixel in the radiomap for all directions, and we then calculate the cosine similarity between $\mathcal{G}(\tilde{\mathbf{y}}(i))$ and $\mathcal{G}(\mathbf{R}_{MBI})(i))$ calculated by $L_{Gradient}$.

- **Phase 2 - Estimation of Local Detail:**
In the second phase, our aim is to enhance the details in the radiomaps. Similarly to Phase-1, L_{MSE} and L_{TV} are still considered for L_1 and L_2, respectively. In addition to conventional empirical loss, we further use the similarity between downsampled observations, that is, L_{Geo} and L_{HPF} are the MSE loss terms for geometry-based and frequency-based downsampled radiomaps. In addition, to ensure the quality of generated radiomap, we mimic the multi-scale structural similarity index [24] in image processing for the second phase.
The MS-SSIM loss function is defined by

$$L_{\mathrm{SSIM}} = 1 - \mathrm{SSIM}(\mathbf{y}, \tilde{\mathbf{y}}). \tag{9.12}$$

- **Loss of Two-phase Training:**
With the aforementioned loss terms, our two-phase training can be summarized as

$$L'_G = \begin{cases} \lambda_1 L_G + \lambda_2 L_{\mathrm{MSE}} + \lambda_3 L_{\mathrm{TV}} + \lambda_4 L_{\mathrm{Gradient}}, & \epsilon < \tau \\ \lambda_1 L_G + \lambda_2 L_{\mathrm{MSE}} + \lambda_3 L_{\mathrm{TV}} + & \\ \lambda_5 L_{\mathrm{SSIM}} + \lambda_6 L_{\mathrm{Geo}} + \lambda_7 L_{\mathrm{HPF}} & \epsilon \geq \tau \end{cases}$$

where λ_i are the weights to control each loss term.

9.3.1.6 Some Experimental Results

We next present the experimental results of the RME-GAN compared to conventional RME approaches. In this case study, we tested the performance of different approaches in the RadioMapSeer dataset [25], compared to conventional learning approaches, such as RadioUnet and autoencoder (AE). We also compare

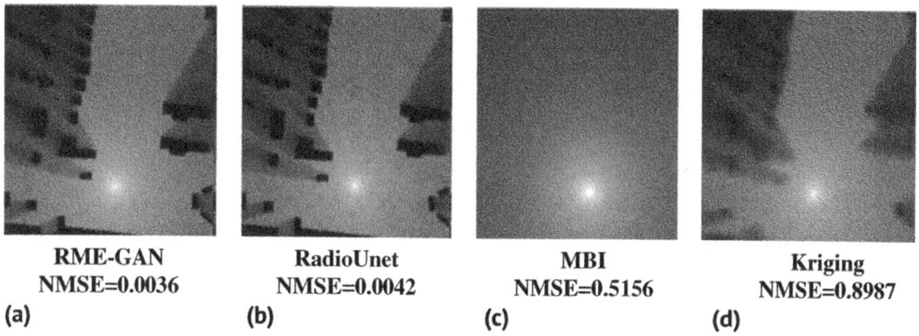

Figure 9.10 Visualization results of radiomap estimation in biased sampling setup: (a) RME-GAN; (b) RadioUnet; (c) MBI; (d) Kriging. The color version of this figure is available on Wiley.com.

it with traditional model-based methods. The visualization results are shown in Figure 9.10, where our proposed methods show superior performance in different setups. The visualization and numerical results demonstrate the efficiency of the physics-inspired GenAI for wireless communication applications.

9.3.1.7 Discussion

RME is only one of the potential applications for physics-inspired GenAI. Other promising scenarios for GenAI, also include the channel state information estimation, channel gain estimation, direction-of-arrival (DOA) estimation and localization. All of these applications are characterized by a demand to address the trade-off between data statistics and wireless physical principles.

Despite some successes, there are still several urgent challenges to be addressed. The first is the lack of realistic datasets. Most existing datasets are simulated data, which is usually captured by several well-defined empirical models. However, in practical applications, the models can be too complex to be formulated by mathematical equations. Incorrect model information may lead to artifacts in GenAI results. Another challenge is to make a more efficient integration of neural networks and physics models. Potential solutions include deep neural networks [26] and physics feedback for GenAI models. Other concerns also include distributed training that integrates different models [27].

9.3.2 Trustworthy Generative AI for Distributed Wireless Communications

In the second case study, we will discuss how to apply GenAI to distributed learning in edge computing, which shall be both communication-efficient and privacy-preserving. In particular, we will focus on a framework, namely PS-FedGAN [28] for FL, as an example.

9.3.2.1 Federated Learning: Gradients Versus Distributions

As introduced in Section 9.2.1, FL [29] is an emerging framework for distributed and collaborative learning while preserving user privacy. Instead of direct sharing of raw data, FL usually shares the model parameters or gradients to enable decentralized training. Despite successes, existing FL methods still suffer from handling nonindependent identically distributed (non-IID) heterogeneous data among different participating FL clients (users). The problem of handling data heterogeneity remains an interesting research topic. Moreover, multiple rounds of gradient sharing may lead to additional communication overhead and privacy concerns.

To address these challenges, generative learning models offer new opportunities. As shown in Figure 9.11, the key to distributed learning is to learn the distribution of data for each of the local clients. Thus, objectives can be achieved if one can transmit local data distribution without direct data sharing, which GenAI can contribute to great efforts. With training from a collection of data samples, GenAI is able to capture the local data distribution and regenerate the synthetic data for downstream tasks. Instead of gradient-based or data-based distributed learning, the GenAI-based FL has no restrictions on the non-IID data distribution and efficiently processes the heterogeneous dataset. However, as a new field in FL, GenAI-based approaches need to address privacy and efficiency concerns. First, although GenAI-based FL does not require raw data sharing, the generator can generate synthetic data with the same labels or memberships, which may lead to membership information leakage. In addition, the full sharing of the GenAI model can result in additional communication overheads. Consequently, addressing both privacy concerns and communication efficiency emerges as a key challenge in advancing the robustness and applicability of GenAI-based FL methods.

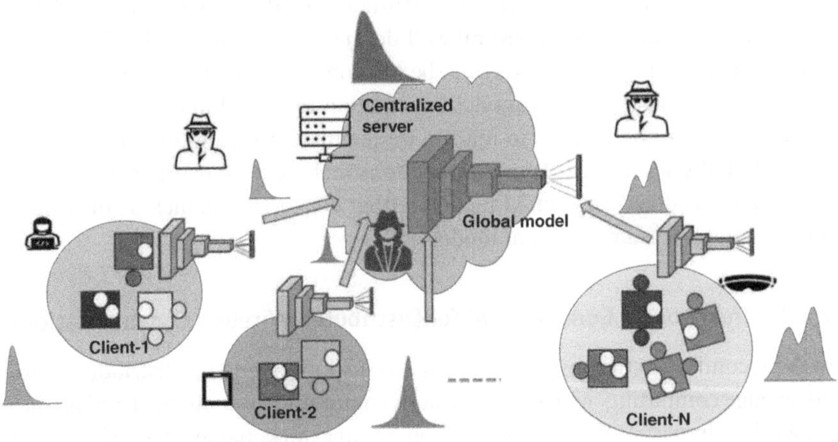

Figure 9.11 Federated learning based on sharing data distributions.

To address these concerns, one potential method is split learning, which trains different parts of GenAI models in suitable locations to avoid full model sharing and sensitive information passing. In the following parts, we will provide an example of split training of GenAI-based FL, called PS-FedGAN [28] with the GAN model as the backbone.

9.3.2.2 Training Schemes of PS-FedGAN

As shown in Figure 9.12, in PS-FedGAN, each client trains a local GAN model with a personalized generator and discriminator. Then, the discriminator is shared to the server, where the server learns a generator from the received discriminator. After training the learned generator on the server, synthetic data samples are generated to train a global neural network, such as classification or prediction. Finally, the global downstream task NN is transmitted back to the local clients for inference. Now, the challenge is how to split-train the GAN model efficiently from a partially shared discriminator.

To capture the underlying data distributions among different local clients, we initiate a GAN model on the server, including a generator (G_s) and a discriminator (D_s) to be updated by each client u. As illustrated in Figure 9.13, we consider a training process consisting of three steps, that is, two steps of D training and one step of G training. We divide the dual-step training D into the server and client sides. On the server side, the server initiates a discriminator D_u and then shares the D_u model with the corresponding client u. Subsequently, the client end performs a forward pass (FP) on D_u using a single batch of real local data I_u. Gradients and loss are calculated and then shared with the server on the network. Without direct generator sharing, privacy and efficiency can be enhanced. Upon receiving the updated information D_u from the sensor u, the server completes the training of D_u. It generates a noise vector $\mathbf{z}$ that passes through G_u. Finally, synthetic data

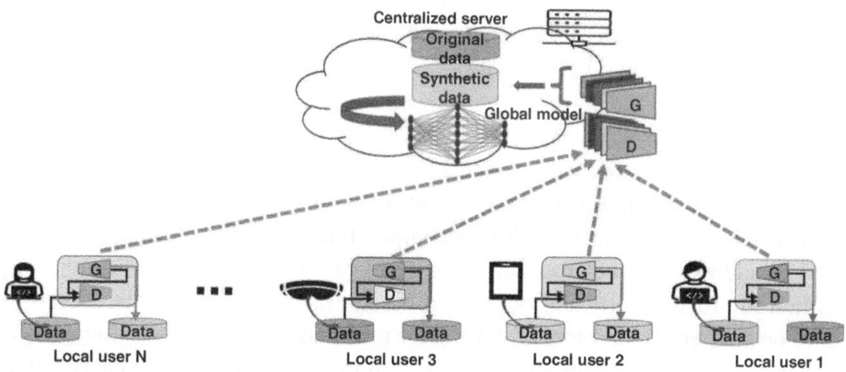

Figure 9.12 Overall structure of PS-FedGAN.

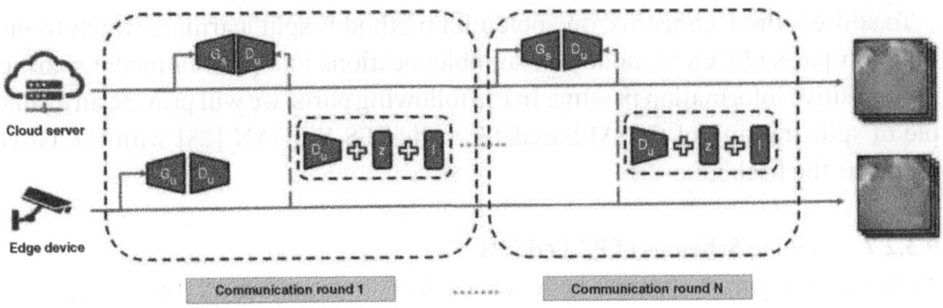

Figure 9.13 Training of PS-FedGAN by partially GAN model sharing. The color version of this figure is available on Wiley.com.

$G_u(\mathbf{z})$ is generated from the generator. $G_u(\mathbf{z})$ is then sent to D_u to calculate the corresponding gradients and loss, which are further combined with the gradient updates of the previous step and backpropagated by D_u for the parameter update. A convergent generator G_s will be obtained after several rounds of communication. In this process, we can create a synthetic dataset using trained G_u, with which we aim to design corresponding unsupervised or semi-supervised algorithms to implement specific downstream learning tasks.

9.3.2.3 Some Experimental Results

We now present some experimental results to validate the privacy preservation and efficiency of GenAI-based FL.

We consider the reconstruction attack with a multiplicative factor $r \in [0, 1]$. We assume that attackers $\mathcal{A}_1$ and $\mathcal{A}_2$ have the same architecture as G_u and the same weights and bias terms as G_u in every layer of the deep learning network except the first. For the first layer, $\mathcal{A}_1$ are set to $w_{a11} = rw_1$ and $b_{a11} = b_1$ for weights and bias, while $\mathcal{A}_2$ are set to $w_{a21} = w_1$ and $b_{a21} = rb_1$. Although the models of attackers have just a slight difference from the true parameters, the trained generator from the partially share discriminator fails to regenerate good-quality synthetic data samples, as shown in Figure 9.14.

9.3.2.4 Discussion

GenAI-based FL has recently attracted significant attention for distributed learning in wireless communications. In addition to PS-FedGAN, some other works, such as UFed-GAN to tackle a limited computational resource [30] and PFL-GAN [31] to address data heterogeneity, are developed for different applications. To further improve the efficiency of GenAI-based FL, one processing direction is to develop efficient GenAI model sharing for multimodal FL. Since existing GenAI sharing cannot protect privacy within authorized clients, additional research efforts are expected to increase data privacy and fairness.

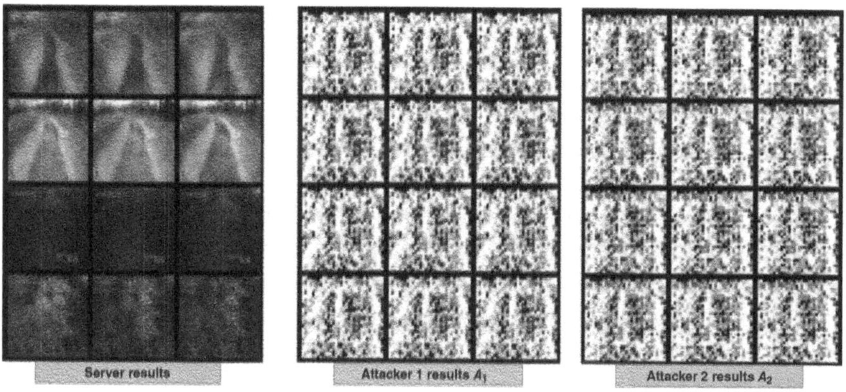

Figure 9.14 Divergence of the attacker: $r = 0.9999$ for $\mathcal{A}_1$ and $\mathcal{A}_2$. The color version of this figure is available on Wiley.com.

9.4 Summary

In next-generation wireless communication systems, generative AI is expected to play an increasingly important role due to its power to capture data distribution and generate high-fidelity data samples. In this chapter, we first present the existing challenges in applying GenAI in wireless communications, covering different aspects of technologies, ethics, and regulation. To provide a general guide for the application of GenAI in wireless communication applications, we present two case studies for physics-inspired GenAI for RME and split training of GenAI for FL. With the development of efficient GenAI algorithms and the improvement of hardware for computing, GenAI shall expect great success in wireless communications.

References

1 Anutusha Dogra, Rakesh Kumar Jha, and Shubha Jain. A survey on beyond 5G network with the advent of 6G: Architecture and emerging technologies. *IEEE Access*, 9: 67512–67547, 2020.

2 Di Wu, Xiang Sun, and Nirwan Ansari. An FSO-based drone assisted mobile access network for emergency communications. *IEEE Transactions on Network Science and Engineering*, 7(3): 1597–1606, 2019.

3 Songyang Zhang, Brian Choi, Feng Ouyang, and Zhi Ding. Physics-inspired machine learning for radiomap estimation: Integration of radio propagation models and artificial intelligence. *IEEE Communications Magazine*, 62(8): 155–161, 2024.

4 Jun Wang, Rong Li, Jian Wang, Yi-qun Ge, Qi-fan Zhang, and Wu-xian Shi. Artificial intelligence and wireless communications. *Frontiers of Information Technology & Electronic Engineering*, 21(10): 1413–1425, 2020.

5 Abdulkadir Celik and Ahmed M. Eltawil. At the dawn of generative AI era: A tutorial-cum-survey on new frontiers in 6G wireless intelligence. *IEEE Open Journal of the Communications Society*, 2024.

6 Yueling Zhou, Achintha Wijesinghe, Yibo Ma, Songyang Zhang, and Zhi Ding. TIRE-GAN: Task-incentivized generative learning for radiomap estimation. *IEEE Wireless Communications Letters*, 14(5): 1401–1405, 2025.

7 George Em Karniadakis, Ioannis G Kevrekidis, Lu Lu, Paris Perdikaris, Sifan Wang, and Liu Yang. Physics-informed machine learning. *Nature Reviews Physics*, 3(6): 422–440, 2021.

8 Rui Guo, Tianyao Huang, Maokun Li, Haiyang Zhang, and Yonina C. Eldar. Physics-embedded machine learning for electromagnetic data imaging: Examining three types of data-driven imaging methods. *IEEE Signal Processing Magazine*, 40(2): 18–31, 2023.

9 An Liu, Zhe Huang, Min Li, Yubo Wan, Wenrui Li, Tony Xiao Han, Chenchen Liu, Rui Du, Danny Kai Pin Tan, Jianmin Lu, Yuan Shen, Fabiola Colone, and Kevin Chetty. A survey on fundamental limits of integrated sensing and communication. *IEEE Communications Surveys & Tutorials*, 24(2): 994–1034, 2022.

10 Mohammad Rasouli, Tao Sun, and Ram Rajagopal. FedGAN: Federated generative adversarial networks for distributed data. *arXiv preprint arXiv:2006.07228*, 2020.

11 Davis Blalock, Jose Javier Gonzalez Ortiz, Jonathan Frankle, and John Guttag. What is the state of neural network pruning? *Proceedings of Machine Learning and Systems*, 2: 129–146, 2020.

12 Jianping Gou, Baosheng Yu, Stephen J Maybank, and Dacheng Tao. Knowledge distillation: A survey. *International Journal of Computer Vision*, 129(6): 1789–1819, 2021.

13 Guojun Wu, Xin Zhang, Ziming Zhang, Yanhua Li, Xun Zhou, Christopher Brinton, and Zhenming Liu. Learning lightweight neural networks via channel-split recurrent convolution. In *Proceedings of the IEEE/CVF Winter Conference on Applications of Computer Vision*, pages 3858–3868, 2023.

14 Achintha Wijesinghe, Songyang Zhang, Suchinthaka Wanninayaka, Weiwei Wang, and Zhi Ding. Diff-GO: Diffusion goal-oriented communications with ultra-high spectrum efficiency. In *2024 IEEE International Conference on Communications Workshops (ICC Workshops)*, pages 1079–1084, 2024.

15 Anirban Chakraborty, Manaar Alam, Vishal Dey, Anupam Chattopadhyay, and Debdeep Mukhopadhyay. A survey on adversarial attacks and defences. *CAAI Transactions on Intelligence Technology*, 6(1): 25–45, 2021.

16 Ziying Yang, Jie Zhang, Wei Wang, and Huan Li. Invisible threats in the data: A study on data poisoning attacks in deep generative models. *Applied Sciences*, 14(19): 8742, 2024.

17 Lucas Bourtoule, Varun Chandrasekaran, Christopher A Choquette-Choo, Hengrui Jia, Adelin Travers, Baiwu Zhang, David Lie, and Nicolas Papernot. Machine unlearning. In *2021 IEEE Symposium on Security and Privacy (SP)*, pages 141–159. IEEE, 2021.

18 Suzhi Bi, Jiangbin Lyu, Zhi Ding, and Rui Zhang. Engineering radio maps for wireless resource management. *IEEE Wireless Communications*, 26(2): 133–141, 2019.

19 ALTAIR. Radio coverage planning for heterogeneous wireless networks, including 5G. https://web.altair.com/winprop-telecom. Online; accessed 5 March 2024.

20 Songyang Zhang, Tianhang Yu, Brian Choi, Feng Ouyang, and Zhi Ding. Radiomap inpainting for restricted areas based on propagation priority and depth map. *IEEE Transactions on Wireless Communications*, 23(8): 9330–9344, 2024.

21 Mehdi Mirza. Conditional generative adversarial nets. *arXiv preprint arXiv: 1411.1784*, 2014.

22 Songyang Zhang, Achintha Wijesinghe, and Zhi Ding. RME-GAN: A learning framework for radio map estimation based on conditional generative adversarial network. *IEEE Internet of Things Journal*, 10(20): 18016–18027, 2023.

23 Songyang Zhang, Tianhang Yu, Jonathan Tivald, Brian Choi, Feng Ouyang, and Zhi Ding. Exemplar-based radio map reconstruction of missing areas using propagation priority. In *GLOBECOM 2022 - 2022 IEEE Global Communications Conference*, pages 1217–1222, 2022.

24 Zhou Wang, Eero P Simoncelli, Alan C Bovik. Multiscale structural similarity for image quality assessment. *In the thirty-seventh Asilomar conference on signals, systems & computers*, 2003 volume 2, pages 1398–1402, IEEE, 2003.

25 Ron Levie, Çağkan Yapar, Gitta Kutyniok, and Giuseppe Caire. Radiounet: Fast radio map estimation with convolutional neural networks. *IEEE Transactions on Wireless Communications*, 20(6): 4001–4015, 2021.

26 Vishal Monga, Yuelong Li, and Yonina C Eldar. Algorithm unrolling: Interpretable, efficient deep learning for signal and image processing. *IEEE Signal Processing Magazine*, 38(2): 18–44, 2021.

27 Dong Yang, Yue Wang, Songyang Zhang, Yingshu Li, and Zhipeng Cai. Physics-inspired distributed radio map estimation. *arXiv preprint arXiv:2502.00319*, 2025.

28 Achintha Wijesinghe, Songyang Zhang, and Zhi Ding. PS-FEDGAN: An efficient federated learning framework with strong data privacy. *IEEE Internet of Things Journal*, 11(16): 27584–27596, 2024.

29 Brendan McMahan, Eider Moore, Daniel Ramage, Seth Hampson, and Blaise Aguera y Arcas. Communication-efficient learning of deep networks from decentralized data. In *Artificial Intelligence and Statistics*, pages 1273–1282. PMLR, 2017.

30 Achintha Wijesinghe, Songyang Zhang, Siyu Qi, and Zhi Ding. UFED-GAN: Secure federated learning over wireless sensor networks with unlabeled data. In *2024 IEEE International Conference on Communications Workshops (ICC Workshops)*, pages 1048–1053. IEEE, 2024.

31 Achintha Wijesinghe, Songyang Zhang, and Zhi Ding. PFL-GAN: When client heterogeneity meets generative models in personalized federated learning. *arXiv preprint arXiv:2308.12454*, 2023.

10

Future Research Directions

Nam H. Chu[1,2,], Diep N. Nguyen[2], Dinh Thai Hoang[2], Octavia A. Dobre[3], Dusit Niyato[4], and Petar Popovski[5]*

[1] *School of Information and Technology, Crown Institute of Higher Education, NSW, Australia*
[2] *School of Electrical and Data Engineering, University of Technology Sydney, NSW, Australia*
[3] *Faculty of Engineering and Applied Science, Memorial University, Newfoundland, St. John's, Canada*
[4] *College of Computing and Data Science, Nanyang Technological University, Singapore*
[5] *Department of Electronic Systems, Aalborg University, Aalborg, Denmark*

*Corresponding author: hoainam.chu@uts.edu.au

10.1 Introduction

Thanks to its abilities, particularly the creativity and innovation, generative AI (GenAI) has been emerging as a revolutionary driver in wireless communications, effectively tackling various challenges that traditional approaches have struggled to overcome. GenAI also enables new paradigms of automation, adaptability, and efficiency. It demonstrates the potential across various layers in wireless communication system, from optimizing the physical layers to advancing application-level functionalities, to assist the system to meet its key performance metrics, as shown in Figure 10.1. In previous chapters, we have explored the fundamentals of GenAI, and its applications and challenges in wireless communication systems, across the PHY, MAC, network, and application layers. These discussions highlight the GenAI's ability to optimize network operations, enhance resource allocation, and enable novel communications, for example, semantic communications. However, achieving this envision requires mitigating various challenges hindering the application of GenAI, ranging from technical aspects like scalability to

Generative AI for Communications Systems: Fundamentals, Applications, and Prospects,
First Edition. Edited by Diep N. Nguyen et al.

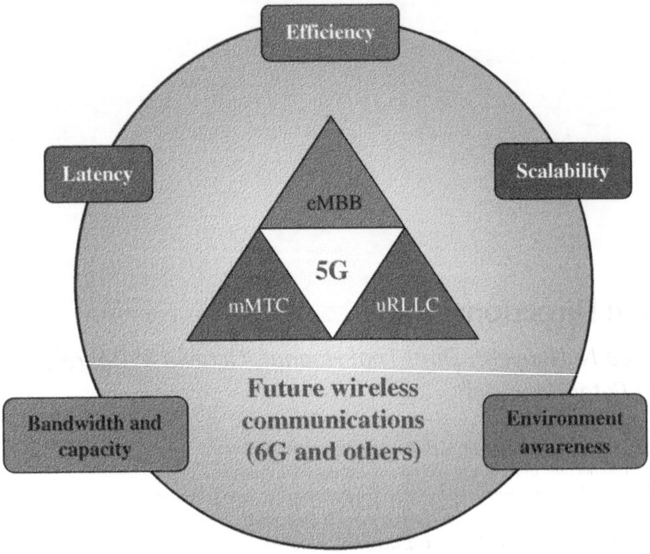

Figure 10.1 Key performance metrics of future wireless systems.

ethical and societal impacts. Additionally, as with any rapidly evolving technologies, there are still many aspects to be explored, discovered, and innovated.

Thus, this chapter aims to set the vision of the future of GenAI in wireless communication systems, exploring emerging trends, technologies, and research frontiers that will shape the next generation of AI-enabled wireless communications. In summary, key objectives of this chapter include:

- Highlight emerging foundation research frontiers, such as the development of innovative generative AI models and their integration with technologies like IoT, blockchain, and quantum computing to drive the next wave of AI-powered wireless communications.
- Showcasing real-world case studies that exemplify the transformative potential of GenAI in communications.
- Encouraging cross-disciplinary research initiatives to address technical, ethical, and societal challenges.

10.2 Emerging Foundational Research Frontiers

The potential of GenAI in wireless communications has led to the emergence of new foundation research frontiers that pioneer the next wave of innovation. These frontiers are driven by the need for more efficient, scalable, and robust GenAI models that can effectively address the unique challenges in wireless communications, for example, distributed and dynamic environments. In this section, we will explore two key research directions: dedicated GenAI models and the fusion of GenAI with emerging technologies, expected to significantly impact the applications of GenAI in wireless communications.

10.2.1 Dedicated GenAI Models for Communication Systems

As wireless communication environments are highly dynamic, distributed, and consist of heterogeneous dedicated entities (e.g., eNodeB and RAN networks), developing dedicated GenAI models for this domain is crucial for ensuring the effectiveness of GenAI. Traditional GenAI models, such as generative adversarial networks (GANs) and variational Autoencoders (VAEs), have been widely used in various applications, ranging from image and video generation, natural language processing, to data imputation and synthesis. However, they are often developed for general-purpose tasks and require customization (e.g., fine-tuning) to meet unique characteristics of wireless communication system [1, 2]. Thus, the development of dedicated GenAI models tailored for wireless communications is crucial in advancing this field. However, to achieving it, dedicated models need to effectively address several major challenges as follows.

Challenges and Opportunities.

- ***Scalability and Efficiency:*** As future wireless systems (e.g., 6G) are expected to be ultra-dense, serving heterogeneous connectivity and device characteristics, GenAI models' operations may face unprecedented operational challenges [3]. For example, they must handle exponentially increasing computational and communication demands from a massive number of devices, and may operate under resource limitations on edge devices [4]. Thus, dedicated GenAI models must be optimized for low latency, energy efficiency, and scalability to handle dynamic network environments effectively.
- ***Domain-specific Features:*** GenAI models developed for wireless communication systems should incorporate domain-specific characteristics, for example, channel state information (CSI), interference, and mobility predictions [5–7]. For example, network traffic prediction accuracy can be improved significantly by leveraging the spatial-temporal correlations in traffic. In addition, GenAI-driven semantic communication networks have been explored [8], aiming to intelligently and automatically support artificial intelligence generated content (AIGC) services.
- ***Robustness and Adaptability:*** Since wireless networks typically face challenges like interference, noise, and adversarial activities, dedicated GenAI models should exhibit robustness to these factors to ensure their success in dynamic environments with heterogeneous network conditions [9].

Although some studies (e.g., [10]) show that GenAI models can adaptively and effectively work in wireless environments, these models are vulnerable to adversarial attacks [11].

The above challenges can be mitigated by adopting several approaches. First, leveraging customizable modular architectures can allow models quickly adapt to various tasks (e.g., wireless channel modeling and mitigating interference) in different environments, for example, line-of-sight (LOS) and non-line-of-sight (NLOS) [12]. Second, integrating GenAI with mobile edge computing is a promising solution to reduce latency and enhance scalability and efficiency [13]. Additionally, leveraging lightweight GenAI models (e.g., quantized models) can further complement this approach by reducing computational overhead at the edge [14]. Third, combining generative capabilities with reinforcement learning for enhancing decision-making in complex scenarios, for example, optimizing resource allocation in dynamic network environments or managing interference in densely populated wireless networks [15].

10.2.2 Fusion of GenAI and Emerging Technologies

The rapid advancement of emerging technologies, for example, 6G, IoT, quantum computing, and blockchain, is expected to shaping the future of wireless communications, as shown in Figure 10.2. Thus, the fusion of GenAI with emerging technologies promises a transformative opportunity for advancing wireless communication systems in various aspects, such as performance, reliability, and scalability. It also lays the foundation for intelligent, autonomous, and secure wireless communication systems. Promising integration are outlined as follows.

- **GenAI and Quantum Computing:** The integration of GenAI and quantum computing represents a groundbreaking fusion of two disruptive technologies. Specifically, quantum computing (with its ability to handle complex computations at an unprecedented scale) and GenAI (with its capacity for creativity, prediction, and optimization) can complement each other in unique ways. For example, GenAI models can leverage quantum circuits to generate high-dimensional and complex datasets, and encode/decode high-dimensional data for compression and anomaly detection in real time [16–20]. Thus, it can open new opportunities to shape the future wireless communication across domains (such as cryptography and optimization) at various layers in the systems.
- **GenAI and 6G technologies:** The 6G perspectives from ITU/IMT and the latest 3GPP release are evolving to incorporate enhanced capabilities beyond the current 5G technologies. Release 21 of 3GPP is expected to produce the first set of 6G technical specifications, which will be submitted for IMT-2030 before 2030 [21]. This release will mark the formal beginning of 6G

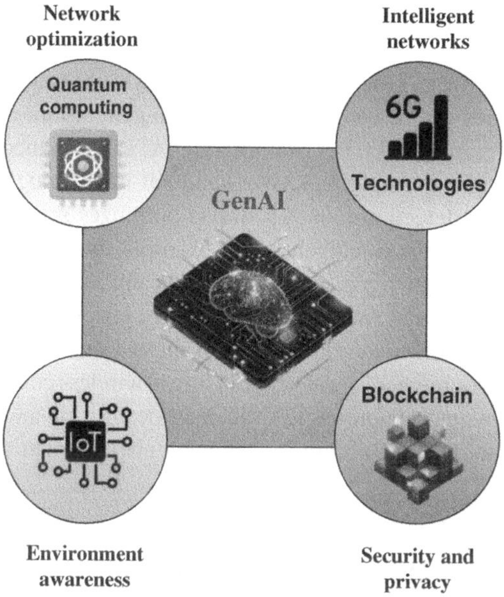

Figure 10.2 The fusion of GenAI with emerging technologies.

standardization, incorporating elements such as wireless, AI, and the Internet of Everything (IoE) [22]. Thus, GenAI can open a new era of intelligent networks. This aligns with the potential of GenAI to enable new use cases and applications within 6G technologies, such as Joint Communication and Sensing (JCAS), Zero or Near-Zero Energy communication, cognitive radio, and Network Slicing) [23]. Additionally, GenAI promises the shift from traditional transmission to semantic-based communication by interpreting the meaning of information. It can also simulate realistic and adaptive wireless channels, enhancing system design and optimizing system performance. In this way, GenAI-enabled Massive MIMO and beam-forming can be a promising solution to meet 6G specifications. Last but not least, fusing GenAI with network slicing in 6G can bring a next generation of intelligent network management.

- **GenAI and Blockchain:** Since Blockchain provides a decentralized and secure framework for wireless communication systems, integrating with GenAI offer various benefits [24, 25]. For instance, GenAI can enhance smart contract generation by automating the creation of context-aware smart contracts, thus reducing human error and development time [26, 27]. In addition, GenAI models can analyze transaction patterns and identify potential fraud or malicious activities in Blockchain-based system [28, 29]. Furthermore, it can optimize consensus algorithms, therefore improving speed and scalability of blockchain systems.

- **GenAI and the Internet of Things (IoT):** IoT systems, which is characterized by massive, heterogeneous device connectivity and data generation, is likely to benefit significantly from GenAI. For example, GenAI can enhance IoT communication protocols, for example, LoRaWAN and NB-IoT, through traffic prediction and anomaly detection [30, 31]. It also optimizes sensor data collection and control actuators intelligently, for example, synthesizing real-time traffic pattern to adjust signal timings dynamically [32]. As a result, it can enable next generation of smart IoT applications, for example, traffic management and smart home. Moreover, the fusion of GenAI and IoT can also enhance the environment awareness of the future wireless communications by improving sensing capabilities [33], such as localization and motion detection, thereby enabling more accurate and reliable environmental monitoring.

10.3 Enhancing Generative AI Models for Wireless Communication Systems

While the foundational research frontiers emphasize the development of dedicated GenAI models and their fusion with emerging technologies, this section shifts the focus to the practical enhancements required to adapt GenAI models for addressing challenges in wireless communication systems. The unique mobility, dynamic characteristics of wireless environments, and the rapid evolution of wireless systems (driven by the advent of emerging technologies like 6G, edge computing, cognitive radio technologies) demand GenAI models that are not only highly efficient but also robust, adaptable, and scalable. However, enhancing these models is challenging since they need to meet intensive demand of the high-quality service while addressing critical concerns such as energy efficiency and security. This section will explore key areas for enhancing GenAI in wireless communication systems.

10.3.1 Model Optimization and Generalization

Currently, the performance of GenAI models in wireless environment hinders their application due to the diverse and dynamic network conditions as well as resource-constrained and heterogeneous devices in the network. Thus, it is essential to focus on optimizing GenAI models for computational efficiency while ensuring their generalization across various wireless network environments. By addressing these areas, we can develop efficient, robust, and scalable GenAI models.

- **Model Optimization:** Model optimization aims to improve the computational efficiency and deployment flexibility of GenAI models to meet unique demands and characteristics of wireless communication systems. This is crucial for real-time applications (e.g., dynamic and adaptive resource allocation) in wireless communications, where low-latency and resource constraints are significant challenges. In this context, several effective strategies can be used to optimize GenAI models. These involve refining GenAI models to enhance performance while sustaining scalability and robustness [34]. In practice, compact models can be developed, tailored for implementing on edge devices with resource constrain [35, 36]. For example, advanced techniques like knowledge distillation and transfer learning can create smaller models from large, pretrained models while maintaining their performance [37]. Other techniques like weight quantization and pruning can be used to reduce size and complexity of GenAI models [34]. Additionally, designing models that are optimized for specific hardware accelerator (e.g., GPU or specialized AI chips) is also a promising solution [38, 39].
- **Model Generalization:** Model generalization ensures that GenAI models can adapt to and quickly well-perform in diverse scenarios and applications. This is particularly critical in wireless communication systems but a challenging task since they encompass heterogeneous devices, environments, and network protocols. To deal with this, model generalization can be improved by designing models that are capable of addressing tasks across multiple layers, for example, physical, MAC, and network layers. For instance, a GenAI model can generalize channel estimation and resource allocation simultaneously [40]. In addition, other existing approaches can also be applied to improve model generalization. These include enhancing training datasets with diverse and realistic synthetic data generated by GenAI models. For example, GenAI models can be used to generate channel state information (CSI) dataset for training wireless communication models [40]. In addition, incorporating adversarial training and regularization techniques improve the model robustness against unforeseen data.

10.3.2 Energy Efficiency

Energy efficient is one of the top considerations when deploying GenAI models in wireless communication, especially in the resource-constrains environments like IoT networks and edge computing platforms where devices have limited computational power, battery, and memory. The reason is that the increasing complexity of GenAI models has led to a significant rise in their computational and energy demand, posing challenges for real-time and sustainable operations in wireless environments [41]. Thus, improving energy efficiency makes large-scale AI systems feasible for deployment in densely connected wireless environment.

This subsection will explore strategies for improving the energy efficiency of GenAI models and their applications in wireless environments with limited resources.

Optimizing the energy efficiency of GenAI models involves minimizing their computational requirements without compromising performance. Several approaches can be applied to achieve this balance:

- **Energy-aware Model Architectures:** Developing sparse, low complexity neural network or hybrid architectures can reduce energy consumption. For example, lightweight transformer models optimized for IoT gateways can be used to predict traffic passing through them [42, 43]. Additionally, techniques like early exit strategies and dynamic neural networks allow models to adjust their computational load based on the complexity of the input, conserving energy when full model capacity is unnecessary [44, 45].
- **Dynamic Power Scaling:** GenAI-based systems can incorporate adaptive algorithms such as Dynamic Voltage and Frequency Scaling (DVFS) and carbon-aware computing to scale computational demand based on task complexity, device power availability, and energy source carbon intensity [46, 47]. For instance, transformer models can utilize early exit strategies for simpler inputs, while carbon-aware scheduling algorithms can shift computationally intensive GenAI workloads to times and locations with abundant renewable energy [48].
- **Battery-aware AI Deployment:** Model inference can be scheduled based on battery level of IoT or edge devices to extend its operational lifetime. For example, noncritical tasks can be postponed during low-battery periods.
- **Low-power Hardware Acceleration:** Optimizing GenAI models for specialized hardware, for example, dedicated AI ASICs, Tensor Processing Units (TPUs), Vision Processing Units (VPUs), and GPU-based edge computing platforms, can reduce energy consumption significantly while maintaining acceptable inference performance [49]. Currently, notable low-power, high-performance modules include Google's Edge TPUs (4 TOPS at 2W) [50], Intel's Neural Compute Stick 2 (4 TOPS with Myriad X VPU) [51], and NVIDIA's Jetson Nano (472 GFLOPS at 5-10W) [52], which can enable the deployment of optimized AI inference models, though with constraints on model size, precision, and framework compatibility.

10.3.3 Generative AI for Spectrum Management

The ever-increasing demand for wireless communication services has led to a significant shortage of available spectrum, making efficient spectrum management a challenging task. In this context, GenAI has emerged as a promising solution that can effectively address these challenges by enabling dynamic, adaptive, and intelligent spectrum management. This section will explore how GenAI can revolutionize spectrum management, focusing on key areas: dynamic spectrum access and cognitive radio networks.

- **Dynamic Spectrum Access (DSA)** is the ability of wireless devices to dynamically access and utilize available spectrum, aiming to reduce interference, congestion, and improve spectrum efficiency. GenAI can enhance DSA by effectively predicting spectrum availability to support real-time, intelligent decision-making. Specifically, GenAI-based approaches first predict spectrum availability with high accuracy by analyzing both historical usage patterns and wireless environment conditions [53, 54]. Then, it can dynamically adjust resource allocation based on current network demand, ensuring efficient spectrum utilization. In addition, GenAI models can also generate scenarios to simulate interference/congestion conditions and then adapt system parameters proactively, thereby improving interference avoidance [54].
- **Cognitive Radio Networks (CRNs)** rely on the ability of devices to sense, learn, and adapt their parameters according to their surrounding environment to optimize spectrum utilization. The rapid evolution of wireless communication with a massive number of heterogeneous devices makes existing solutions in CRNs ineffectively. Thus, GenAI can empower CRNs with advanced learning, creativity, and decision-making capabilities. For example, GenAI can improve spectrum sensing by synthesizing complex interference scenarios used for training CRNs to detect idle spectrum more effectively [6, 55]. In addition, GenAI can also analyze user behavior and traffic patterns to dynamically optimize spectrum utilization. Moreover, with the ability of processing various input types (e.g., text and image) [6], GenAI model can create adaptive spectrum access policies based on regulatory constraints and real-time conditions [56]. As above mentioned, GenAI can enable the creation of end-to-end models that unify sensing, decision-making, and resource allocation processes, leading to robust, adaptable, an intelligent CRNs.

10.3.4 AI-driven Network Management and Orchestration

The increasing complexity of wireless communication systems leads to a significant demand for efficient and adaptive network management and orchestration, which cannot be addressed by traditional approaches effectively. This is the area where the distinct abilities (e.g., creativity) of GenAI shines. Advances in GenAI can unlock new possibilities for managing and orchestrating networks in real time, which is critical in the rapid evolving wireless communication systems. They enable greater autonomy, efficiency, and adaptability across areas in the wireless systems. This section will sketch several key aspects where GenAI-enable solutions can enhance wireless network management and orchestration.

- **Self-organizing Networks (SONs)** are networks that can automatically configure, optimize, and heal themselves, reducing the need for manual activities. Here, GenAI enhances SONs by enabling intelligent decision-making based on real-time collected data. Specifically, GenAI-based SONs can predict network conditions, simulate scenarios, and take actions to sustain network performance under varying conditions. For example, a GenAI-empower system can predict potential faults in hardware components and then redirects traffic to backup nodes before failures occur [57].
- **Policy-driven Management** plays a vital role in navigating complex rules, regulations, and supporting decision-making in modern networks. Generally, it offers a structured approach to enforce rules and configurations. By integrating GenAI, policy-driven management systems can evolve into adaptive frameworks, which are capable of real-time decision-making and dynamic responses based on the current network states, application demands, regulations, and organization goals. For instance, these management systems can leverage GenAI to translate high-level objectives into actionable rules and ensure consistent performance, security, and service quality [58]. By doing so, GenAI-based systems automate traditionally static process, therefore enabling networks to respond intelligently to changes in operation conditions and threats.
- **Adaptive QoS Management** can dynamically adjust network parameters to meet the specific performance requirements of different applications and services. In this context, GenAI brings advanced prediction and resource optimization capabilities to maintain consistent service delivery, even when network conditions vary. For example, GenAI can predict network congestions in advance, enable proactive actions like using load balancing or route re-optimization [59, 60]. In addition, GenAI-based approaches can automatically adjust parameters like throughput and priority tailoring them accordingly to different QoS profiles, for example, video streaming, online gaming, or emergency communications [58].
- **QoE Enhancement** can improve the end-user's perceived quality of services like user satisfaction, which goes beyond the technical parameters. By integrating GenAI, QoE management can predict individual user preferences, for example, streaming resolution, and then adjust services parameters accordingly. Similar to DL-based approaches (e.g., [61, 62]), GenAI can also be used to predict and identify degradation in service (e.g., buffering or lag) to address issues quickly. In addition, user feedback can be integrated into GenAI models, enabling these models to continuously self-improve.

10.3.5 Security and Privacy Concerns

The integration of GenAI into wireless communication opens promising opportunities to revolutionize wireless networks. On the other hand, it also introduces significant security and privacy concerns, including AI-generated vulnerabilities, data confidentiality, and the robustness of training process. Addressing these challenges is crucial for ensuring safe and secure AI-driven wireless communication networks. Notable challenges are as follows.

- **AI-generated Attack Detection:** GenAI can not only mitigate but also create sophisticated and unforeseen security threats to the wireless communication systems, making AI-generated attack detection an emerging challenge. For example, GenAI models can perform spectrum spoofing where it generates signals mimicking legitimate transmission to disrupt network operations [63]. They can also generate signals that deceive signal detection algorithms, leading to the compromise of network security. Furthermore, GenAI models can be trained to bypass AI-based detector. As such, mitigating AI-generated attacks is challenging, requiring a combination of technical strategies, regulatory measures, and collaborative framework. For instance, regulations for validating AI-based system should be enforced to prevent exploitation in future [64]. Additionally, explainable AI techniques should be considered to understand decision-making process, making it easier to identify vulnerabilities.
- **Privacy-preserving AI Models:** In any AI-driven system, privacy of sensitive data is always crucial, especially in distributed and decentralized environments like IoT and edge networks. If not implemented correctly, trained GenAI models can be exploited to reveal sensitive data by advanced techniques like using another AI models. Thus, GenAI should be incorporated with privacy-preserving techniques. For instance, differential privacy mechanisms can be used to ensure that user data cannot be differentiated individually, making it unable to be reconstructed [65]. Notably, GenAI models can be used to anonymize sensitive data and create synthetic data that preserves the original statistical property without compromising user privacy. Furthermore, distributed training approaches, similar to Federated Learning, can preserve user privacy by removing the need of sending user data to center servers.
- **Secure AI Model Training:** As training process highly impacts the operation of trained model, ensuring the security and integrity of GenAI model training process is critical to mitigating the risks of model corruption, privacy breach, and unauthorized access. Several approaches can be leveraged to train GenAI models securely. They include identifying and filtering malicious data from training datasets, thereby protecting models from poisoning attacks that attempt to compromise their performance. In addition, utilizing

blockchain in distributed training process can ensure immutability and verifiability. By doing so, together with authentication mechanism, it can provide a secure foundation for model training, ensuring only authorized entities participate in the process.

10.4 Practical Case Studies

This chapter concludes with an exploration of real-world implementations of GenAI in wireless communication systems. Through these specific use cases and on-going deployments of GenAI in wireless communication systems, this section aims to emphasize the tangible benefits and challenges from GenAI integrations. These use cases can illustrate the applicability of GenAI across various aspects of the industry. For example, Verizon has implemented GenAI-based Customer Service to enhance customer service satisfaction by predicting the purpose of calls and efficiently connect them to suitable agents. Handling approximately 170 million calls annually, GenAI accurately predicts customer's intention 80% of the time, thus improving customer's satisfaction and reducing waiting time [66]. Differently, T-Mobile and DeepSig focus on integrating GenAI to address technical challenges. The remainder of this section will overview notable ongoing projects at T-Mobile and DeepSig.

10.4.1 AI-powered Network Optimization by T-Mobile

T-Mobile, one of the leading telecommunication providers in the United States, has pioneered the usage of GenAI in optimizing its networks. In partnership with OpenAI, a leading GenAI service provider, and other industry leaders, including NVIDIA, Ericsson, and Nokia, T-Mobile aims to transform its traditional network infrastructure into highly automated and self-optimized system, revolutionizing mobile networking and customer service.

- **AI-RAN Innovation Center:** In September 2024, T-Mobile established an AI-based Radio Access Network (AI-RAN) Center, aiming to design and drive the future of mobile communications [67]. Specifically, the main objectives of this initiative include enhancing network performance, driving innovation in mobile network technologies, and creating multipurpose networks by integrating AI (including GenAI). The AI-RAN platform is designed to automate RAN operations, leading to more efficient network management while reducing operational cost.
- **T-Mobile's IntentCX Platform:** T-Mobile has collaborated with OpenAI to develop IntentCX [68]. This is a GenAI platform aiming to deliver personalized services by analyzing real-time customer data as well as the current

condition of T-Mobile's network and service. By doing so, IntentCX can enable T-Mobile to anticipate and address customer needs proactively, leading to improved customer satisfaction.

10.4.2 DeepSig's Generative AI for Wireless Communications

DeepSig is a US technology company who specialize in integrating AI/ML into wireless communication systems. Aiming to innovate how wireless communications are designed and implemented, DeepSig employs GenAI to optimize 5G vRAN infrastructure and 6G networks. One of the areas that DeepSig focuses is leveraging GenAI for modeling realistic wireless environments. These models simulate various complex physical factors, such as electromagnetic wave propagation, noise, and hardware effects and impairments, that directly impact wireless performance. Then, data generated from these tools is used for system design, testing, and performance optimization of wireless systems such as OmniPHY5G receiver and RF sensing solution in OmniSIG [69]. Accelerating performance in OpenRAN is another area focused by DeepSig, who is the first commercial recipient of the grant of the NTIA Public Wireless Supply Chain Innovation (PWSCI) Fund to leverage and extend GenAI tools within 5G OpenRAN. These tools aim to facilitate critical testing and evaluation of the air interface in real-world scenarios, which are not captured by the standard 3GPP test scenarios.

10.5 Conclusion

The future of GenAI in wireless communications is highlighted by exiting possibilities and critical challenges that demand collaborative research and innovation across fields. The integration of GenAI into the wireless communication stack, from the physical to application layers, offers potential to revolutionize how networks operate and adapt in real time. The transformative capabilities of GenAI, for example, automating complex network tasks like optimizing resource allocations and enhancing spectrum efficiency, promise to shape the future of wireless communications. However, realizing these opportunities requires overcoming several foundational research frontiers. They include developing specialized GenAI models dedicated to wireless communication, which are enhanced for energy efficiency and seamless integration with next-generation network architectures, for example, 6G and beyond. Additionally, GenAI and emerging technologies (e.g., quantum computing, blockchain, and semantic communication) can complement others to introduce both opportunities for innovation as well as technical challenges that must be overcome.

The practical use case explored in this chapter demonstrates the feasibility and transformative impact of GenAI in wireless communication systems, highlighting its potential to enhance operational efficiency and user satisfaction. However, security and privacy remain pivotal concerns as they are not the main focus of current projects. Therefore, it necessitates research on AI-generated attack detection, privacy-preserving models, and security training methodologies.

In conclusion, the future research directions for GenAI in wireless communications demand a balanced focus on innovation and ethical considerations. By overcoming these challenges and leveraging emerging opportunities, GenAI can revolutionize wireless communication network into a more intelligent and efficient systems, meeting the ever-increasing demand from user and service providers.

References

1 Lin, Y., Zhang, R., Huang, W., Wang, K., Ding, Z., So, D. K., & Niyato, D. Empowering large language models in wireless communication: A novel dataset and fine-tuning framework. *arXiv preprint arXiv:2501.09631*, 2025.

2 Zhang, S., Cheng, G., Huang, X., Li, Z., Wu, W., Song, L., & Shen, X. Split fine-tuning for large language models in wireless networks. *arXiv preprint arXiv:2501.09237*, January 2025. Submitted on 16 Jan 2025.

3 Muhammad Sajjad Akbar, Zawar Hussain, Muhammad Ikram, Quan Z Sheng, and Subhas Chandra Mukhopadhyay. On challenges of sixth-generation (6G) wireless networks: A comprehensive survey of requirements, applications, and security issues. *Journal of Network and Computer Applications*, 233: 104040, 2025.

4 Abdulraqeb Alhammadi, Ibraheem Shayea, Ayman A. El-Saleh, Marwan Hadri Azmi, Zool Hilmi Ismail, Lida Kouhalvandi, and Sawan Ali Saad. Artificial intelligence in 6G wireless networks: Opportunities, applications, and challenges. *International Journal of Intelligent Systems*, 2024: 1–27, 2024. https://doi.org/10.1155/2024/8845070. https://onlinelibrary.wiley.com/doi/10.1155/2024/8845070.

5 Nguyen Van Huynh, Jiacheng Wang, Hongyang Du, Dinh Thai Hoang, Dusit Niyato, Diep N. Nguyen, Dong In Kim, and Khaled B. Letaief. Generative AI for physical layer communications: A survey. *IEEE Transactions on Cognitive Communications and Networking*, 10(3): 706–728, 2024. https://doi.org/10.1109/TCCN.2024.3384500.

6 Jiacheng Wang, Hongyang Du, Yinqiu Liu, Geng Sun, Dusit Niyato, Shiwen Mao, Dong In Kim, and Xuemin Shen. Generative AI-based secure wireless sensing for ISAC networks. *IEEE Transactions on Information Forensics and Security*, 2025.

7 Jiawei Shao, Jingwen Tong, Qiong Wu, Wei Guo, Zijian Li, Zehong Lin, and Jun Zhang. WirelessLLM: Empowering large language models towards wireless intelligence. *arXiv preprint arXiv:2405.17053*, 2024.

8 C. Liang, C., Du, H., Sun, Y., Niyato, D., Kang, J., Zhao, D., & Imran, M. A. Generative AI-driven semantic communication networks: Architecture, technologies and applications. *IEEE Transactions on Cognitive Communications and Networking*, 2024. https://doi.org/10.1109/TCCN.2024.3435524.

9 Abdulkadir Celik and Ahmed M. Eltawil. At the dawn of generative AI era: A tutorial-cum-survey on new frontiers in 6G wireless intelligence. *IEEE Open Journal of the Communications Society*, 5: 2433–2489, 2024.

10 Mauro Castelli, Luca Manzoni, Tatiane Espindola, Aleš Popovič, and Andrea De Lorenzo. Generative adversarial networks for generating synthetic features for Wi-Fi signal quality. *PLOS One*, 16(11): e0260308, 2021.

11 Zikun Liu, Changming Xu, Emerson Sie, Gagandeep Singh, and Deepak Vasisht. Exploring practical vulnerabilities of machine learning-based wireless systems. In *20th USENIX Symposium on Networked Systems Design and Implementation (NSDI 23)*, pages 1801–1817, 2023.

12 Coline Devin, Abhishek Gupta, Trevor Darrell, Pieter Abbeel, and Sergey Levine. Learning modular neural network policies for multi-task and multi-robot transfer. In *2017 IEEE International Conference on Robotics and Automation (ICRA)*, pages 2169–2176. IEEE, 2017.

13 Minrui Xu, Hongyang Du, Dusit Niyato, Jiawen Kang, Zehui Xiong, Shiwen Mao, Zhu Han, Abbas Jamalipour, Dong In Kim, Xuemin Shen, Victor C. M. Leung, H. Vincent Poor. Unleashing the power of edge-cloud generative AI in mobile networks: A survey of AIGC services. *IEEE Communications Surveys & Tutorials*, 26(2): 1127–1170, 2024.

14 Mart van Baalen and Abhijit Khobare. Quantization: Unlocking scalability for large language models. Qualcomm AI Research Webinar, 2024. https://www.qualcomm.com/news/onq/2024/06/quantization-unlocking-scalability-for-large-language-models. Accessed: August 28, 2025.

15 Geng Sun, Wenwen Xie, Dusit Niyato, Fang Mei, Jiawen Kang, Hongyang Du, and Shiwen Mao. Generative AI for deep reinforcement learning: Framework, analysis, and use cases. *IEEE Wireless Communications*, 2025.

16 Tailong Xiao, Xinliang Zhai, Jingzheng Huang, Jianping Fan, and Guihua Zeng. Quantum deep generative prior with programmable quantum circuits. *Communications Physics*, 7(1): 276, 2024.

17 Jonathan Romero, Jonathan P Olson, and Alan Aspuru-Guzik. Quantum autoencoders for efficient compression of quantum data. *Quantum Science and Technology*, 2(4): 045001, 2017.

18 Kilian Tscharke, Maximilian Wendlinger, Afrae Ahouzi, Pallavi Bhardwaj, Kaweh Amoi-Taleghani, Michael Schrödl-Baumann, and Pascal Debus. Quantum autoencoder for multivariate time series anomaly detection. *arXiv preprint arXiv:2504.17548*, 2025.

19 Robin Frehner and Kurt Stockinger. Applying quantum autoencoders for time series anomaly detection. *Quantum Machine Intelligence*, 7(1): 1–21, 2025.

20 Vishal S Ngairangbam, Michael Spannowsky, and Michihisa Takeuchi. Anomaly detection in high-energy physics using a quantum autoencoder. *Physical Review D*, 105(9): 095004, 2022.

21 Puneet Jain. 3GPP: The road to 6G. https://www.3gpp.org/ftp/Information/presentations/Presentations_2024/03_2024_09_17_Puneet_v03.pdf, 2024. Accessed: Mar. 2025.

22 Harri Pennanen, Tuomo Hänninen, Oskari Tervo, Antti Tölli, and Matti Latva-Aho. 6G: The intelligent network of everything. *IEEE Access*, 13:1319–1421, 2024.

23 CableLabs. Up next in mobile connectivity: The technologies that will usher in 6G, 2024. https://www.cablelabs.com/blog/mobile-technologies-6g-era. Accessed: Jan. 2025.

24 Zhonghua Zhang, Xifei Song, Lei Liu, Jie Yin, Yu Wang, and Dapeng Lan. Recent advances in blockchain and artificial intelligence integration: Feasibility analysis, research issues, applications, challenges, and future work. *Security and Communication Networks*, 2021(1): 9991535, 2021.

25 Ziyi Zhou, Oluwakayode Onireti, Hao Xu, Lei Zhang, and Muhammad Imran. AI and blockchain enabled future wireless networks: A survey and outlook. *Distributed Ledger Technologies: Research and Practice*, 3(3): 1–30, 2024.

26 Xiang Hui and Catherine Tucker. Decentralization, blockchain, artificial intelligence (AI): Challenges and opportunities. *Journal of Product Innovation Management*, 42(5): 947–957, 2023.

27 Hanouf Al Ghanmi, Sabreen Ahmadjee, and Rami Bahsoon. Evaluating explanations needs in blockchain smart contracts to reconcile surprises. *ACM Transactions on Software Engineering and Methodology*, 2025.

28 Moetez Abdelhamid, Layth Sliman, Raoudha Ben Djemaa, and Guido Perboli. A review on blockchain technology, current challenges, and AI-driven solutions. *ACM Computing Surveys*, 57(3): 1–39, 2024.

29 Chibuzo Obi-Okoli, Olamide Jogunola, Bamidele Adebisi, and Mohammad Hammoudeh. Machine learning algorithms to detect illicit accounts on Ethereum blockchain. In *Proceedings of the 7th International Conference on Future Networks and Distributed Systems*, pages 747–752, 2023.

30 Francesca Cuomo, Domenico Garlisi, Alessio Martino, and Antonio Martino. Predicting LoRAWAN behavior: How machine learning can help. *Computers*, 9(3): 60, 2020.

31 Fatima Salahdine, Shobhit Aggarwal, and Asis Nasipuri. Short-term traffic congestion prediction with deep learning for lora networks. In *SoutheastCon 2022*, pages 261–268. IEEE, 2022.

32 Huan Yan and Yong Li. Generative AI for intelligent transportation systems: Road transportation perspective. *ACM Computing Surveys*, 2025.

33 Nassim Sehad, Lina Bariah, Wassim Hamidouche, Hamed Hellaoui, Riku Jantti, and Merouane Debbah. Generative AI for immersive communication: The next frontier in internet-of-senses through 6G. *IEEE Communications Magazine*, 2024.

34 Jing Liu, Bohan Zhuang, Peng Chen, Chunhua Shen, Jianfei Cai, and Mingkui Tan. Single-path bit sharing for automatic loss-aware model compression. *IEEE Transactions on Pattern Analysis and Machine Intelligence*, 45(10): 12459–12473, 2023.

35 Fred Hohman, Mary Beth Kery, Donghao Ren, and Dominik Moritz. Model compression in practice: Lessons learned from practitioners creating on-device machine learning experiences. In *Proceedings of the 2024 CHI Conference on Human Factors in Computing Systems*, pages 1–18, 2024.

36 Zhuoran Zhao, Kamyar Mirzazad Barijough, and Andreas Gerstlauer. Deepthings: Distributed adaptive deep learning inference on resource-constrained IoT edge clusters. *IEEE Transactions on Computer-Aided Design of Integrated Circuits and Systems*, 37(11): 2348–2359, 2018.

37 Geoffrey Hinton, Oriol Vinyals, and Jeff Dean. Distilling the knowledge in a neural network. *arXiv preprint arXiv:1503.02531*, 2015.

38 Xinyi Zhang, Weiwen Jiang, Yiyu Shi, and Jingtong Hu. When neural architecture search meets hardware implementation: From hardware awareness to co-design. In *2019 IEEE Computer Society Annual Symposium on VLSI (ISVLSI)*, pages 25–30. IEEE, 2019.

39 Yutaka Yoshihama, Kenichi Yadani, and Shota Isobe. Hardware-aware zero-shot neural architecture search. In *2023 18th International Conference on Machine Vision and Applications (MVA)*, pages 1–5. IEEE, 2023.

40 Lina Bariah, Qiyang Zhao, Hang Zou, Yu Tian, Faouzi Bader, and Merouane Debbah. Large generative AI models for telecom: The next big thing? *IEEE Communications Magazine*, 62(11): 84–90, 2024.

41 Berk Ciloglu, Gorkem Berkay Koc, Afsoon Alidadi Shamsabadi, Metin Ozturk, and Halim Yanikomeroglu. Strategic demand-planning in wireless networks: Can generative-AI save spectrum and energy? *IEEE Communications Magazine*, 2025.

42 Shikhar Tuli and Niraj K Jha. Edgetran: Device-aware co-search of transformers for efficient inference on mobile edge platforms. *IEEE Transactions on Mobile Computing*, 23(6): 7012–7029, 2023.

43 Fang Liu, Zimeng Fan, Wei Hu, Dian Xu, Min Peng, Jing He, and Yanxiang He. Vision transformer-based overlay processor for edge computing. *Applied Soft Computing*, 156: 111459, 2024. https://doi.org/10.1016/j.asoc.2024.111459.

44 Surat Teerapittayanon, Bradley McDanel, and Hsiang-Tsung Kung. Branchynet: Fast inference via early exiting from deep neural networks. In *2016 23rd international conference on pattern recognition (ICPR)*, pages 2464–2469. IEEE, 2016.

45 Yizeng Han, Gao Huang, Shiji Song, Le Yang, Honghui Wang, and Yulin Wang. Dynamic neural networks: A survey. *IEEE Transactions on Pattern Analysis and Machine Intelligence*, 44(11): 7436–7456, 2021.

46 Manjari Gupta, Lava Bhargava, and Indu Sreedevi. Dynamic voltage frequency scaling in multi-core systems using adaptive regression model. In *2020 Fourth International Conference on I-SMAC (IoT in Social, Mobile, Analytics and Cloud)(I-SMAC)*, pages 1201–1206. IEEE, 2020.

47 Radovanović, A, Koningstein, R., Schneider, I., Chen, B., Duarte, A., Roy, B., Xiao, D., Haridasan, M., Hung, P., Care, N., Talukdar, S., Mullen, E., Smith, K., Cottman, M., & Cirne, W. Carbon-aware computing for datacenters. *IEEE Transactions on Power Systems*, 38(2): 1270–1280, 2022.

48 Haseena Rahmath P, Vishal Srivastava, Kuldeep Chaurasia, Roberto G Pacheco, and Rodrigo S Couto. Early-exit deep neural network-a comprehensive survey. *ACM Computing Surveys*, 57(3): 1–37, 2024.

49 Tianming Gao and Jozsef Suto. Acceleration of image classification and object tracking by the intel neural compute stick 2 with power efficiency evaluation on raspberry pi 4b. *Sensors*, 25(6): 1794, 2025.

50 Artificial. 10 questions you should know about Google Edge TPU. *Medium*, 2021. https://medium.com/@deve321/1-what-is-the-edge-tpu-a7dc7c5272dd.

51 carmelito. Intel neural compute stick 2 - review. *element14 Community*, 2021. https://community.element14.com/products/roadtest/rv/roadtest_reviews/954/intel_neural_compute_3.

52 Moe Long. NVIDIA jetson nano launches: New NVIDIA AI computer platform released. *Electromaker*, 2019. https://www.electromaker.io/blog/article/nvidia-jetson-nano-launches-for-ai.

53 Lixing Yu, Yifan Guo, Qianlong Wang, Changqing Luo, Ming Li, Weixian Liao, and Pan Li. Spectrum availability prediction for cognitive radio communications: A DCG approach. *IEEE Transactions on Cognitive Communications and Networking*, 6(2): 476–485, 2020.

54 Abiodun Gbenga-Ilori, Agbotiname Lucky Imoize, Kinzah Noor, and Paul Oluwadara Adebolu-Ololade. Artificial intelligence empowering dynamic spectrum access in advanced wireless communications: A comprehensive overview. *AI*, 6(6): 126, 2025.

55 Shuhang Zhang, Shuai Jiang, Wanjie Lin, Zheng Fang, Kangjun Liu, Hongliang Zhang, and Ke Chen. Generative AI on spectrum net: An open benchmark of multiband 3D radio maps. *IEEE Transactions on Cognitive Communications and Networking*, 2024.

56 Minrui Xu, Dusit Niyato, Jiawen Kang, Zehui Xiong, Shiwen Mao, Zhu Han, Dong In Kim, and Khaled B Letaief. When large language model agents meet 6G networks: Perception, grounding, and alignment. *IEEE Wireless Communications*, 2024.

57 Emilia Mikołajewska, Dariusz Mikołajewski, Tadeusz Mikołajczyk, and Tomasz Paczkowski. Generative AI in AI-based digital twins for fault diagnosis for predictive maintenance in industry 4.0/5.0. *Applied Sciences*, 15(6): 3166, 2025.

58 Dimitrios Michael Manias, Ali Chouman, and Abdallah Shami. Towards intent-based network management: Large language models for intent extraction in 5G core networks. In *2024 20th International Conference on the Design of Reliable Communication Networks (DRCN)*, pages 1–6. IEEE, 2024.

59 Feibo Jiang, Yubo Peng, Li Dong, Kezhi Wang, Kun Yang, Cunhua Pan, Dusit Niyato, and Octavia A Dobre. Large language model enhanced multi-agent systems for 6G communications. *IEEE Wireless Communications*, 2024.

60 Chengming Hu, Hao Zhou, Di Wu, Xi Chen, Jun Yan, and Xue Liu. Self-refined generative foundation models for wireless traffic prediction. *arXiv preprint arXiv:2408.10390*, 2024.

61 Yu-Ting Lin, Eduardo Mucelli Rezende Oliveira, Sana Ben Jemaa, and Salah Eddine Elayoubi. Machine learning for predicting QoE of video streaming in mobile networks. In *2017 IEEE International Conference on Communications (ICC)*, pages 1–6. IEEE, 2017.

62 Monalisa Ghosh, Dr Chetna Singhal, and Rushikesh Wayal. DESVQ: Deep learning based streaming video QoE estimation. In *Proceedings of the 23rd International Conference on Distributed Computing and Networking*, pages 19–25, 2022.

63 Yi Shi, Kemal Davaslioglu, and Yalin E Sagduyu. Generative adversarial network in the air: Deep adversarial learning for wireless signal spoofing. *IEEE Transactions on Cognitive Communications and Networking*, 7(1): 294–303, 2020.

64 Davinder Kaur, Suleyman Uslu, Kaley J Rittichier, and Arjan Durresi. Trustworthy artificial intelligence: A review. *ACM Computing Surveys (CSUR)*, 55(2): 1–38, 2022.

65 Cynthia Dwork. Differential privacy. In *International Colloquium on Automata, Languages, and Programming*, pages 1–12. Springer, 2006.

66 Reuters. Verizon uses GenAI to improve customer loyalty, 2024. https://www.reuters.com/technology/artificial-intelligence/verizon-uses-genai-improve-customer-loyalty-2024-06-18/. Accessed: Jan. 2025.

67 T-Mobile. T-Mobile launches AI-enabled ran innovation center with NVIDIA, 2023. https://www.t-mobile.com/news/business/t-mobile-launches-ai-ran-innovation-center-with-nvidia. Accessed: Jan. 2025.

68 T-Mobile. T-Mobile launches intentCX with openAI, 2023. https://www.t-mobile.com/news/business/t-mobile-launches-intentcx-with-openai. Accessed: Jan. 2025.

69 DeepSig. Generative AI at DeepSig and its role in shaping the future of wireless communications. https://www.deepsig.ai/generative-ai-at-deepsig-and-its-role-in-shaping-the-future-of-wireless-communications/, 2023. Accessed: Jan. 2025.

Index

Generative AI for Communications Systems: Fundamentals, Applications, and Prospects, First Edition. Edited by Diep N. Nguyen et al.

b

q

r

u

v

Printed and bound by CPI Group (UK) Ltd, Croydon, CR0 4YY

17/09/2026

14964333-0004